Hamed Taherdoost

Innovation Through Research and Development

Strategies for Success

Hamed Taherdoost
University Canada West
Vancouver, BC, Canada

ISSN 1860-4862　　　　　　　　ISSN 1860-4870　(electronic)
Signals and Communication Technology
ISBN 978-3-031-52567-4　　　　ISBN 978-3-031-52565-0　(eBook)
https://doi.org/10.1007/978-3-031-52565-0

© The Editor(s) (if applicable) and The Author(s), under exclusive license to Springer Nature Switzerland AG 2024
This work is subject to copyright. All rights are solely and exclusively licensed by the Publisher, whether the whole or part of the material is concerned, specifically the rights of translation, reprinting, reuse of illustrations, recitation, broadcasting, reproduction on microfilms or in any other physical way, and transmission or information storage and retrieval, electronic adaptation, computer software, or by similar or dissimilar methodology now known or hereafter developed.
The use of general descriptive names, registered names, trademarks, service marks, etc. in this publication does not imply, even in the absence of a specific statement, that such names are exempt from the relevant protective laws and regulations and therefore free for general use.
The publisher, the authors, and the editors are safe to assume that the advice and information in this book are believed to be true and accurate at the date of publication. Neither the publisher nor the authors or the editors give a warranty, expressed or implied, with respect to the material contained herein or for any errors or omissions that may have been made. The publisher remains neutral with regard to jurisdictional claims in published maps and institutional affiliations.

This Springer imprint is published by the registered company Springer Nature Switzerland AG
The registered company address is: Gewerbestrasse 11, 6330 Cham, Switzerland

Paper in this product is recyclable.

Signals and Communication Technology

Series Editors

Emre Celebi, Department of Computer Science, University of Central Arkansas, Conway, AR, USA

Jingdong Chen, Northwestern Polytechnical University, Xi'an, China

E. S. Gopi, Department of Electronics and Communication Engineering, National Institute of Technology, Tiruchirappalli, Tamil Nadu, India

Amy Neustein, Linguistic Technology Systems, Fort Lee, NJ, USA

H. Vincent Poor, Department of Electrical Engineering, Princeton University, Princeton, NJ, USA

Antonio Liotta, University of Bolzano, Bolzano, Italy

Mario Di Mauro, University of Salerno, Salerno, Italy

This series is devoted to fundamentals and applications of modern methods of signal processing and cutting-edge communication technologies. The main topics are information and signal theory, acoustical signal processing, image processing and multimedia systems, mobile and wireless communications, and computer and communication networks. Volumes in the series address researchers in academia and industrial R&D departments. The series is application-oriented. The level of presentation of each individual volume, however, depends on the subject and can range from practical to scientific.

Indexing: All books in "Signals and Communication Technology" are indexed by Scopus and zbMATH

For general information about this book series, comments or suggestions, please contact Mary James at mary.james@springer.com or Ramesh Nath Premnath at ramesh.premnath@springer.com.

To my wonderful wife, my life's co-author and forever collaborator

Preface

Innovation is the lifeblood of progress, and at its core lies the transformative power of research and development (R&D). In the rapidly evolving landscape of contemporary business practices, where technology constantly pushes boundaries and global markets shift toward knowledge-driven economies, the pivotal role of R&D cannot be overstated. It is within this dynamic and ever-shifting realm that our journey begins. *Innovation Through Research and Development; Strategies for Success* is your guide into the world of R&D, an exploration that spans from foundational principles to the intricacies of crafting effective strategies, project management, and performance evaluation.

Innovation Through Research and Development; Strategies for Success embarks on a comprehensive journey into the world of research and development (R&D) within the dynamic landscape of contemporary business practices. This book is a comprehensive exploration of research and development (R&D) within the context of contemporary business practices and innovation. In today's rapidly evolving landscape, characterized by constant technological advancement and a global shift toward knowledge-driven economies, the importance of R&D in fostering innovation and facilitating the sustainability of organizations cannot be overstated. With the goal of equipping both seasoned professionals and newcomers to the field with valuable insights, this textbook has been meticulously structured into 14 distinct chapters, each serving as a key piece of the intricate puzzle that is R&D.

This book commences this scholarly journey by laying the foundational framework of R&D, introducing fundamental concepts and principles inherent to the field. From there, this book navigates the intricate domain of R&D and its critical role in shaping the destiny of businesses of all scales. In the pursuit of excellence, this book delves into the intricacies of crafting effective R&D strategies and their seamless integration into the broader organizational vision. A parallel exploration examines the financial aspects of R&D, delving into funding and budgeting considerations to ensure the effective and efficient allocation of resources. In the quest to execute successful R&D projects, this book provides a detailed overview of methodologies and best practices, guiding the reader from the conceptual phase through to project completion.

Furthermore, the complexities of the R&D process are examined in depth, offering insights into the intricate steps and stages that underlie every research and development endeavor. Moreover, this textbook explores the professional dimensions of R&D as a career path, elucidating the requisite skills and the diverse opportunities that this field offers. In tandem, this book addresses the art of assembling and cultivating high-performance R&D teams, showcasing the advantages of diverse talent utilization. Recognizing the importance of sound project management, this book delves into the realm of R&D project management, from inception to successful execution. The exploration extends to the critical domain of performance management within R&D, underscoring the need for continuous improvement and optimization in the innovation process. Furthermore, this book focuses on the assessment of success in R&D projects, offering a comprehensive framework for the evaluation of key performance indicators.

This book comprehensively navigates the multifaceted landscape of R&D, addressing issues such as risk management and risk mitigation, the utilization of technological tools and resources for optimizing R&D capabilities, and the imperative of safeguarding intellectual property and adhering to copyright principles in R&D. My intent is to impart a holistic understanding of the intricacies and intricacies that underlie R&D endeavors. Through the amalgamation of real-world examples, practical insights, and established best practices, this text aims to guide individuals and organizations toward the harnessing of R&D as a catalyst instrument for innovation, change, and organizational success.

In *Innovation Through Research and Development; Strategies for Success*, you will embark on a transformative journey into the heart of R&D. The power to innovate and succeed is now in your hands, armed with insights, strategies, and best practices.

Vancouver, BC, Canada   Hamed Taherdoost

Acknowledgment

I would like to express my heartfelt gratitude to all those who supported and inspired me throughout the journey of creating this book.

First and foremost, I want to thank my parents, whose unwavering support, love, and belief in my abilities have been a driving force throughout my life. Your sacrifices and encouragement have been the foundation of my journey.

To my dear children, Hamta and Kiasha, your love, laughter, and the joy you bring to my life are the fuel that powers my passion for research and development. Your presence has been a constant reminder of the importance of innovation in shaping a better future. You are my motivation and my ultimate source of pride.

A special thank you to my beloved wife, Mitra, whose unwavering love, encouragement, and patience were the pillars that sustained me during the long hours and countless revisions. Your support and understanding were invaluable.

To my PhD and MSc supervisors, Prof. Shamsul Sahibuddin and Prof. Maslin Masrom, who have guided me throughout my studies and opened the doors to research, your wisdom and guidance have been instrumental in shaping both my work and my thinking.

I would also like to extend my appreciation to my dedicated colleagues at Hamta Business Corporation and Q Minded | Quark Minded Technology Inc. Your enthusiasm, expertise, and collaborative spirit enriched the ideas within this book, and I am grateful for the camaraderie we share.

My heartfelt thanks go to my colleagues at University Canada West, especially Assoc. Prof. George Drazenovic, for all his support and encouragement. I also want to acknowledge Dr. Stephanie Chu, who consistently supports scholarly activities at UCW.

Lastly, I extend my thanks to all the readers and supporters of this book. Your interest in the field of research and development fuels the pursuit of innovative strategies for success. Your feedback and engagement continue to inspire me in my work.

Contents

1	**Fundamentals of R&D**		1
	1.1 Definition of R&D		2
	1.2 Research Dimensions		3
	1.3 R&D Elements		3
		1.3.1 Basic Research	4
		1.3.2 Applied Research	7
		1.3.3 Development	7
	1.4 R&D Methods		8
		1.4.1 Research Methods	8
		1.4.2 Apply Now	11
		1.4.3 Research Steps	12
		1.4.4 Development Methods	13
		1.4.5 Development Steps	14
		1.4.6 The Distinction Between R&D and Other Projects	15
	1.5 R&D Challenges		16
		1.5.1 The Functional Tensions Challenge	17
		1.5.2 Globalization Challenge	18
		1.5.3 Attracting Talent Challenge	18
		1.5.4 Managing Risks Challenge	19
		1.5.5 Managing Scientific Freedom Challenge	20
		1.5.6 Funding R&D Expenditure Challenge	20
	References		21
2	**The Role of R&D in Business**		23
	2.1 The Strategic Importance of R&D		24
		2.1.1 R&D and Competitive Advantage	26
		2.1.2 Aligning R&D with Business Goals	27
	2.2 R&D's Impact on Innovation and Competitiveness		29
	2.3 R&D Budgeting and Investment		31
		2.3.1 Balancing Short-Term and Long-Term R&D Investments	34
		2.3.2 ROI Analysis for R&D Projects	35

	2.4	Managing R&D Teams and Intellectual Property		36
	2.5	Measuring and Assessing R&D Performance		39
		2.5.1	Key Performance Indicators (KPIs) for R&D	40
		2.5.2	Benchmarking R&D Success	41
		2.5.3	Evaluating the Efficiency of R&D Processes	42
	2.6	The Future of R&D in Business		44
	References			46
3	**R&D Strategy and Planning**			47
	3.1	Significance of R&D in Modern Businesses		48
	3.2	Defining R&D Strategy		48
	3.3	Setting R&D Goals and Objectives		49
		3.3.1	The Role of SMART Objectives in R&D	50
		3.3.2	Developing R&D Goals That Reflect Market Needs and Competitive Dynamics	51
	3.4	Environmental Scanning and Market Research		52
		3.4.1	Techniques for Gathering and Analyzing Market Data	52
		3.4.2	Using Competitive Intelligence to Inform R&D Decisions	55
	3.5	Resource Allocation and Budgeting		56
	3.6	Technology Roadmaps		58
		3.6.1	Creating Technology Roadmaps to Guide R&D Efforts	58
		3.6.2	Balancing Short-Term and Long-Term Technology Development	59
	3.7	Project Selection and Prioritization		59
		3.7.1	Criteria for Project Selection	60
		3.7.2	Decision-Making Processes and Tools for Project Prioritization	61
	3.8	Collaboration and Partnerships		62
	3.9	Risk Management and Contingency Planning		64
	3.10	Monitoring and Evaluation		65
		3.10.1	Implementing Key Performance Indicators (KPIs) for R&D Projects	65
		3.10.2	Regular Monitoring and Evaluation of R&D Progress	65
		3.10.3	Adjusting Strategies Based on Performance Data	66
	References			67
4	**R&D Funding and Budgeting**			69
	4.1	The Basics of R&D Funding		69
		4.1.1	Sources of R&D Funding	70
		4.1.2	Advantages and Disadvantages of Different Funding Sources	73
	4.2	Budgeting for R&D		74
		4.2.1	Setting R&D Budget Goals and Objectives	75
		4.2.2	Creating a Budgeting Process	75
		4.2.3	Factors to Consider When Budgeting	76
		4.2.4	Allocating Funds for Different R&D Projects	80

	4.3	Managing R&D Budgets	81
		4.3.1 The Role of Financial Reporting in R&D Budget Management	82
		4.3.2 Making Budget Adjustments When Necessary	83
		4.3.3 Avoiding Common Budgeting Pitfalls	83
	4.4	Ensuring ROI	85
		4.4.1 Measuring the Success of R&D Projects	85
		4.4.2 Calculating ROI for R&D Initiatives	86
		4.4.3 Factors Impacting ROI in R&D	87
		4.4.4 Strategies for Improving ROI	88
	4.5	Government Grants and Incentives	88
	4.6	Private Sector Investment	89
		4.6.1 Attracting Venture Capital and Angel Investors	89
		4.6.2 Corporate Partnerships and Strategic Investors	90
		4.6.3 Equity Financing and Its Impact on Ownership	91
	4.7	Future Trends in R&D Funding and Budgeting	91
		4.7.1 Emerging Funding Sources and Methods	92
		4.7.2 Technological and Economic Trends Shaping R&D Financing	92
	References		93
5	**How to Carry Out an R&D Project**		**95**
	5.1	Project Approach, Alignment, and Fit	96
	5.2	Complementary Discipline Components	98
	5.3	Facilitate the Kind of R&D Project	99
	5.4	Systems Engineering and Project Management Documents	101
	5.5	Project Measurements and Baseline	104
	5.6	Active Management	105
	5.7	Seize Opportunities and Reduce Risks	106
	5.8	Support and Facilitate the R&D Group	107
	References		109
6	**R&D Process**		**111**
	6.1	Importance of R&D Process	111
	6.2	Stage 1: Idea Generation	112
	6.3	Stage 2: Feasibility Assessment	115
		6.3.1 Technology Assessment	116
	6.4	Stage 3: Project Planning	119
	6.5	Stage 4: Concept Development	123
	6.6	Stage 5: Testing and Evaluation	125
	6.7	Stage 6: Scaling and Production	127
	6.8	Stage 7: Commercialization	128
	6.9	Stage 8: Post-Launch Evaluation	131
	6.10	Challenges and Common Pitfalls	133
	References		133

7	R&D as a Job.		135
	7.1	Job Attributes	136
	7.2	Career Pathways.	137
	7.3	Hierarchies with Dual and Triple Levels	138
	7.4	Work with R&D and Its Heterogeneity	140
		7.4.1 Non-routine Work	140
		7.4.2 Uncertainty in Activities	141
		7.4.3 The Work in R&D: Engagements and Interactions with the Environment.	141
		7.4.4 The Job with a Certain Amount of Occupational and Autonomy Norms	142
	7.5	Decentralization and Centralization.	142
	7.6	Career Conflict and Design	143
	7.7	Future of R&D Jobs.	146
		7.7.1 How Long-Lasting Is This Concentrated Focus and Heightened Pressure, Both Privately and Communally?	147
		7.7.2 How Will Relocation, Globalization, Outsourcing, and Open Innovation Affect R&D Work Future?	147
		7.7.3 What Effects Will the Digital Revolution Have on R&D Work?	149
	References.		150
8	Team Building in R&D Projects		153
	8.1	R&D and HRM: Complex Relationships	154
		8.1.1 Project Management and HRM Adaptation in R&D.	154
		8.1.2 Integration and Recruitment	155
	8.2	Relationship of the Researcher to Peers and Management	157
	8.3	Creating Teams	158
	8.4	The R&D Team Leadership.	160
		8.4.1 Project Management for R&D.	161
		8.4.2 The Role of Leadership	162
		8.4.3 The Management Role.	163
	8.5	Workforce Management.	164
		8.5.1 Staff Diversity and Selection	165
		8.5.2 Activities for Building Teams	165
		8.5.3 Alignment of Objectives	166
		8.5.4 Workforce Sharing.	167
		8.5.5 Evolution of Team	167
	8.6	Improving the Productivity of R&D	168
		8.6.1 Alignment of Organizations.	169
		8.6.2 Environment of Learning.	170
		8.6.3 Responsibility and Accountability.	175
	8.7	Managing Projects for Certain Specialties.	176
		8.7.1 Scientist or Researcher R&D.	177
	References.		178

9 R&D Project Management 179
9.1 Project Management's Importance 180
9.2 Definitions and Controls for Project Management 181
9.3 Traditionally Managed Projects 182
9.4 Adaptive Project Management 183
 9.4.1 Portfolio, Project, and Program 184
 9.4.2 Managing Stakeholders 186
9.5 Risks of Project 187
9.6 Complexity of Project 188
9.7 Change and Project Management 191
9.8 Project Life Cycle for R&D 191
9.9 Navigating R&D Challenges: The Way Forward 193
References 194

10 Managing Performance in R&D Projects 197
10.1 Performance in R&D 198
 10.1.1 A Hard-to-Define Concept 198
 10.1.2 Particular Management Challenges in R&D 200
 10.1.3 The Challenges of Performance 201
10.2 The Sensitive Matter of Measurement 202
 10.2.1 Traditional Indications 202
 10.2.2 Considering the Different Stakeholders 203
 10.2.3 Limitations of Technological Sophistication 204
 10.2.4 R&D Departments' Financial Management 204
10.3 Project Management for Innovation 205
 10.3.1 Project Economic Evaluation: The Two Methods 205
10.4 Enhancing Team Collaboration for R&D Success 210
10.5 Leveraging Technology for Efficient R&D Performance 212
References 214

11 Success Evaluation in R&D 215
11.1 Success Definition 215
11.2 Crucial Foundations 217
 11.2.1 Project Manager 217
 11.2.2 Communications 217
 11.2.3 Basis for the Project 218
 11.2.4 Success Definitions in Various Fields 219
11.3 Measurement Types 221
 11.3.1 Metrics Versus Measures 222
11.4 The Analysis of Variance and Trend 224
 11.4.1 Technical Performance 225
 11.4.2 Schedule Performance 226
 11.4.3 Budgetary Performance 228
 11.4.4 Risk Management 230
 11.4.5 Resources Allocation 231
 11.4.6 Outcomes 231

11.5 Various Project Kinds' Measurements.................... 232
    11.5.1 Measures for Applied and Basic Research............ 232
    11.5.2 Development Measures 235
    11.5.3 Innovation Measures 237
    11.5.4 Measures for Production 237
11.6 R&D Performance Measurement......................... 239
    11.6.1 Documentation.................................. 239
    11.6.2 Schedule and Budget 240
    11.6.3 Outcomes....................................... 241
    11.6.4 Configuration and Change Management 242
    11.6.5 Risk Management 242
    11.6.6 Reviews 243
    11.6.7 Quality 244
    11.6.8 Leadership 245
    11.6.9 Communications 245
References... 246

12 Risk Management in R&D Projects 247
12.1 Risk Management 248
    12.1.1 Processes of Risk Management 249
    12.1.2 A Project's Level of Risk........................... 251
    12.1.3 A Program's Level of Risk......................... 255
    12.1.4 A Portfolio's Level of Risk 255
12.2 R&D Risk Management................................... 257
    12.2.1 Commercial Risks 257
    12.2.2 Technical Risks 258
    12.2.3 Regulatory Risks 258
    12.2.4 Organizational Risks 258
12.3 R&D-Specific Risk Factors 259
    12.3.1 Inherent Uncertainty in R&D....................... 259
    12.3.2 Long Project Timelines 260
    12.3.3 Rapid Technological Advancements 261
    12.3.4 Resource Constraints 262
    12.3.5 Intellectual Property Risks........................ 263
12.4 Risk Assessment Tools and Techniques..................... 264
    12.4.1 Probability and Impact Analysis 265
    12.4.2 Risk Matrices................................... 265
    12.4.3 Monte Carlo Simulation.......................... 266
    12.4.4 Expert Judgment 266
    12.4.5 Benchmarking 267
12.5 Reward Strategy and Risk 268
References... 269

13 R&D Tools and Technologies 271
13.1 Evolution of R&D Tools and Technologies.................. 271
    13.1.1 Milestones and Breakthroughs in R&D Tools.......... 272
    13.1.2 How R&D Tools Have Transformed over Time 274

	13.2	Key Categories of R&D Tools and Technologies	274
		13.2.1 Laboratory Equipment	274
		13.2.2 Computational Tools	276
		13.2.3 Communication and Collaboration Tools	279
	13.3	Impact of Tools and Technologies on R&D	280
		13.3.1 Accelerating Research	281
		13.3.2 Enhancing Precision and Accuracy	282
		13.3.3 Data-Driven Decision-Making	283
		13.3.4 Facilitating Global Collaboration	283
	13.4	Innovations in R&D Tools	284
		13.4.1 Advanced Technologies on the Horizon	284
		13.4.2 Breakthroughs in Data Management and Storage	286
		13.4.3 Novel Methods for Research and Experimentation	287
	13.5	Role of R&D Tools in the Coming Years	288
	References		289
14	**R&D Intellectual Property and Copyright**		**291**
	14.1	Understanding Intellectual Property in R&D	291
		14.1.1 Types of IP	292
	14.2	Copyright Basics	295
		14.2.1 What Can Be Copyrighted?	296
		14.2.2 Rights Conferred by Copyright	297
		14.2.3 Duration of Copyright Protection	297
	14.3	Copyright in the R&D Process	298
		14.3.1 Copyright Ownership in R&D	298
		14.3.2 Work for Hire and Copyright	299
		14.3.3 Fair Use and Research	300
		14.3.4 Licensing and Permissions	301
		14.3.5 Open Access and R&D	301
	14.4	Intellectual Property and Collaboration in R&D	302
	14.5	Protecting R&D Through Copyright	303
		14.5.1 Strategies for Copyright Protection	303
		14.5.2 Copyright Notice	304
		14.5.3 Digital Rights Management (DRM)	304
	14.6	Intellectual Property Enforcement and Infringement	305
	14.7	Emerging Trends and Challenges	306
		14.7.1 Digital and Online R&D	306
		14.7.2 AI and Machine-Generated Works	307
		14.7.3 Open Source and Collaboration	308
		14.7.4 Future of IP in R&D	310
	References		311
Index			**313**

Chapter 1
Fundamentals of R&D

This chapter comprehensively explores the research and development (R&D) landscape. It begins by defining R&D and distinguishing its core components. This chapter then delves into the diverse dimensions of research, from basic to applied and experimental development, elucidating their distinct roles. It further dissects the essential elements that underpin R&D processes, emphasizing the synergy between creativity, innovation, technology, and scientific inquiry. Readers are introduced to various R&D methods, from empirical research to theoretical modeling, showcasing their significance in knowledge advancement.

Additionally, this chapter addresses the multifaceted challenges faced in R&D, spanning resource allocation, risk management, regulatory compliance, and global competition. This holistic overview equips readers with a firm grasp of R&D's fundamental principles, fostering a deeper appreciation for its pivotal role in driving innovation and progress in diverse industries. Companies regularly invest money in specific investigative projects to generate findings that may aid in developing new products or methods of doing things, as well as efforts to improve already-existing goods or procedures. These actions fall under the category of R&D. R&D is crucial for ensuring future development and keeping a product relevant in the marketplace.

> **Learning Objectives**
> - Know what research and development (R&D) are all about
> - Describe the most important parts of R&D
> - Name the main types of R&D
> - Understand the most important steps in R&D
> - Learn how to do R&D projects

1.1 Definition of R&D

Research is a systematic and organized process of inquiry that involves the exploration, investigation, and analysis of a particular subject or topic in order to generate new knowledge, expand understanding, or find solutions to existing problems. It often encompasses a series of structured steps, from formulating research questions and hypotheses to collecting and analyzing data. Research can take various forms, including scientific experiments, surveys, literature reviews, and observational studies, and it can be conducted in a wide range of fields, from science and technology to social sciences and humanities. The primary goal of research is to contribute to the accumulation of reliable and valuable information, fostering progress, innovation, and informed decision-making in various domains of human knowledge and endeavor.

According to the National Science Foundation (NSF), research is classified as follows (Science & Engineering Indicators, 2008):

- *Basic Research*: The goal of basic research is to

  learn or understand more about the subject being studied, without any specific applications in mind.

To accommodate industrial goals, the National Science Foundation changed this definition for the industry sector to indicate that basic research advances scientific knowledge.

However, it may be in prospective or existing commercial interest areas.

- *Applied Research*: The goal of applied research is to

  Gain knowledge or comprehension to determine how to address a particular, identified need.

In the business world, applied research is research that is done

to find new scientific knowledge with specific commercial goals for services, processes, or products.

- *Development*: Development is the

  Utilization of research-based information or understanding to create beneficial techniques, systems, equipment, or materials

This includes the development and design of processes and prototypes.

In the Organization for Economic Co-operation and Development (OECD) book [1], the OECD discusses the following research activities:

Basic research is the theoretical or experimental study done without a particular application in mind. This kind evaluates rules, theories, and hypotheses using connections, structures, and attributes. Research results are frequently published or shared with colleagues. Long-term effort is not made to improve society or the economy, nor are the discoveries applied to real-world challenges or passed on.

Fundamental research builds a knowledge base for future opportunities or challenges.

Applied research is also original research done to discover something new. It is, however, mostly aimed at a specific practical goal. Through applied research, ideas are turned into things that can be done.

Experimental development is systematic work that builds on real-world and research experience to make big improvements, install new services, systems, and processes, or make new devices, products, and materials for those already put or made in place.

Many of these things are part of R&D. The OECD [1] says that R&D is

> Systematic creative labor performed to grow the store of information about society, culture, and man, and the utilization of this stock of knowledge to generate novel applications.

Science Indicators categorizes R&D activities as efforts in engineering and science in the following ways so that they can be defined in a way that is easy and useful to understand (Fig. 1.1).

1.2 Research Dimensions

Harvey Brooks (1968, p. 46) [2] compiled a comprehensive list of study classifications and dimensions (Fig. 1.2):

Brooks (1968, p. 57) [2] proposes three kinds of research organizations that might be utilized for planning: academic research, scientific, institutional research, and mission-oriented research (Table 1.1):

1.3 R&D Elements

R&D is a way to look into a subject to learn more about it and make changes to what is already known. Researchers usually define areas of study, look at what is already known about set-up experiments and the area of interest, and try to figure out what the results mean to change existing theories or develop new ones. Sometimes, they start by looking at what is already known about an area of interest to help them develop new ideas. Research can also confirm what has already been found or be used to solve problems. Even though research is a broad field, this section outlines the main elements of R&D that can be used to show how project management can be used to get good results [3].

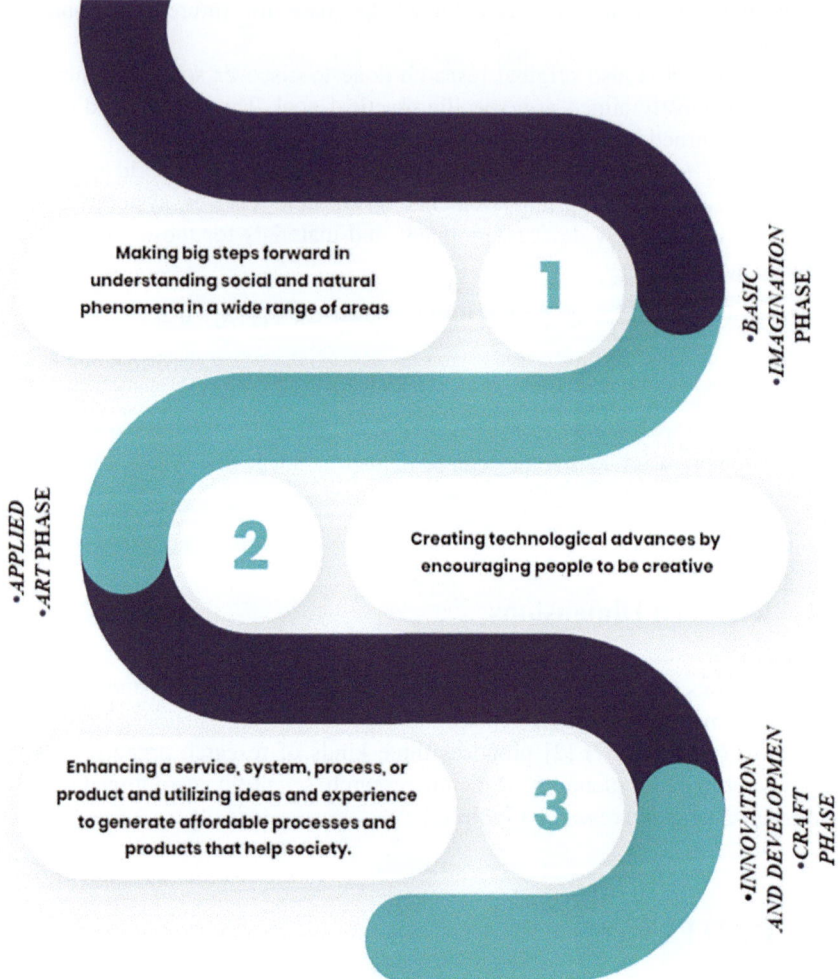

Fig. 1.1 R&D activities categorization by Science Indicators

1.3.1 Basic Research

Basic research, often called the "imagination" phase of research, is pivotal in expanding the frontiers of human knowledge. During this phase, researchers delve into the unknown, driven by curiosity and a thirst for discovery. Their primary aim is not to achieve a predefined outcome but rather to unravel the mysteries of the universe. In this research stage, groundbreaking ideas take shape, novel ways of thinking are tested, and hypotheses are born. It is an intellectual exploration into uncharted territory with boundless possibilities. Researchers at this juncture embark on journeys fueled by questions, not answers. These questions serve as beacons

1.3 R&D Elements

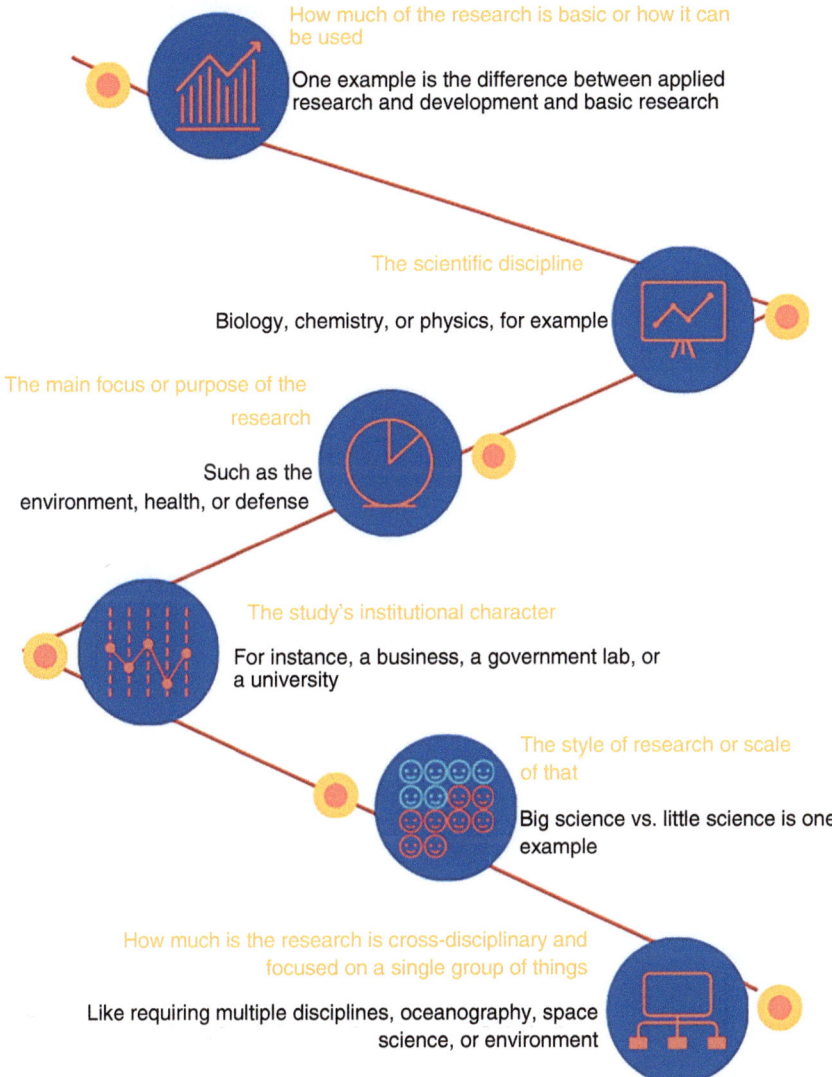

Fig. 1.2 List of research categories and dimensions coming up by Harvey Brooks

guiding their pursuit of new insights. Crucially, during basic research, there is no predetermined blueprint for applying the acquired knowledge. The emphasis is on uncovering fundamental truths and principles that may lay the groundwork for future breakthroughs.

Ideas generated during this phase are like raw materials awaiting refinement and application. To further illustrate the significance of basic research, consider the concept of oriented basic research. This subtype of basic research is intentionally conducted to generate a foundational knowledge base that can be harnessed in

Table 1.1 Three broad kinds of research organizations suggested by Harvey Brooks

Sorts of research organizations	Description
Academic	In university academic departments, research associates or students often do fundamental research under the direction of university professors who also function as teachers
Scientific Institutional	This includes groups whose goals are mostly scientific, like making progress in molecular biology or high-energy physics. These organizations research by following a plan that can be changed as new information comes out in their field
Mission-Oriented	"Mission" refers to an organization's long-term ambitions, not a technical goal. The research laboratories undertake applied and fundamental research and may give manufacturing or operational assistance. While their research is sophisticated and important, it serves the organization's objectives and aims, not science itself

Fig. 1.3 R&D movement

subsequent development activities and applied research endeavors. It provides the scaffolding upon which practical innovations can be constructed. Research studies typically fall into two categories: applied or basic research. The former is characterized by a direct focus on practical applications and immediate problem-solving, while the latter is centered around exploration and discovery without specific utilization in mind. The journey of development and research is intricate and dynamic, as depicted in Fig. 1.3. While the figure may present a linear progression, the reality is far more complex. Research endeavors can fluidly transition between phases. For example, a project initially intended for applied research may evolve into early development driven by unforeseen insights and opportunities.

Conversely, a basic research project may remain within its initial category without moving to the next stage. Flexibility is key in this process. If a prototype or experiment fails to yield the expected results, researchers can revisit a prior phase to reevaluate their approach. The outcomes of each stage dictate the subsequent course of action, emphasizing the importance of adaptability and continuous learning. Basic research is the cornerstone of scientific progress, where ideas are born and tested without predefined outcomes. It is fertile ground from which future innovations and applications may emerge, shaping the research and development trajectory.

1.3.2 Applied Research

Applied research, the following level in R&D, may be compared to the art phase since its primary objective is to remove a question from the domain of speculation and create something. It involves gathering more data to satisfy a recognized void or necessity. This is sometimes referred to as strategic research. Fundamental or targeted fundamental research is the initial stage in applied research. The next phase is to demonstrate how the concepts operate to achieve a goal, fulfill a need, or close a knowledge gap. Ideas might lead to studies designed to show and test their validity during this phase. It is also possible to create rough prototypes or test ideas in other methods that do not require the actual world. As in basic research, actions at this level may involve debate, peer reviews, publishing, study, and other endeavors that characterize the topic.

In contrast to being dispersed, these acts have a clear emphasis. This is a highly creative phase with high risk and low experimental validation success probability. When activities in this phase are successful, they are prepared to move on to the "craft" phase, where the emphasis is more on how an idea functions, fits, and looks [4].

1.3.3 Development

The next part of development, experimental development, or the R&D process, can be considered the "craft" part. Its goal is to make things like software, methods, systems, tools, and materials. Development, on the other hand, is focused on making something better than it is now. This can be done by changing an existing service, system, process, or product or by making something new. This development is directed by technicians, engineers, scientists, and researchers who rely on their expertise and research-based knowledge.

Rarely used by itself, the term "development" may have a variety of meanings in many professions. When used in technology and science, it is usually grouped as "R&D." Economic development, organizational development, and leadership development are all common things in the business world.

The term is also used in cultural fields, such as in the development of music art, composition, and even character development in the acting world. People often use "development" to discuss social maturity, growth, community, rural, social, regional, or personal.

There are clear rules about what R&D is and what it is not. In particular, the Frascati Manual [5] is a collection of knowledge that clearly and carefully explains what R&D is and what it is not. This handbook compiles global data on R&D and innovation, so the definitions are solid and consistent. This useful resource illustrates how the rest of the world perceives each occupation. Typically, "R&D" refers to creating new services, processes, systems, materials, components, and products. R&D is a methodical effort using novel, untested concepts in a new field. A commercial, social, technological, or scientific advantage is typically anticipated from a development endeavor, although it may only occur for a while or even decades.

1.4 R&D Methods

In the ever-evolving realm of research and development (R&D), the efficacy of methods employed is paramount. This chapter delves into the intricate world of R&D methods, shedding light on the diverse approaches, techniques, and tools that fuel the process of innovation. Here, we explore the methodologies that underpin successful R&D endeavors, providing a comprehensive understanding of the scientific and systematic approaches that drive progress and transformation. With the right methods at one's disposal, R&D projects can yield not only groundbreaking results but also the competitive advantage that organizations seek in a knowledge-driven era.

1.4.1 Research Methods

There are many different research methods, and many are based on what is most commonly done in a certain field of discipline or study. How people argue about the best methods for each situation depends on many things. Each of these research methodologies is supported by voluminous evidence and study. To comprehend this material, however, you can handle these topics without extensive knowledge. It is important to know, though, that these different ways of looking at a phenomenon try to explain it or learn more about it. They either use qualitative, quantitative, or a mix of the two. They also show whether something is subjective or objective. So that everything is clear, let us talk more about each of these approaches.

A *quantitative study* poses a specific issue that may be addressed via numerical or statistical data collection. Beginning with a broad query about a topic, qualitative research collects evidence through pictures or words, which are then subjectively analyzed. Some research projects work better in an environment that is more

1.4 R&D Methods

subjective and qualitative, like arts and crafts. Some scientific research projects work better in a more objective and quantitative environment. Subjectivity is a point of view that is not based on concrete evidence. Objectivity is a result that is not based on the researcher's interpretations [6]. Figure 1.4 shows how different methods, whether objective or subjective, quantitative or qualitative, usually fit with different kinds of research.

An explanatory study examines an existing occurrence or hypothesis and attempts to understand x and its causes. Exploratory research seeks the appropriate research questions. On the other hand, an explanation study attempts to determine why something is as it is without altering the variables. Formative research is adding shape or form to an idea or concept. These two forms of research are often done in the social sciences. Constructive research aims to give solutions to the stated issues, while empirical research aims to evaluate whether or not the proposed solutions are practical [7].

The exploratory study involves looking into something that needs to be better understood. Researchers or other interested people explore to find answers to specific questions or to narrow down the range of questions they might have. This helps them ensure they ask the "right" questions or the ones that could have the most impact. Sometimes, these research questions lead to more formal research projects. These sorts of fundamental research follow the pattern of asking questions to determine a research objective or how to conduct a study. This study does not necessarily result in a clear route but is often used to generate research topics or ideas. This very innovative process is driven by inquisitive individuals [8].

A mixed method may be utilized when a research issue is too complex to solve with a single kind of data or when a secondary analysis using a different approach is required. To fulfill the demands of the study, mixed methods use a variety of approaches for analyzing and collecting both qualitative and quantitative data. The data analysis and collection can be done at the same time. For example, two surveys

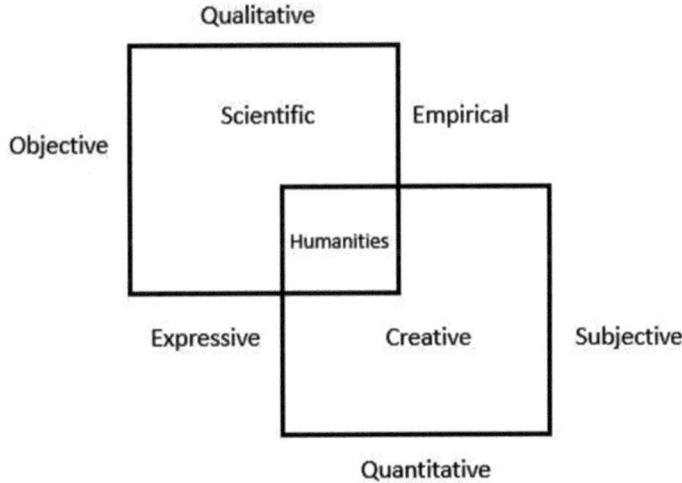

Fig. 1.4 Methods used for various study categories [3]

can be made and sent out simultaneously, and the results can be looked at together from different points of view. They may also be investigated and collected in a certain order utilizing quantitative and qualitative approaches set up in sophisticated ways to answer research questions that should be answered in a specific order [9].

Whether the study is qualitative and subjective, quantitative and objective, or a combination of the two, project management approaches may be used to assist. Many researchers, for instance, believe that the greatest approach to achieving the maximum degree of creativity is to have adequate time to study however they want. This is an essential creative thinking often required while generating or evaluating various ideas. It may also be useful when a study plan or approach cannot be expressed in words. When a researcher attempts to determine what is conceivable, having sufficient unstructured time to generate and evaluate many ideas typically leads to breakthrough concepts; thus, R&D should support this. Utilizing project management does not restrict exploratory research; it may assist in safeguarding it. However, you should try to set aside enough free time for this to happen. The bulk of researchers explore diverse ideas and products using well-established research methodologies. Even though they are not the only techniques for doing research, they are the most effective means of demonstrating how adopting project management approaches to conduct research activities may make them more helpful and aligned. Among these techniques:

- *The scientific method* is a quantitative way to explore.
- *The method of Socrates* is a method that is both exploratory and descriptive.
- *Mixed methods* combine qualitative and quantitative methods and methods for discovering and explaining.

Research can also go in different directions. Artistic and creative research, often called "practice-based research," does not use measurable, fact-based, or objective results. Instead, it relies on subjective things like judgments and interpretations. This means these tasks only need a little statistical analysis or a clear, objective problem statement. Creative research starts with a question. Well-thought-out questions lead to well-thought-out experiments, and the answers stand independently. The classical Socratic method of research, which is usually thought of as an inquisition style and has been linked to the Greek philosopher Socrates (469–399 BCE), uses creative deconstructive questioning to find the answer to a question. It requires questioning what is assumed and known right now. The modern Socratic technique employs a hypothesis as a plausible response. This response may initiate a sequence of tests or experiments designed to confirm or refute the hypothesis. When writing up the findings, putting up testing, and thoroughly determining the scope, project management techniques may aid both forms of Socratic questioning. In any situation, project management tools for change control may assist in demonstrating how research is progressing [10].

1.4.2 Apply Now

If the reader understands what activities are covered in applied and basic research and if an activity is best done using the Socratic, Scientific, or combined method, they can apply the suitable project management structure. The complex decision-making process faced by researchers when selecting research techniques, activities, and project management structures is depicted visually in Fig. 1.5 by a network of interconnected flowcharts. The crucial decision of study type—either "Basic" or "Applied"—starts the flowchart. The "Activities" phase then follows, when researchers choose between "Exploratory" and "Explanatory" research activities depending on the nature of their study. Researchers now have a choice between two unique approaches: the "Socratic Method" and the "Scientific Method." The former entails qualitative and arbitrary evaluations and gives the option to select either a "Classical" or "Contemporary" project management strategy. While the "Scientific Method" offers the option of either "Exploratory" or "Explanatory" activities, setting the corresponding project management structure, it tends toward quantitative and objective measurements.

There are many ways to set up an experiment. They also depend on the question to be answered by the research. However, here is a quick list of the different kinds of tests and experiments:

- An interview or a question (including surveys)
- Empirically, by looking at things
- Sampling (stratified, systematic, and random)
- Case studies or pilot
- Simulation and modeling

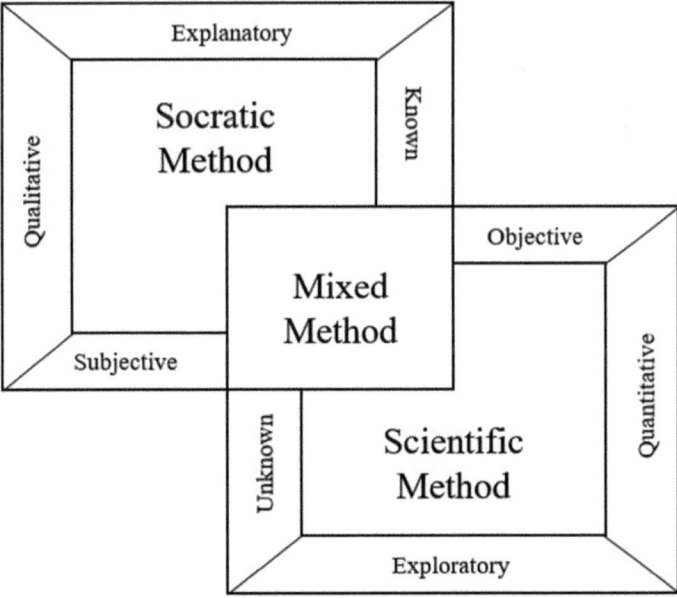

Fig. 1.5 Flowchart of research methods [3]

1.4.3 Research Steps

Organizing research tasks into meaningful groups is essential to effectively apply project management techniques in research. This organization helps streamline the research process and ensures that key activities are appropriately managed. Three common research methodologies—Socratic, Scientific, and mixed methods—can all benefit from such task grouping. Moreover, it is important to note that these tasks are often executed iteratively and flexibly, as illustrated in Fig. 1.6 [11].

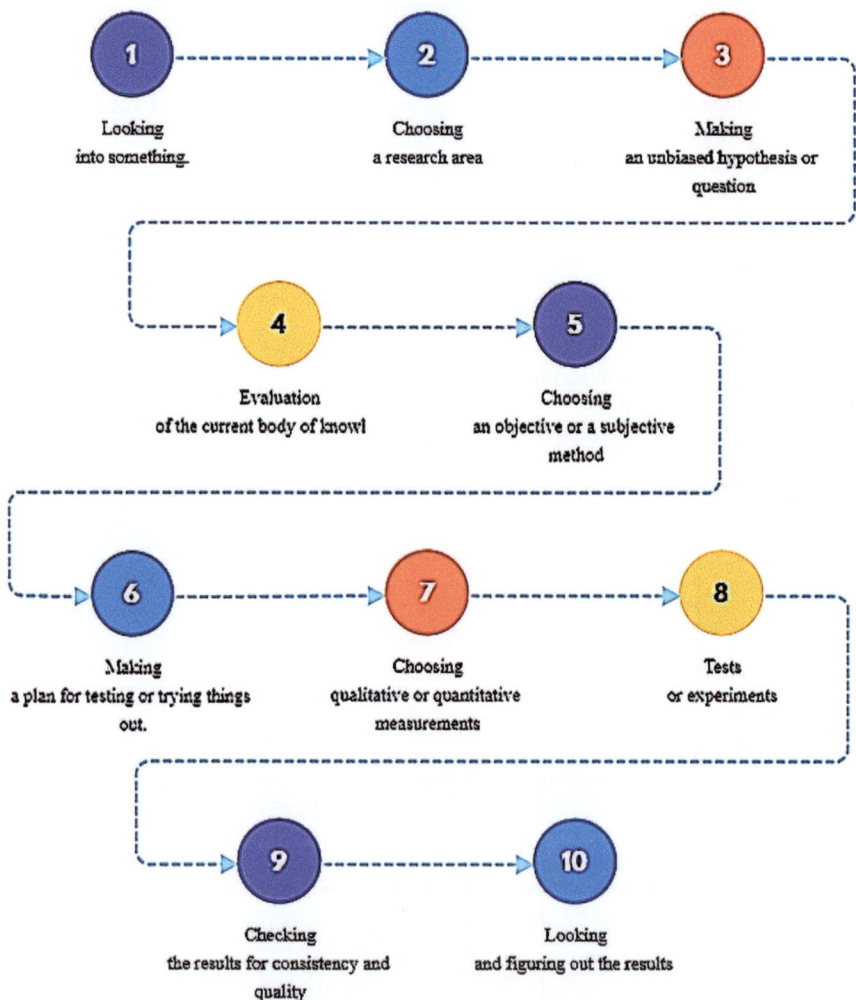

Fig. 1.6 Research: step by step

Indeed, R&D activities encompass various tasks and processes for innovation and improvement. These tasks are essential in R&D to determine feasibility and functionality, as stated in Ref. [3].

Building Models or Prototypes. Creating scaled-down or preliminary versions of products or systems to test concepts and ideas before full-scale development.

Designing or Re-designing Processes. Developing efficient and effective workflows and procedures to enhance productivity and product quality.

Running and Making Simulations. Utilizing computer-based simulations and modeling to simulate real-world scenarios and assess the performance of various solutions.

Conducting Experiments. Carrying out controlled experiments to gather data and test hypotheses, often involving variables and controlled conditions.

Testing, Validating, and Verifying Designs. Rigorously evaluating prototypes or designs to ensure they meet specified requirements and perform as expected. This involves validation to confirm they address real-world needs and verification to ensure they meet design specifications.

These activities are fundamental in the R&D process, pushing the boundaries of knowledge and technology and ultimately creating new or improved products, processes, and innovations.

1.4.4 Development Methods

The Frascati Manual [12] says that operational systems development, manufacturing development, demonstration (test), engineering, validation, and applied and basic research are all included in the R&D category. Even though it might be difficult to distinguish between research and development, there is a continuum of work done in the development area. The earlier an action is in the development process, the less is known about how to bring about the desired change. The later it is in the development process, the closer it is to being ready for implementation. It is sometimes difficult to distinguish between R&D and preproduction. Most of the time, "preproduction" refers to nonexperimental tasks to prepare for production runs. People often mix preproduction and R&D because preproduction tasks, like operational systems development and manufacturing development, seem like R&D tasks.

Table 1.2 depicts the spectrum of activities between late and early development. This approach assists in determining whether a development activity is ready to go to the next level, which might be implementation or preproduction. When deciding where to draw the lines, a helpful metric is evaluating whether the process, model, or prototype has matured enough for verification and testing.

Once a successful testing phase has concluded and the findings have been validated to match what was anticipated, it should be ready to proceed into an implementation or production environment and no longer be called R&D. In practice, the distinction between the two may be blurred. An R&D effort might result in a design

Table 1.2 The path between early and late development

Items	Early development	Late development
Techniques and methods	Undefined	Established
Prototype and simulations	None	Developed
Question/theory	Unclear	Clear
Validation/verification	None	Completed
Replication readiness	Not ready	Ready

that has been demonstrated and tested to work. Most of the time, preproduction follows predefined standards and procedures to create a duplicate of the final prototype. R&D services and products are often moved into the preproduction phase before being tested and ready. As a result, during setup, a severe design flaw may emerge that should be addressed before creating many component duplicates. Changes that are required are implemented outside of the traditional R&D environment. People working on the design of the component and any adjustments that need to be made should collaborate closely. Although it is optional to be able to distinguish between preproduction and R&D to employ project management methods, it is crucial to understand the purpose of each activity.

In process implementation and development, the same thing can happen. A process can be proven and tested to work. However, a flaw is found when the first operations are implemented. This happens sometimes because a test environment cannot always match the real world. Sometimes, this happens because stakeholders who were not known before find a missing part of the process that affects them. After the R&D phase, a change should be made, which could mean reengineering the design or just making a small change to the process. The manager in charge of putting operations into place will decide if it needs to go back into the R&D process or if it can be changed in the operational environment.

Whatever the kind of project, it is critical to identify the intended output and the boundary between R&D and the next life cycle phase to split the work into manageable work packages. So that forward pricing is accurate, it is also important to know how much work and money goes into R&D. Forward pricing means estimating the total cost of development for future bids [13].

1.4.5 Development Steps

At the R&D stage, people usually do things like peer reviews, make prototypes, and talk to each other. These activities start to be run more formally, but the project manager is still in charge of keeping track of progress and being responsible for it. Most of the time, this creative process is managed by putting numbers on how well the hypotheses are being proven and the prototypes that come from those proofs. If everything goes well in this phase, the project is ready to move on to the implementation or preproduction phase. To employ project management techniques in R&D,

1.4 R&D Methods

activities should be organized logically, much as in the research stages, so they can be mapped to project management methods. Some activities can be put into general groups from the beginning to the end of R&D. As with research, these things move flexibly iteratively and often return to where they started. This flexible loop-back method should be made possible by the project management methods that will be used (Fig. 1.7) [3].

1.4.6 The Distinction Between R&D and Other Projects

The enormous number of tries and unknowns to manage in R&D initiatives and their indirect, roundabout evolution truly sets them apart from other projects. R&D demands a high level of invention and an accelerated pace of projected change due to the ever-evolving landscape of technology and innovation. Uncertainties,

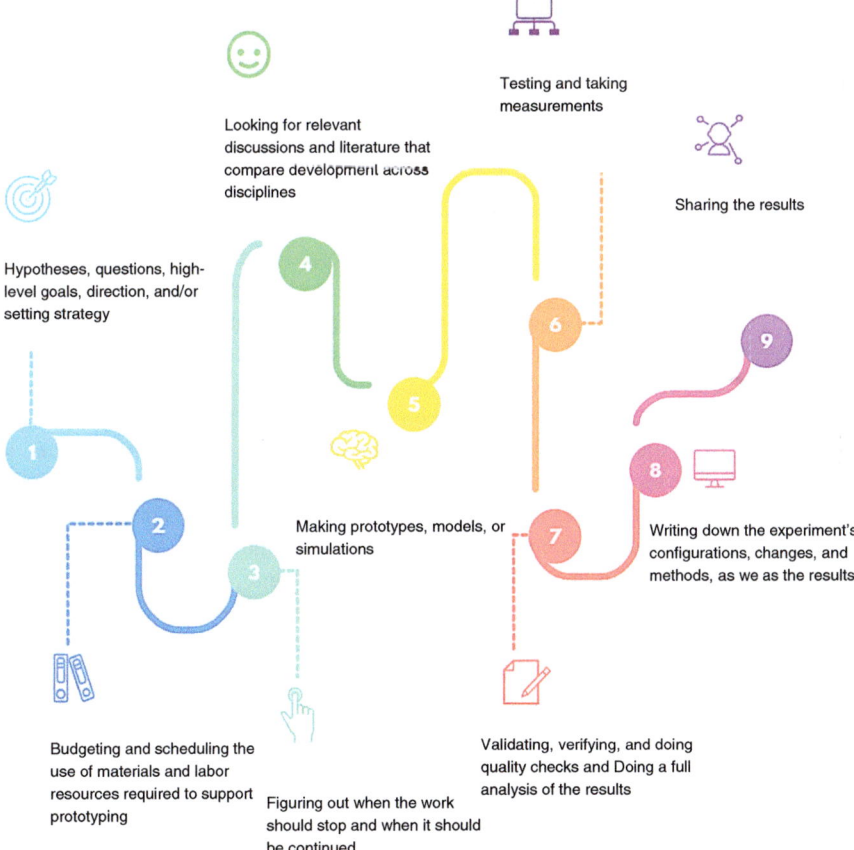

Fig. 1.7 Development: step by step

however, should not be viewed as insurmountable obstacles but rather as opportunities for effective management. Recognizing these unique aspects is crucial as it enables the deployment of tailored project management strategies that not only navigate the complexities but also amplify and accentuate the chances of success in R&D projects. Such an approach often yields superior results compared to a hands-off, laissez-faire approach, which may need to be revised to manage the dynamic nature of R&D initiatives.

Specific R&D characteristics can serve as crucial control points in managing these endeavors effectively. For instance, acquiring information in R&D often demands a purposeful and methodical approach, particularly when experimentation is necessary. To ensure that the obtained answers are meaningful, it is essential to specify and, to some extent, control the factors being assessed. Typically, research in the R&D domain requires a thorough, comprehensive, and impartial inquiry, untainted by the investigator's prior beliefs about the expected outcome. Furthermore, R&D should be presented to permit others to replicate the findings using the same techniques, thereby establishing credibility and providing a solid platform for future research and development [14]. This emphasis on transparency and replicability reinforces the scientific rigor of R&D and fosters an environment of collaborative innovation, enhancing the overall success of these groundbreaking projects.

1.5 R&D Challenges

R&D has played a pivotal role in significantly improving the quality of life, especially over the past five decades. Vital sectors such as the environment, information technology, healthcare, transportation, and communication have all reaped substantial benefits from the relentless pursuit of innovation and R&D activities in many laboratories and research institutes worldwide. However, it is important to acknowledge that sustaining and initiating a research and development program comes with its fair share of challenges. As Pavitt astutely noted in Ref. [15], "In many areas, it is unknown before the event who will be participating in the innovation race, where the starting and finishing lines are, and what the race includes." This profound statement encapsulates the inherent unpredictability and complexity associated with R&D endeavors. Numerous publications have delved into different facets of the unique challenges of R&D initiatives. For example, von Zedtwitz and colleagues, in their work referenced as [16], have identified ten impediments to effective innovation management. These impediments illuminate the intricacies of navigating the innovation landscape and managing R&D efforts in a competitive and rapidly evolving world.

Furthermore, a recent study by UNESCO, cited as [17], sought to quantify and pinpoint R&D difficulties that are particularly pertinent to developing countries. This research underscores the global nature of R&D challenges and highlights the need for tailored strategies to foster innovation and technological advancement in regions with varying resources and capabilities. While the impact of R&D on our lives is undeniable, it is essential to recognize and address the multifaceted challenges that researchers, innovators, and organizations encounter when embarking on the research and development journey. By understanding and surmounting these challenges, we can continue to harness the transformative power of R&D for the betterment of society and the world.

1.5.1 The Functional Tensions Challenge

The inherent conflict between two competing components is one of the greatest hurdles to completing a standard R&D project on time and within budget. On the one hand, there is a softer, more creative R&D culture, but on the other hand, there is a much more concentrated approach to project management. The project manager is entrusted with serving as a mediator between representatives of varied cultures whose natures are influenced by diametrically opposing influences. A businessperson is motivated by monetary gain, but a researcher is driven by personal ingenuity and curiosity.

This tension is substantiated by empirical evidence as well. Von Zedtwitz et al. [15] interviewed managers from 18 multinational businesses in the late 1990s. They identified 11 tension-inducing barriers between business management and R&D. Managing human resources and knowledge was a major obstacle. R&D managers supervised very creative knowledge-based resources. In contrast, scientists were more committed to their field than the rigor of project management methods. Moreover, there were creative tensions and goal conflicts between global program objectives and local effectiveness, as well as between project management techniques and R&D hierarchy. A third location of friction between discipline and creativity was uncovered.

Intriguingly, the authors assert that these tensions and challenges are not intrinsically negative. Establishing an integrated program and project management in R&D was one of their solutions for overcoming these tensions. The conflicts above, which were identified in the late 1990s, are still prevalent, and the idea to expand the use and understanding of project management in R&D remains relevant. A study by Alik [14] revealed the requirement for a project manager who could mediate between R&D professionals and the external nonacademic world to address functional concerns. These functional stresses provide a difficult task to solve. Here, the regulation of intellectual pride, result-oriented motivation, and human emotions are discussed. As shown by the empirical study, one method is to enhance the knowledge and understanding of project management techniques among R&D managers. In addition, project managers should be made aware of the complexity and additional risks R&D managers confront. When R&D leadership places project managers and R&D managers in shared facilities and offices, the likelihood of attaining harmony increases.

1.5.2 Globalization Challenge

Globalization is undeniably a significant "megatrend" that will continue to shape the world in the coming decades. Its impact on various aspects of society, including R&D, is profound. This impact is a complex interplay of advantages and disadvantages. Over the past decade, as multinational firms expanded their R&D efforts globally, they had high expectations for the outcomes of their international product and research development divisions. However, challenges have emerged in the context of global R&D programs, as identified by the CREST Working Group [18].

Globalization can sometimes clash with the original motivations for R&D. Companies may need to work on their global expansion goals and core values or objectives. Protecting intellectual property becomes more challenging in a globalized R&D landscape. Different regions may have varying levels of intellectual property protection, potentially leading to IP theft or disputes. While globalization enables the sharing of knowledge and expertise worldwide, it can result in imbalances in the flow of knowledge. Certain regions may become knowledge hubs, while others face knowledge deficits. There has been a noticeable shift in FDI in R&D from traditional powerhouses like the United States and Europe to other regions, notably Asia. This relocation reflects the changing global economic landscape.

In response to these challenges, companies such as General Electric have adopted strategies like "reverse innovation." This approach involves developing product designs in underdeveloped nations, often to address the needs of emerging markets. By doing so, companies can tap into local expertise and market insights, ultimately fostering innovation and competitiveness on a global scale. While globalization offers opportunities for R&D expansion and collaboration globally, it also presents challenges related to motivation, intellectual property, knowledge flow, and shifting investment patterns. Adapting to these challenges and embracing innovative strategies like "reverse innovation" can help organizations thrive in an increasingly interconnected and competitive world.

1.5.3 Attracting Talent Challenge

In the realm of R&D firms, recruiting candidates with the requisite competencies and skills poses a dual challenge. First, R&D roles demand domain-specific knowledge, making it unsurprising that research laboratories tend to hire a higher proportion of individuals with PhDs compared to other departments. However, to enhance the likelihood of generating groundbreaking product concepts, it is imperative for businesses to also identify and nurture creative qualities throughout the entire hiring process. Engaging high-potential project leaders in overseeing the various stages of an R&D program is equally crucial. While recruiting qualified professionals and scientists may be relatively straightforward, it proves considerably more challenging to identify and attract innovative thinkers and high-performers who can thrive in the dynamic and creative environments that R&D settings often embody. The second hurdle lies in the "competence extinction" [19]. In situations characterized by significant technological uncertainty, businesses aiming to pioneer highly innovative or untested concepts (similar to the "dot-com bubble" companies of the past) may initially need to recruit individuals with well-established competencies. These competent employees provide the necessary expertise to navigate complex terrain. However, the inherent unpredictability of such ventures can lead to unforeseen challenges, potentially resulting in workforce layoffs.

Balancing the need for domain-specific knowledge and technical expertise with the imperative for innovation and adaptability is a central challenge in R&D hiring. Organizations must be prepared to strike this delicate equilibrium, recognizing that the ideal candidate for R&D roles possesses the know-how and the creative flair required to push the boundaries of knowledge and drive forward transformative initiatives.

1.5.4 Managing Risks Challenge

R&D is inherently defined as exploring new opportunities, often in uncharted territory, and inevitably involves several risks. These risks encompass a range of uncertainties, including ambiguous outcomes, uncertain timescales, and unclear costs. However, there are other, less obvious hazards associated with R&D endeavors, one of which is highlighted by Trott [19] as "appropriability risk." Appropriability risk revolves around the vulnerability of a newly created product to imitation by competitors. This risk is intricately linked to intellectual property management, typically safeguarded through copyright, trademarks, and patents. In industries where intellectual property protection is more susceptible to risks, such as fast-moving consumer goods, companies often lean toward strategies like line extensions and cumulative technologies that can protect against appropriation.

Furthermore, integrating diverse organizational cultures, including service, development, and research, can introduce significant risk factors. Effective risk management in R&D requires a collaborative approach that involves product managers, accountants, engineers, designers, and scientists. Through this collaborative effort, an organization can tackle quality, cost, and time risks as a unified team.

It is essential to recognize that R&D risk is not isolated but intricately tied to an enterprise's broader risk profile. The dangers, opportunities, and threats a company faces in the market are closely interlinked with those encountered by the R&D team. As such, organizations must adopt a holistic approach to risk management, considering the intricate relationship between R&D initiatives and the overall health and success of the business and the economic landscape.

1.5.5 Managing Scientific Freedom Challenge

The cultural dynamics within R&D, especially between scientists and management, run deep and present a unique challenge. This challenge stems from the perception that formal control measures, such as project planning and cost control, can stifle free thinking and hinder the "spirit of creativity" essential in scientific pursuits. While fostering an environment that nurtures creativity is crucial, it is also imperative to acknowledge that R&D departments operate within finite resource constraints and need to adhere to specific cycle times. This cultural friction could lead to a division within R&D, resulting in the emergence of two distinct organizations. On the one hand, there is the assumption that engineering should be the domain of scientists, while development should be reserved for researchers. However, this distinction is often arbitrary unless supported by genuine experience and knowledge. Fortunately, R&D managers are increasingly cognizant of the imperative for R&D success in a fiercely competitive market. There is a recognition that scientific inquiries often require a certain amount of time to yield meaningful results. Therefore, the primary objective of addressing this cultural challenge is to find a delicate balance—enhancing the effectiveness of R&D initiatives by applying project management processes and tools while safeguarding the spirit of innovation and allowing sufficient time to pursue breakthroughs. In essence, the challenge is to create an environment where project management practices complement rather than hinder the creative and innovative endeavors of R&D scientists. It is about finding the right equilibrium between structured management and the unhindered pursuit of scientific discovery, all while recognizing the need for efficiency and competitiveness in today's rapidly evolving world.

1.5.6 Funding R&D Expenditure Challenge

Developing a robust business case can be a formidable challenge when initiating an R&D activity, particularly in basic research. The difficulty arises from the inherent uncertainty and long-term nature of many research endeavors, making it challenging to predict the outcomes and quantify the potential returns accurately. In contrast, an initial cost–benefit analysis for creating a new product within an existing product line, based on prior experience, can be relatively straightforward to construct. This is because it involves incremental improvements or modifications to existing offerings, and historical data and market trends can provide valuable insights into potential outcomes.

The challenge of securing the necessary financing for R&D projects is a significant one. Funding for R&D activities often comes from various sources, as below.

Risk-Taking Investors. Investors are willing to take on the inherent risks of R&D projects, expecting significant returns if successful.

Trust Funds. Some R&D initiatives may be supported by endowments or trust funds established for research purposes, particularly in academic and nonprofit sectors.

Individual Businesses. Companies may allocate internal funds to finance R&D projects as part of their innovation strategies, especially when the R&D aligns with their long-term goals.

Industries. Trade associations and industry consortiums may pool resources to fund collaborative R&D efforts that benefit the entire sector.

Governments. Government agencies, such as grants from national research bodies or incentives for R&D investments, play a vital role in supporting research initiatives. Governments sometimes act as venture capitalists, directly investing in promising R&D projects.

Venture Capitalists. Specialized venture capital firms focus on funding early-stage start-ups and high-risk, high-reward R&D ventures.

Each funding source comes with its expectations, requirements, and risk profiles. Successfully securing financing for R&D projects often involves crafting a compelling business case demonstrating the potential for innovation, market impact, and a clear path to achieving desired outcomes. It also requires effective communication to convey the value and feasibility of the R&D initiative to potential investors or funding organizations.

References

1. (OECD), O.f.E.C.-o.a.D, *The Measurement of Scientific and Technical Activities* (OECD Publication, Paris, 1993)
2. H. Brooks, Definitions, concepts, themes, in *Managing the Environment: Report*, National Academies, United States vol. 2, (1968), p. 21. https://books.google.ca/books?id=BTcrAAAA YAAJ&dq=brooks+1967+applied+sciences&lr=&source=gbs_navlinks_s
3. L.M. Wingate, *Project Management for Research and Development: Guiding Innovation for Positive R&D Outcomes* (CRC Press, Boca Raton, 2014)
4. H. Borgdorff, *The Debate on Research in the Arts*, vol 2 (Kunsthøgskolen i Bergen, Bergen, 2006)
5. F. Manual, *Organisation for Economic Co-operation and Development* (Organisation for Economic Co-operation and Development, Paris, 1993)
6. K. McCusker, S. Gunaydin, Research using qualitative, quantitative or mixed methods and choice based on the research. Perfusion **30**(7), 537–542 (2015)
7. R.W. Scapens, Researching management accounting practice: The role of case study methods. Br. Account. Rev. **22**(3), 259–281 (1990)
8. J. Iden, L.B. Methlie, G.E. Christensen, The nature of strategic foresight research: A systematic literature review. Technol. Forecast. Soc. Chang. **116**, 87–97 (2017)
9. J. Brannen, Combining qualitative and quantitative approaches: An overview, in *Mixing Methods: Qualitative and Quantitative Research*, (Routledge, Abingdon, 2017), pp. 3–37
10. L.M. Wingate, Bounding the creative spaces, in *Project Management for Research and Development*, (Auerbach Publications, 2014), pp. 148–201
11. J. Passmore, A. Fillery-Travis, A critical review of executive coaching research: A decade of progress and what's to come. Coaching: Int. J. Theory Res. Pract. **4**(2), 70–88 (2011)

12. F. Manual, *Proposed Standard Practice for Surveys on Research and [43_TD $ DIFF] [37_TD $ DIFF] Experimental Development* (Organisation for Economic Co-Operation and Development, Paris, 2002)
13. E. Baydoun et al., An overview of research and development in academia, in *Higher Education in the Arab World: Research and Development*, (Springer, Cham, 2022), pp. 13–37
14. J. Allik, Quality of Estonian science estimated through bibliometric indicators (1997–2007). Proc. Estonian Acad. Sci. **57**(4), 225 (2008)
15. K. Pavitt, What we know about the strategic management of technology. California management review, **2**(3), 17–26 (1990)
16. M. Von Zedtwitz, O. Gassmann, R. Boutellier, Organizing global R&D: Challenges and dilemmas. J. Int. Manag. **10**(1), 21–49 (2004)
17. UNESCO, U., *Measuring R&D: Challenges faced by developing countries*. UNESCO Institute for Statistics PO Box, 2010. 6128
18. C. W. Group, Internationalisation of R&D–facing the challenge of globalisation: approaches to a proactive International policy in S&T(Brussles: CREST, 2007)
19. Trott, P., *Innovation management and new product development* (Pearson education, 2008)

Chapter 2
The Role of R&D in Business

This chapter delves into the critical importance of research and development (R&D) in contemporary business strategies. R&D is not merely a functional unit within a company but the driving force behind sustainable growth, innovation, and competitive advantage. This chapter begins by elucidating the strategic significance of R&D, emphasizing its role as a catalyst for business expansion. It explores how R&D efforts directly influence innovation and competitiveness, discussing real-world examples where innovation arising from R&D has transformed industries. Managing R&D investments is vital for success, and this chapter dissects the intricacies of budget allocation, balancing short-term and long-term commitments, and evaluating return on investment in R&D projects. Effective R&D management extends beyond financial considerations. This chapter examines the critical aspects of building high-performing R&D teams and safeguarding intellectual property, offering insights into collaborative practices and protection strategies. Measuring R&D performance is another essential facet, exploring key performance indicators, benchmarking, and evaluating process efficiency. Finally, this chapter ventures into the future of R&D in business, predicting emerging trends and digital-age adaptations while preparing businesses for disruptive technologies. It provides a comprehensive overview of how businesses can harness R&D to fuel sustainable growth, adaptation, and innovation in an ever-evolving marketplace.

Learning Objectives

- Understanding the strategic significance of R&D
- R&D investment management
- Building high-performing R&D teams and intellectual property protection
- Measuring R&D performance
- Future trends in R&D and business adaptation

2.1 The Strategic Importance of R&D

R&D commands a pivotal and formalized role within the corporate framework, signifying a critical enabler of strategic direction for contemporary enterprises. In this section, we embark on a comprehensive exploration of the multifaceted and strategic significance of R&D in shaping the trajectories of modern businesses. At its core, R&D embodies the essence of innovation, the crucible where ideas are conceived, refined, and ultimately transmuted into pioneering products, services, and solutions. For businesses, this innovation-centric paradigm fundamentally influences their ability to remain competitive and, crucially, to adapt with agility to the shifting dynamics of the marketplace. The strategic importance of R&D is unmistakable, offering businesses the capability to respond to external change and actively steer it.

In the realm of competition, R&D emerges as the clandestine weapon in an enterprise's arsenal. Companies that make strategic R&D investments are strategically poised to discern emerging market trends and anticipate evolving consumer needs. Armed with this foresight, they can conceive and deliver cutting-edge products and services, establishing themselves as vanguards within their industries. The strategic advantage engendered by R&D transcends mere market presence; it is, fundamentally, about setting the pace. R&D transcends transient interests; it encompasses a long-term perspective. It serves as the engine behind sustainable growth and resilience. By perpetually refining existing products and optimizing processes, R&D equips businesses to weather industry disruptions and challenges while preserving relevance and adaptability. These investments extend beyond sustaining current operations; they are the foundations for future revenue streams.

Furthermore, the strategic import of R&D extends to facilitating exploration and expansion. The fruits of breakthrough technologies, innovative solutions, and distinctive products pave the way to unexplored markets and global growth prospects. The strategic role of R&D is not confined to upholding the existing order; it represents the impetus for continuous growth and diversification.

The strategic importance of R&D lies at the heart of innovation, engendering competitiveness, assuring sustainable growth, and affording businesses the means to perennially succeed. Enterprises that embrace R&D as a strategic imperative are equipped to navigate the intricate terrain of the ever-evolving business landscape. They are positioned to seize untapped opportunities and maintain a pioneering stance within their respective industries.

The pivotal role of R&D in driving business growth has been extensively acknowledged within academic and corporate discourse. This section elucidates the nexus between R&D endeavors and the amplification of business growth, providing a comprehensive academic perspective on the subject.

Innovation Propagation
One of the central mechanisms through which R&D fuels business growth is the propagation of innovation. R&D departments act as the crucible for innovative

concepts, converting nascent ideas into tangible products or services. This innovation cycle is fundamental for business growth, as it empowers companies to meet existing market demands more effectively and enables them to create entirely new market segments. Such innovative products and services catalyze revenue expansion and bolster a company's market position [1].

Competitive Advantage

R&D plays a pivotal role in fostering competitive advantage. Firms that invest strategically in R&D often find themselves at the vanguard of technological advancements and market trends. This positions them as early movers in identifying and catering to evolving customer needs, thereby staying ahead of competitors. The competitive advantage emanating from R&D is grounded in its ability to offer products or services that are superior in quality, cost-effectiveness, or innovation, thereby capturing a larger market share.

Market Expansion

The strategic application of R&D enhances the existing product portfolio and broadens the scope for market expansion. By developing new products, tapping into unexplored market segments, or penetrating global markets, businesses can experience exponential growth. R&D investments open doors to diversification and enhance the brand's footprint, contributing to revenue diversification and broader customer reach.

Operational Efficiency

R&D can optimize internal processes and systems, leading to improved operational efficiency. Companies can streamline their operations, reduce costs, and enhance productivity by researching and developing innovative methods, technology, or automation solutions. This efficiency positively impacts profit margins and, consequently, business growth.

Enhanced Stakeholder Value

Businesses that demonstrate a strong commitment to R&D often garner increased investor and stakeholder confidence. A robust R&D strategy signals a company's dedication to continuous improvement, which, in turn, can enhance stock prices, attract investment, and boost market capitalization.

In conclusion, the academic exploration of R&D as a driver of business growth underscores its multifaceted impact on revenue expansion, competitive advantage, market diversification, operational efficiency, and stakeholder value. A strategic commitment to R&D not only signifies a dedication to innovation but is also demonstrative of a proactive approach to navigating the dynamic terrain of contemporary business landscapes. Academic research in this field underscores the vital importance of R&D in ensuring enterprises' sustainable and prosperous growth.

2.1.1 R&D and Competitive Advantage

The dynamic synergy between R&D and competitive advantage is a topic of significant scholarly interest and practical import within strategic management. This section scrutinizes the intricate and pivotal connection between R&D activities and the attainment of a competitive advantage, elucidating the multifaceted dynamics at play. At the core of the R&D and competitive advantage nexus lies the intrinsic ability of R&D to foster innovation. R&D functions as an innovation engine, continually generating novel ideas, products, and processes. Such innovations, whether incremental improvements or disruptive transformations, bestow organizations with a distinctive competitive edge. Companies can outpace their rivals and elevate their market positioning by consistently introducing new or improved offerings.

The proactive nature of R&D initiatives equips companies with the agility to swiftly respond to evolving market conditions and shifting consumer preferences. Firms that commit to R&D investments can adapt their product or service portfolios to align with emerging trends, preserving their relevance in the eyes of consumers. This adaptability is paramount in securing a competitive advantage as it empowers organizations to seize opportunities and mitigate threats with precision. R&D investments can culminate in the development of cost-effective processes or technologies. This innovation-driven cost efficiency can significantly reduce production costs and enhance supply chain operations, ultimately translating into competitive pricing. Companies that leverage R&D to optimize operations often enjoy a distinct cost advantage, which can be converted into augmented profit margins and a competitive foothold in the market.

R&D investments extend to brand differentiation. Businesses that consistently introduce innovative products or services through R&D foster a robust brand identity associated with quality and innovation. This brand equity can be a formidable competitive advantage, as customers are likelier to choose products or services from a trusted, innovative brand. In several cases, the perceived value of a brand can become a decisive factor in market competition, enabling a premium pricing strategy [2].

The examination of the R&D and competitive advantage relationship underscores its dynamic and multifaceted character. R&D activities, through their capacity for innovation, market responsiveness, cost efficiency, brand differentiation, and potential barriers to entry, contribute significantly to attaining and preserving a competitive advantage. An intricate understanding of this relationship informs strategic decision-making, underscoring the strategic significance of R&D in the contemporary landscape of business competition.

2.1.2 Aligning R&D with Business Goals

Aligning R&D with the strategic goals of a business is a crucial endeavor in the contemporary business landscape. The efficacy of this alignment can significantly impact a company's success and market positioning. At the heart of this alignment lies a clear and purposeful strategic focus. By identifying and articulating overarching business objectives, companies establish a framework within which R&D activities can operate. Whether the goals revolve around expansion, cost efficiency, market share, or technological leadership, aligning R&D efforts with these goals ensures that the organization's resources are concentrated on endeavors that actively contribute to realizing these objectives (Fig. 2.1).

Effective alignment between R&D and business goals is not unilateral; it demands robust cross-functional collaboration. Business leaders, marketing teams, and finance departments must work with R&D units to ensure that the products or innovations developed are in sync with market demands, financial constraints, and customer preferences. This collaboration serves as a two-way street, allowing business leaders to better understand the technological possibilities and limitations while enabling R&D teams to remain closely attuned to market needs. It also permits real-time adjustments to R&D strategies to ensure alignment with dynamic business objectives.

Assessment of alignment is an essential step in the process, often facilitated through establishing Key performance indicators (KPIs) and metrics. These quantitative measures enable businesses to gauge the effectiveness of their R&D initiatives in contributing to desired outcomes. KPIs encompass innovation success rates, time-to-market for new products, or return on investment (ROI) for R&D projects. This data-driven approach allows for a systematic evaluation of alignment and provides insights for refining R&D strategies to better meet business objectives.

Equally important is the adoption of an adaptive strategy. In today's rapidly changing business landscape, business goals may evolve in response to shifting market conditions, technological advancements, or competitive pressures. An adaptive R&D strategy enables businesses to pivot and refocus R&D efforts in response to these changing dynamics. This flexibility positions organizations to protect their competitive edge, seize emerging opportunities, and navigate potential challenges effectively.

Resource allocation plays a pivotal role in the alignment process. It is incumbent upon businesses to ensure that sufficient financial, human, and technological resources are allocated to R&D projects that align most closely with the defined business objectives. Strategic resource allocation ensures that R&D efforts have the requisite support to deliver outcomes that align with strategic business goals. The prioritization of R&D projects according to their alignment with business objectives is a crucial facet of resource allocation [3].

Fig. 2.1 Aligning R&D with business goals

The alignment of R&D with business goals is an essential and dynamic aspect of modern strategic management. It necessitates a well-defined strategic focus, fostering cross-functional collaboration, employing KPIs and metrics, an adaptive strategy, and resource allocation. When R&D initiatives align with overarching business objectives, the organization is better equipped to achieve its growth targets, bolster its competitive position, and adapt to the ever-evolving business landscape. This alignment represents a cornerstone of sustained success in today's highly competitive and ever-changing business environment.

2.2 R&D's Impact on Innovation and Competitiveness

R&D is a cornerstone of innovation, driving businesses forward in highly competitive markets. R&D serves as the crucible where novel ideas are born and transformed into tangible innovations, shaping the trajectory of companies. It continually generates new concepts and refines existing ones, keeping businesses at the forefront of technological advancements and market trends. This catalytic role of R&D is the lifeblood of business evolution.

One of the primary impacts of R&D is its capacity to enhance market responsiveness. Companies strategically investing in R&D can swiftly adapt their product or service offerings to align with emerging trends and changing consumer preferences. This responsiveness is crucial for maintaining a competitive edge, allowing companies to seize new opportunities and mitigate threats effectively.

Moreover, R&D plays a pivotal role in fostering competitive advantage. Companies committed to R&D investments often find themselves at the forefront of technological advancements and market insights. This positions them as early movers in identifying and addressing evolving customer needs, enabling them to secure a strong competitive position. The competitive advantage arising from R&D lies in offering products or services that are superior in quality, cost-effectiveness, or innovation, capturing a larger market share.

In highly competitive markets, product differentiation is key to success. R&D is instrumental in creating unique and distinctive products or services that stand out in crowded marketplaces. Through innovative solutions, advanced features, and superior quality, businesses can set themselves apart from rivals and command premium pricing. In essence, R&D's impact on innovation and competitiveness is profound, and it serves as a strategic cornerstone for businesses striving to excel in competitive landscapes, fostering market responsiveness, creating competitive advantages, enabling product differentiation, and enhancing productivity.

R&D is the epicenter of transformative innovations redefining industries and driving progress. The groundbreaking products and services it yields are the core of R&D's impact. These innovations span a wide spectrum, from technological breakthroughs like smartphones and electric vehicles to design enhancements that change how we live and work. R&D does not stop at products; it extends to innovative processes that optimize manufacturing, supply chain management, and operational procedures. By continually refining these processes, R&D increases efficiency, cost savings, and heightened productivity. Concepts like just-in-time inventory systems and agile project management methodologies are prime examples of how R&D's influence transcends products and extends to the mechanisms underpinning businesses.

Technological progress owes much to R&D. Cutting-edge technologies like 5G telecommunications, artificial intelligence, and renewable energy sources like solar panels and wind turbines have all emerged from dedicated research efforts. These innovations have disrupted existing industries and given birth to entirely new ones, reshaping the global technological landscape. In healthcare, R&D is a pivotal driver of medical breakthroughs. From life-saving medications to advanced medical devices, it has revolutionized patient care. Innovations such as gene therapy, precision medicine, and minimally invasive surgical techniques have drastically improved healthcare outcomes, providing countless individuals hope and quality of life [4].

R&D also plays a central role in sustainability. In an era of growing environmental concerns, R&D efforts are focused on creating eco-friendly materials, energy-efficient technologies, and renewable energy solutions. Innovations like electric vehicles, sustainable building materials, and waste-to-energy technologies are testaments to R&D's commitment to addressing pressing environmental challenges and promoting a more sustainable future for all.

R&D stands as a formidable driver of market disruption, wielding the power to reshape industries and redefine business landscapes. The cornerstone of this transformative capacity lies in the development of innovative solutions. R&D teams are the architects of game-changing innovations in new products, services, or technologies. These innovations challenge conventional market dynamics, offering fresh approaches and superior value propositions that can quickly erode the foundations of established market players. Market disruption often finds its epicenter in the emergence of disruptive technologies. R&D plays a pivotal role in conceiving and developing these groundbreaking technologies. The smartphone revolution that altered the telecommunications and consumer electronics industries is a prime example, as is the disruptive potential of blockchain technology in redefining financial systems. R&D's capacity to drive technological innovation is fundamental to initiating and sustaining market disruptions.

However, R&D's role in market disruption is not limited to technological innovation. It extends to challenging existing business models. Companies with robust R&D capabilities often explore new ways of conducting business, offering fresh value delivery systems. E-commerce giants like Amazon redefined the retail industry by changing how consumers shop, challenging the traditional brick-and-mortar model. R&D drives these shifts, allowing businesses to experiment with alternative strategies and business paradigms.

Consumer-centric innovation, stemming from R&D efforts, is a significant force behind market disruption. Businesses that invest in understanding and meeting consumer needs can introduce products and services that directly address those demands. This approach enables them to gain market share quickly and redefine industry norms. Ride-sharing services like Uber and streaming platforms like Netflix emerged as direct responses to consumer needs, leading to transformative disruptions in the transportation and entertainment sectors. R&D also equips businesses to extend their global reach and compete internationally. Innovations from R&D efforts can transcend geographic boundaries, challenging established markets globally. Companies that leverage R&D to create products or services with universal appeal can enter new markets and provoke disruptive changes, impacting industries far beyond their domestic borders.

R&D occupies a central role in inciting market disruption. Its capacity to foster innovation, introduce disruptive technologies, challenge prevailing business models, and focus on consumer-centric solutions mark it as the driving force behind transformative change. Businesses investing in R&D position themselves as agents of market disruption and agile and resilient entities capable of navigating external disruptions. The strategic commitment to R&D has the potential to transform businesses and entire industries, emphasizing its pivotal role in shaping the future of markets and economies. Table 2.1 includes the company name, industry, key R&D innovation, and the transformation outcome for each case study.

2.3 R&D Budgeting and Investment

R&D budgeting and investment are integral to a company's innovation strategy. Strategic alignment, risk management, resource allocation, flexibility, performance evaluation, investment in talent, and ongoing monitoring are key elements in optimizing R&D budgets. Companies that master the art of R&D budgeting and investment are better positioned to drive innovation, remain competitive, and shape their industries' future (Fig. 2.2). Strategic alignment with a company's objectives forms the foundation for R&D budgeting and investment, ensuring resources are directed toward endeavors that contribute to the company's long-term vision. This alignment is essential for setting clear R&D objectives, whether developing new products, enhancing existing ones, or exploring new technologies. Balancing risk and

Table 2.1 Case studies demonstrating the transformative power of R&D in companies

Company name	Industry	Key R&D innovation	Transformation outcome
Apple Inc.	Technology	Development of iPhone and iOS	Transformed into a global tech giant, redefining consumer technology
Tesla, Inc.	Automotive	Electric vehicles and autonomous driving	Revolutionized the automotive industry with sustainable transportation
Amazon.com, Inc.	E-commerce	AWS	Expanded from e-commerce to a dominant player in cloud computing
Google (Alphabet)	Technology	Search engine, Android, and AI research	Dominance in online search, AI-driven services, and tech innovation
Netflix, Inc.	Entertainment	Streaming platform and original content	Redefined entertainment, changing how people consume content
SpaceX	Aerospace	Reusable rockets and Mars colonization	Pioneering commercial space exploration and satellite deployment
Pfizer Inc.	Pharmaceuticals	COVID-19 vaccine development	Critical role in the global pandemic response, life-saving innovation
IBM	Technology	The invention of the first computer and software	Pioneered the computer industry, leading in enterprise solutions

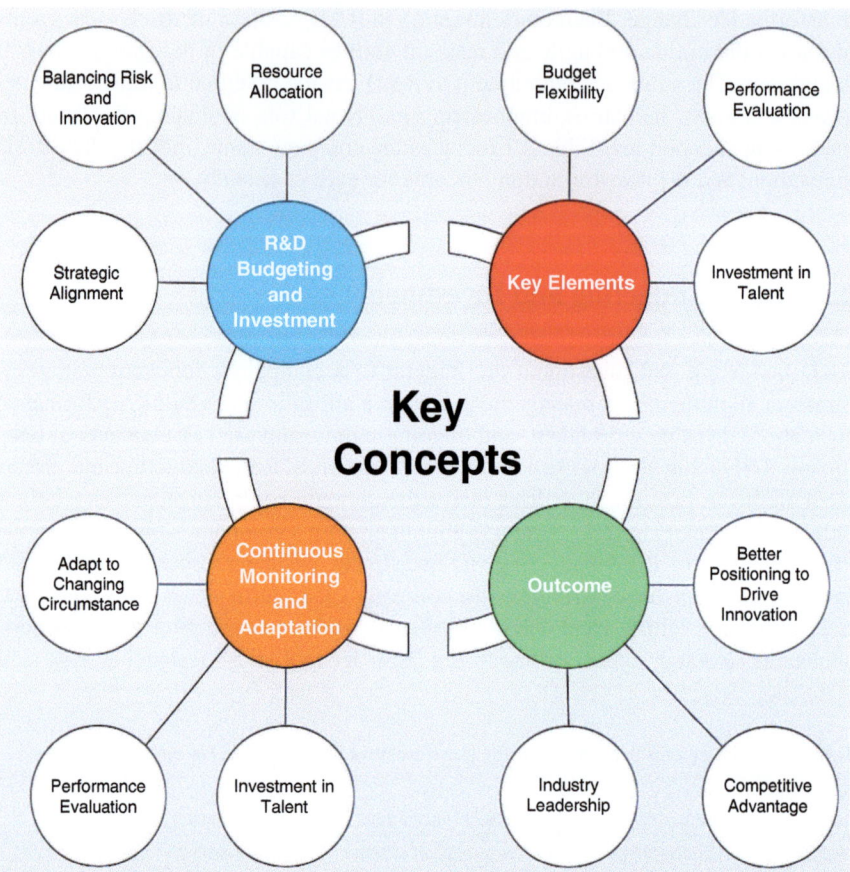

Fig. 2.2 Simplified flowchart illustrating R&D budgeting and investment

innovation is a crucial aspect of R&D budgeting. It requires walking the fine line between stifling innovation through conservative investment and overcommitting to risky ventures. A well-considered budget takes calculated risks, exploring new opportunities while ensuring that investments are managed prudently. Resource allocation is fundamental in R&D budgeting. It involves allocating financial, human, and technological assets to projects that align with the company's strategic goals and offer substantial returns. This ensures that R&D efforts are in harmony with the overarching objectives.

Budget flexibility is essential to adapt to changing circumstances and emerging opportunities. Companies must be prepared to adjust budget allocations as market dynamics evolve. Flexibility enables them to seize unexpected opportunities and reallocate resources from less promising projects. R&D budgeting necessitates the establishment of key performance indicators (KPIs) and metrics for performance evaluation. These measures may include innovation success rates, time-to-market for new products, or return on investment (ROI). Quantitative metrics enable data-driven decisions regarding future budget allocations. Investing in talent is a critical aspect of R&D budgeting. Skilled researchers, engineers, and scientists are essential for driving innovation and achieving strategic objectives. Investing in talent development and retention is a long-term strategy that pays dividends in innovation. Continuous monitoring and adaptation are vital components of R&D budgeting. Regular assessment of project progress ensures alignment with strategic objectives. Companies must be prepared to make necessary adjustments, including resource reallocation or discontinuing projects that no longer serve their best interests.Allocating resources to R&D is a strategic imperative for organizations seeking to maintain a competitive edge and foster innovation. The process begins with strategic alignment, where resources are distributed by a company's broader objectives and R&D goals. This alignment serves as a guiding beacon, ensuring that resource allocation follows a clear path that resonates with the organization's vision. Whether the intent is to pioneer cutting-edge products, enhance existing offerings, or explore innovative technologies, the fundamental principle remains: resources must align with these objectives.

Project prioritization emerges as a pivotal element of resource allocation. Not all R&D projects are of equal importance. Prioritizing projects based on potential return on investment, alignment with strategic objectives, and market demand is essential. High-impact projects merit more substantial resource allocations, while lower-priority initiatives may receive limited resources or be temporarily deferred. Balancing risk and reward is a recurring theme in resource allocation. R&D inherently carries a level of risk, and a well-thought-out strategy balances resource allocation between innovative, high-risk projects and more conservative, lower-risk undertakings. While high-risk projects have the potential for significant rewards, they may also entail substantial costs. Diversifying the portfolio with varying risk profiles is a prudent approach.

Flexibility and adaptability are imperative aspects of resource allocation. Market dynamics are fluid, and R&D projects often face unexpected opportunities or challenges. Being prepared to shift resources from underperforming projects to more promising ones or promptly respond to emerging trends is essential. Resource allocation encompasses a spectrum of assets, including financial resources, human talent, and technological infrastructure. While financial allocation is a focal point in budget planning, allocating human expertise is equally vital. Nurturing a skilled workforce of researchers, engineers, and scientists is paramount in driving innovation.

Resource allocation goes hand in hand with performance metrics. Key performance indicators (KPIs) and performance metrics play a vital role in evaluating the effectiveness of resource allocation, tracking project progress, and guiding data-driven decision-making. These metrics may encompass project success rates, time-to-market, and return on investment (ROI), offering a quantifiable framework to assess resource utilization. The competitive landscape serves as a backdrop for resource allocation. Companies must vigilantly analyze their industry peers' R&D

activities and investments. Staying competitive often necessitates matching or exceeding competitors' resources to maintain a strong market position.

Resource allocation is not a one-time endeavor but an ongoing process. Companies must perpetually review and reassess resource allocation to ensure that projects align with strategic objectives and dynamic market conditions. Regular adjustments are essential to adapt to evolving circumstances and retain agility and responsiveness in the fast-paced world of R&D. In conclusion, allocating resources to R&D is a multifaceted process that requires meticulous planning, strategic alignment, and an astute assessment of risks and opportunities. Companies that master the art of resource allocation are better equipped to drive innovation, maintain their competitive edge, and significantly influence the future of their industries.

2.3.1 Balancing Short-Term and Long-Term R&D Investments

Balancing short-term and long-term R&D investments is a strategic tightrope organizations must navigate to stay competitive and innovative. In the realm of short-term investments, the focus lies on immediate outcomes. These endeavors address pressing market demands and are aimed at incremental improvements, product enhancements, and efficiency gains. They connect the organization to its present market reality, ensuring it remains relevant and competitive. In contrast, long-term R&D investments take on a more visionary perspective. These investments are marked by their focus on breakthrough innovations, disruptive technologies, and the creation of entirely new markets. While long-term R&D projects have the potential to redefine industry standards and secure an organization's future, they also come with higher risks and longer time horizons. Striking the right balance between these two investment horizons is crucial. This balance ensures that short-term and long-term investments are synchronized with the organization's overarching strategic goals. It also prevents fragmentation or conflicts in resource allocation, creating a cohesive approach to R&D investment that supports immediate and long-term objectives. Resource allocation is at the heart of this balancing act. Organizations must carefully allocate financial, human, and technological resources between short-term and long-term R&D projects. Short-term investments may require more immediate financial resources, while long-term projects necessitate sustained commitment and patient capital. Effective risk management is essential, as short-term and long-term R&D investments differ in risk profiles. Short-term projects offer more predictable outcomes, while long-term endeavors are associated with greater uncertainty.

In addition, maintaining a diverse innovation portfolio is key to achieving balance. Such a portfolio allows organizations to engage in incremental innovations for short-term returns and disruptive innovations for long-term growth. Performance metrics and key performance indicators (KPIs) are critical in evaluating the success of these investments. They should be tailored to each investment horizon and reflect the specific objectives and challenges encountered along the way. Balancing short-term and long-term R&D investments is an ongoing process that requires regular review and adjustment to adapt to changing market conditions and emerging opportunities.

2.3.2 ROI Analysis for R&D Projects

Return on investment (ROI) analysis is paramount in R&D projects. It serves as a measure of the financial returns generated by R&D initiatives concerning the initial investment. ROI analysis is pivotal in quantifying the efficiency and effectiveness of R&D endeavors. This objectivity is indispensable for organizations in assessing the economic impact of their innovation initiatives. Within the context of R&D, ROI becomes a vital performance metric, providing a yardstick for evaluating the success and impact of R&D projects. It offers a window into how these endeavors contribute to the organization's financial well-being. This insight is critical for informed decision-making, resource allocation, and strategic planning in innovation.

The first step to embark on ROI analysis is defining and quantifying the financial investment involved in R&D projects. This encompasses various costs, including research and development expenditures, labor, and overhead. Simultaneously, one must delve into quantifying the benefits generated by these projects. Benefits can span a spectrum, from increased revenue and cost savings to the value of intellectual property generated.

Quantifying the benefits generated by R&D projects can be a nuanced endeavor. While certain benefits, such as revenue generated by a new product, are straightforward to measure, others might necessitate a more comprehensive evaluation. Benefits, like enhanced brand reputation or potential market share gains, may not lend themselves to simple numerical quantification.

ROI analysis also must consider the time horizons over which benefits are realized. R&D projects may yield short- and long-term results and a comprehensive ROI assessment should account for the time horizon over which benefits accrue. This temporal aspect ensures that ROI analysis provides a balanced view of the project's financial impact. Furthermore, evaluating risks is intrinsic to ROI analysis for R&D. R&D projects inherently involve uncertainties and potential setbacks. Effective ROI analysis incorporates risk factors to provide a more realistic view of the potential outcomes, enabling organizations to make informed decisions. Comparing ROI across various R&D projects or benchmarking it against industry standards and competitors is a valuable practice. Such comparisons facilitate prioritization, identifying best practices, and allocating resources more efficiently and strategically. It enables organizations to gauge the relative success of their R&D initiatives. An essential consideration in ROI analysis is the recognition of intangible benefits. Only some R&D benefits can be readily quantified in monetary terms. Intangible benefits such as brand equity, technological leadership, or first-mover advantage may be equally valuable but challenging to measure through traditional financial metrics. Finally, ROI analysis is not a one-time exercise but an ongoing process. Regular monitoring and reassessment of the ROI of R&D projects ensure that they align with strategic goals and market dynamics. This continuous monitoring approach keeps organizations agile and responsive to the ever-evolving innovation landscape.

In conclusion, ROI analysis is a linchpin in managing R&D efforts. It provides a structured and data-driven approach to assess the financial impact and success of R&D projects, equipping organizations with valuable insights for decision-making, resource allocation, and long-term strategic planning. By comprehending the intricacies of ROI analysis for R&D, organizations can optimize their innovation initiatives, bolster their competitive position, and secure future growth.

2.4 Managing R&D Teams and Intellectual Property

Effectively managing R&D teams and safeguarding intellectual property are two key components that underpin an organization's innovation and competitiveness. To manage R&D teams successfully, the initial step involves forming well-rounded teams with diverse skills and perspectives. Multidisciplinary teams of researchers, engineers, designers, and subject matter experts foster innovation through exchanging ideas and cross-collaboration. An environment that encourages open communication and collaboration propels innovation, allowing teams to explore new ideas and approaches. Furthermore, having clearly defined R&D objectives is essential for team management. These objectives should closely align with the organization's broader strategic goals, providing a well-lit path for the team's efforts. This shared sense of purpose and direction is pivotal in keeping the team motivated and on track. Adequate resource allocation is equally important, encompassing financial, technological, and human resources. Teams should have access to the necessary tools, technology, and funding to effectively execute their projects, with resource allocation matching the strategic significance of R&D projects.

Effective project management is another critical aspect of managing R&D teams. Establishing project timelines and milestones is essential for keeping R&D initiatives organized and on schedule. This includes defining roles and responsibilities, setting deadlines, and monitoring progress to ensure the team meets its objectives. In the realm of intellectual property, safeguarding innovations is a primary concern. Implementing robust intellectual property protection processes, including patents, trademarks, and copyrights, is vital. In addition to these legal safeguards, clear documentation and confidentiality measures are crucial to prevent unauthorized use and secure innovations.

Intellectual property protection is strengthened by access to legal and compliance expertise. In-house legal teams or partnerships with legal experts specializing in intellectual property law can provide essential guidance to ensure that all innovations are legally protected. Training R&D team members on intellectual property matters also bolster protection efforts. Educating employees on the significance of intellectual property protection, including best practices and ethical considerations, fosters a culture of awareness and compliance. Moreover, collaborating with legal experts or specialized firms is often advantageous. These experts can offer guidance on patent applications, trademark registrations, and enforceable contracts to protect intellectual property rights effectively. Continuous monitoring of intellectual property rights is vital to safeguard innovations. Organizations should be prepared to enforce their rights through legal action when necessary, effectively deterring infringement or unauthorized use. Finally, regular reviews of R&D team performance and intellectual property protection processes are essential. Identifying areas for improvement and adapting to evolving legal and technological landscapes ensure the organization's long-term sustainability and competitiveness in the ever-evolving landscape of R&D and innovation.

Building effective R&D teams is a multifaceted endeavor that is the foundation of successful innovation and growth. One fundamental principle in this process is assembling multidisciplinary teams with diverse expertise. These teams often encompass researchers, engineers, designers, and subject matter experts, creating a dynamic environment where a rich exchange of ideas and perspectives occurs. Multidisciplinary teams are often more adept at finding innovative solutions and approaching challenges from various angles. To maximize the effectiveness of R&D teams, it is vital to establish clear objectives and a shared vision. These objectives should directly align with the organization's strategic goals, ensuring the team comprehends its role in achieving the broader vision. Motivation and focus naturally follow when the team shares a clear sense of purpose and direction, contributing to higher performance. Adequate resource allocation is another critical factor in the R&D team's success. This entails ensuring the team can access the financial, technological, and human resources to execute their projects effectively. Resource allocation should correspond to the strategic importance of the R&D projects. Properly allocated resources help the team meet its goals and drive innovation.

Effective project management is a linchpin for keeping R&D initiatives on track. It involves setting clear project timelines, defining milestones, roles, and responsibilities, and monitoring progress regularly. Effective project management ensures that the team remains organized and that goals are met within predetermined timeframes. Fostering a culture of collaboration and open communication within R&D teams is essential. Innovation flourishes when team members can freely exchange ideas, share insights, and work cohesively. Establishing effective communication channels and platforms is crucial for facilitating this collaborative environment. Strong leadership is another crucial element in building effective R&D teams. Effective leaders inspire, guide, and provide direction to the team. They should set a clear vision, offer support, and remove obstacles hindering progress. Effective leadership contributes to team morale, productivity, and overall success. Creating an innovation-friendly environment is a fundamental practice in building effective R&D teams. This involves encouraging risk-taking and a tolerance for failure. Innovation often involves trial and error, and teams should feel comfortable exploring new ideas and approaches without fear of repercussions. Such an environment fosters creativity and allows for the discovery of breakthrough solutions.

Building effective R&D teams is a multifaceted process that requires careful planning, resource allocation, and cultivating a culture of innovation, collaboration, and strong leadership. When multidisciplinary teams are led by effective leaders, equipped with the necessary resources, and empowered with a clear vision and a culture of innovation, they are well positioned to drive innovation, meet objectives, and contribute to an organization's long-term success.

Intellectual property (IP) is a cornerstone of innovation and competitiveness for organizations. It encompasses various forms, such as patents, trademarks, copyrights, and trade secrets, each offering distinct protections and opportunities. Organizations must adopt a multifaceted approach to ensure the preservation and effective utilization of intellectual property. One crucial aspect of managing intellectual property is safeguarding it. Robust safeguards include filing for patents, registering trademarks, and establishing copyrights. These legal protections grant exclusive rights to the creator or owner, preventing unauthorized use. Alongside these protections, clear documentation and confidentiality measures are pivotal in preventing intellectual property from falling into the wrong hands. Legal expertise in intellectual property law is indispensable for organizations. Legal experts, whether in-house or through collaborations with specialized firms, guide various aspects of IP protection. They assist in patent applications, trademark registrations, and the formulation of contracts and agreements. Their insights and legal knowledge are invaluable in navigating the complex field of IP protection.

Moreover, fostering an innovation-friendly environment encourages the creation and utilization of intellectual property. Embracing risk-taking, tolerating failure, and supporting the exploration of new ideas are all vital components of this culture. By nurturing an environment where innovation is encouraged and protected, organizations stimulate the generation of valuable intellectual property. Effective utilization of intellectual property often involves strategic licensing and partnerships. Licensing agreements grant third parties permission to use specific IP in exchange for royalties or fees, generating additional revenue. Collaborative partnerships allow organizations to pool resources and expertise, facilitating the joint development or commercialization of IP assets.

Safeguarding and leveraging intellectual property is essential for organizations seeking to protect their innovations and maximize their value. By implementing robust safeguards, embracing an innovation-friendly culture, utilizing legal expertise, and engaging in strategic licensing and partnerships, organizations can protect their intellectual property and harness it as a source of competitive advantage and revenue.

Collaborative R&D practices have gained prominence in the contemporary business landscape, enabling organizations to pool resources, expertise, and innovation efforts to tackle complex challenges and drive meaningful innovation. Collaborations can take various forms, from partnerships with external entities to internal cross-functional teams. The benefits of collaborative R&D include the ability to share risks and costs, access complementary skills, and accelerate innovation. To ensure successful collaborations, organizations need to establish clear objectives, define roles and responsibilities, and create an environment conducive to teamwork and knowledge sharing (Table 2.2).

2.5 Measuring and Assessing R&D Performance

Table 2.2 Collaborative R&D models and benefits

Collaborative model	Description	Key benefits
Industry-academia	Partnerships between industries and academic institutions	Access to cutting-edge research, academic expertise, and potential talent pipeline
Public-private	Collaboration between public organizations (e.g., government) and private companies	Shared funding, regulatory support, and access to public resources
Cross-industry	Collaborations across different sectors or industries	Synergy of diverse knowledge, insights into unrelated markets, and expanded innovation potential
Open innovation	Involvement of external contributors such as customers, suppliers, and start-ups	Diverse ideas, faster product development, and market responsiveness
Joint ventures	Formation of a separate entity with shared ownership	Risk and cost-sharing, expanded market presence, and access to new technology
Research consortia	Collaboration among multiple organizations on a shared research agenda	Reduced R&D costs, pooling of expertise, and access to broader research capabilities

Collaborative R&D practices offer organizations a spectrum of models to choose from, each with its unique advantages. These partnerships enhance access to resources, reduce financial burden, and accelerate innovation. However, their success hinges on effective management, clear communication, and a commitment to shared goals. Organizations must carefully assess the suitability of each model for their specific R&D needs and establish robust frameworks for collaboration to reap the full benefits of these practices.

2.5 Measuring and Assessing R&D Performance

Measuring and assessing the performance of R&D activities is essential for organizations committed to innovation and competitiveness. By employing a range of key performance indicators (KPIs), organizations can gain valuable insights into the effectiveness and impact of their R&D efforts.

Research productivity is a quantitative KPI encompassing the number of research projects, publications, or patents produced. It measures the organization's output regarding knowledge and intellectual property generation. Time-to-market is another critical KPI that quantifies the efficiency of bringing innovations to market. Reducing this timeframe is often a key goal in R&D, as it can provide a competitive edge.

R&D expenditure is a vital financial KPI, offering insights into the resources allocated to R&D activities. A high R&D expenditure suggests a significant investment in innovation, but it must be balanced with successful outcomes to ensure efficient resource utilization. The KPI of innovation success rates quantifies the percentage of R&D projects that result in successful innovations. This metric provides an indicator of R&D project quality and effectiveness. Return on investment (ROI) is a financial KPI that measures the financial returns generated by R&D efforts relative to the initial investment. It is a critical metric for assessing the economic impact of R&D projects and ensuring they contribute positively to the organization's financial health. Intellectual property–related KPIs track the number of patents, trademarks, or copyrights filed and granted. These indicators reflect the organization's efforts in securing its innovations. Collaboration success and customer feedback are qualitative KPIs. Collaboration success measures the effectiveness of collaborative R&D initiatives and partnerships, helping organizations gauge the value derived from such endeavors. Customer feedback is a crucial indicator of the market's perception of R&D-driven products or services. Positive feedback reflects successful innovation that meets customer needs and expectations.

Measuring and assessing R&D performance involves a combination of quantitative and qualitative KPIs that collectively offer a comprehensive view of an organization's innovation efforts. These indicators enable data-informed decision-making, allowing organizations to refine their R&D strategies, allocate resources more effectively, and ultimately drive innovation that aligns with their objectives and maintains competitiveness.

2.5.1 Key Performance Indicators (KPIs) for R&D

Key performance indicators (KPIs) are pivotal for organizations seeking to assess the effectiveness and impact of their R&D endeavors. Research productivity is a foundational KPI, measuring the quantity of research projects, publications, or patents produced. This metric serves as a benchmark of an organization's output regarding intellectual property generation and knowledge creation. High research productivity indicates an active and innovative R&D environment.

Time-to-market is another key KPI, quantifying the time it takes to bring new products, innovations, or technologies to the market. A shorter time-to-market is often a strategic imperative, providing organizations with a competitive edge by responding swiftly to changing customer needs and market dynamics. Monitoring this KPI ensures that R&D projects remain synchronized with real-world demands. R&D expenditure is a financial KPI that reflects the financial resources allocated to R&D activities. By tracking R&D expenditure, organizations gain insights into the scale of their investments in innovation. This KPI plays a vital role in resource allocation, ensuring that financial resources are used efficiently to support R&D initiatives that contribute to the organization's objectives and growth. Innovation success rates are a core KPI that quantifies the percentage of R&D projects yielding successful innovations. This metric offers an invaluable measure of the quality and effectiveness of R&D initiatives. By assessing innovation success rates, organizations can refine their project selection and allocation strategies, optimizing the use of resources. Return on Investment (ROI) is a pivotal financial KPI, as it assesses the financial return generated by R&D efforts relative to the initial investment. This metric provides organizations with a clear understanding of the economic impact of their R&D projects. A positive ROI ensures that R&D efforts contribute favorably to the organization's financial health.

Furthermore, intellectual property (IP) metrics are an integral component of R&D KPIs, encompassing the number of patents, trademarks, copyrights, or other forms of IP filed and granted. These metrics are tangible evidence of an organization's efforts to secure its innovations and protect its intellectual property. IP metrics are essential for tracking innovation output and safeguarding valuable assets. Collaboration success, another critical KPI, measures the effectiveness of collaborative R&D initiatives and partnerships. This KPI underscores the importance of successful teamwork and knowledge sharing in R&D activities. It is instrumental in evaluating the value derived from collaborative efforts and underlines the significance of building effective collaborative relationships.

Finally, customer feedback, while qualitative, is a vital KPI that reflects the market's perception of products or services resulting from R&D efforts. Positive customer feedback demonstrates the successful alignment of innovations with customer needs and expectations, while negative feedback provides opportunities for improvement and adjustment. These KPIs collectively offer a comprehensive view of an organization's innovation endeavors, enabling data-driven decision-making to refine R&D strategies, optimize resource allocation, and sustain competitiveness. Understanding and monitoring these KPIs are fundamental for organizations considering R&D a strategic asset.

2.5.2 Benchmarking R&D Success

Benchmarking R&D success is a strategic imperative for organizations. It allows them to evaluate their R&D initiatives compared to industry standards and competitors, providing critical insights into performance. This evaluation process is essential for several reasons. It facilitates assessing an organization's R&D efforts, helping determine how well they align with industry norms, competitors, and historical data. Benchmarking is a powerful tool for identifying best practices and strategies employed by successful entities, which can be adopted to enhance R&D effectiveness.

Key performance indicators (KPIs) are integral to benchmarking R&D success. These KPIs provide a structured approach to assessing R&D performance. Metrics such as research productivity, time-to-market, R&D expenditure, innovation success rates, return on investment (ROI), intellectual property (IP) metrics, collaboration success, and customer feedback are commonly used. Research productivity quantifies the quantity and quality of research projects, publications, and patents, reflecting the organization's intellectual output. Time-to-market assesses the organization's efficiency in bringing R&D-driven innovations to market, indicating its responsiveness to changing market dynamics. R&D expenditure provides insights into the allocation of financial resources, ensuring that investments align with industry standards and competition.

Innovation success rates are critical for measuring the percentage of R&D projects that result in successful innovations, shedding light on the effectiveness and quality of R&D efforts. ROI evaluates the financial return generated by R&D investments relative to the initial outlay, offering a clear financial perspective. IP metrics include the number of secured patents, trademarks, copyrights, and other IP assets, signifying the organization's innovation output and IP protection. Collaboration success evaluates the effectiveness of collaborative R&D initiatives and partnerships, highlighting the value derived from teamwork and knowledge sharing. Customer feedback gauges market perception and customer satisfaction with R&D-driven products or services, providing insights into their impact.

To effectively benchmark R&D success, organizations must follow best practices. The first step is selecting relevant metrics that align with the industry, business objectives, and R&D goals. Identifying peer organizations or competitors with similar R&D profiles ensures meaningful comparisons. Regular evaluation against benchmarks is crucial for ongoing improvement and adaptation of strategies. Benchmarking should encompass internal (comparing R&D performance over time) and external (against competitors and industry standards) perspectives. Collaboration and knowledge sharing with benchmarking partners can drive mutual improvement and innovation. In Fig. 2.3, the process of benchmarking R&D success is outlined, emphasizing the selection of relevant metrics, the identification of peer organizations, the choice of KPIs, data collection and analysis, strategy adaptation, the implementation of best practices, and the commitment to continuous improvement. This visual representation can serve as a useful guide for organizations seeking to engage in the benchmarking process effectively.

2.5.3 Evaluating the Efficiency of R&D Processes

Evaluating the efficiency of R&D processes is a fundamental practice for organizations striving to maximize the impact of their innovation efforts [5]. The significance of this evaluation cannot be overstated. Efficient R&D processes help organizations optimize their resource allocation, which includes financial, human, and technological assets. By identifying inefficiencies and bottlenecks, organizations can ensure that their R&D endeavors are cost-effective and streamlined, thus minimizing unnecessary expenditures. This leads to time and cost reduction in the innovation process, enabling organizations to bring new products and solutions to the market at an accelerated pace, a crucial advantage in the dynamic business landscape.

Moreover, evaluating the efficiency of R&D processes gives organizations a competitive edge. In today's fast-paced business environment, responding to market changes swiftly is essential. Efficient R&D processes enable organizations to adapt

2.5 Measuring and Assessing R&D Performance

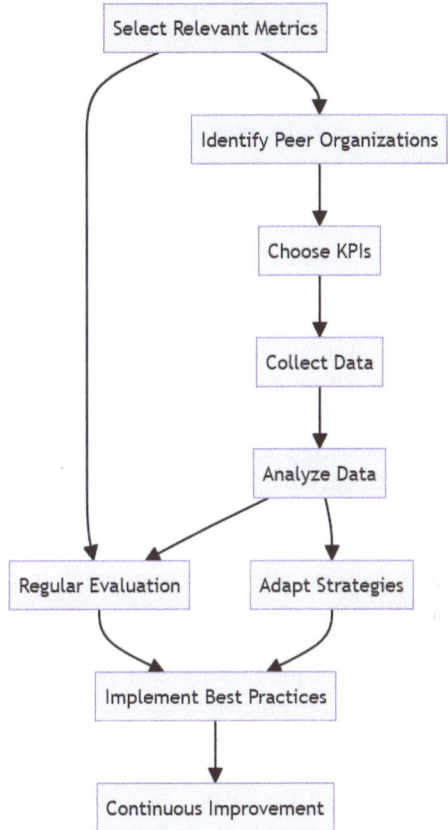

Fig. 2.3 Benchmarking R&D success—a visual guide to effective evaluation and continuous improvement

to evolving circumstances, outpacing competitors and maintaining a strong market position. Additionally, focusing on the efficiency of R&D processes translates into innovations of higher quality that are closely aligned with organizational objectives, ensuring that the organization's innovation efforts are swift and of high value.

To evaluate R&D process efficiency, organizations employ a range of key performance indicators (KPIs). Research time, for instance, measures the time taken to complete research projects, indicating the speed at which knowledge is acquired and innovation is developed. Cost per project evaluates the financial resources allocated to individual R&D projects, providing crucial insight into cost-effective initiatives. Project success rate measures the percentage of R&D projects that culminate in successful innovations, offering insights into the effectiveness of the R&D pipeline. Resource allocation assesses how financial and human resources are allocated to R&D activities, ensuring alignment with organizational goals. Innovation to market time quantifies the time it takes for innovations to progress from concept to market, providing a valuable measure of commercialization efficiency. Finally, collaboration effectiveness evaluates the efficiency and impact of collaborative R&D

initiatives and partnerships, highlighting the success of teamwork and knowledge sharing in the R&D process.

Several methodologies are employed to assess R&D process efficiency. Process mapping is a visual method that helps organizations identify bottlenecks, redundancies, and areas for improvement. Benchmarking involves comparing an organization's R&D processes against industry standards and competitors to identify gaps and opportunities for optimization. Lean and Six Sigma principles are well-established methodologies that can be applied to reduce waste, improve process flow, and enhance overall efficiency. Moreover, the integration of advanced technologies, such as data analytics, machine learning, and AI, can streamline R&D processes and enhance data-driven decision-making.

Organizations aiming to optimize their R&D operations should adhere to best practices. Clear objectives should be established, ensuring that R&D objectives and desired outcomes are well defined and that the evaluation process aligns with these goals. Fostering a culture of continuous improvement within the R&D team is essential, encouraging members to identify and address inefficiencies in the process. Regular assessments using KPIs and methodologies should be conducted, and strategies should be adapted as necessary based on the findings. Collaborative learning is valuable, allowing organizations to collaborate with benchmarking partners, industry peers, and experts to share insights and best practices for optimizing R&D operations. Additionally, investing in technology can significantly enhance the data analysis, decision-making, and process automation aspects of R&D operations, leading to more efficient processes.

In conclusion, evaluating the efficiency of R&D processes is an indispensable practice for organizations aiming to maximize their innovation efforts' impact. By employing relevant KPIs, methodologies, and best practices, organizations can assess and optimize their R&D operations, ensuring the efficient allocation of resources and the timely delivery of high-quality innovations. This approach enhances the organization's competitiveness and contributes to its long-term success in a rapidly evolving business environment.

2.6 The Future of R&D in Business

The future of R&D in business is poised for significant transformation, primarily driven by emerging trends and advancing technologies. Companies will increasingly harness the power of artificial intelligence, machine learning, and data analytics in their R&D endeavors. These cutting-edge technologies will enhance the efficiency and precision of R&D processes, from product design to operational optimization. Open innovation is set to become a prevailing strategy as businesses seek external collaboration for fresh ideas and expertise. Start-ups, research institutions, and other external partners will be pivotal in shaping R&D efforts. Sustainability is rapidly climbing the corporate agenda, leading to R&D investments in eco-friendly products, processes, and technologies. Businesses will explore renewable energy sources, sustainable materials, and methods to reduce waste. Digital transformation remains a priority, and R&D is central to this evolution. Companies will focus on developing digital products, services, and processes to stay competitive in an increasingly digitized landscape. Personalization and customization will continue to gain traction, driven by customer demands. R&D efforts will leverage data and AI to craft solutions that meet individual preferences and needs.

In the healthcare and biotechnology sectors, R&D will continue to drive progress, resulting in breakthroughs like personalized medicine, gene therapy, and pharmaceutical innovations. The pandemic has accelerated digital health solutions, telemedicine, and healthcare R&D. Automation and robotics will find applications in various industries, optimizing manufacturing and logistics processes. Autonomous vehicles, robotic process automation, and smart factories will redefine R&D in these domains. The escalating threat of cyberattacks necessitates ongoing R&D efforts in cybersecurity. Businesses will allocate resources to develop robust, innovative cybersecurity solutions safeguarding their data and systems. Agile methodologies reshape R&D processes, enabling faster development cycles and adaptability to changing market dynamics. Global collaboration in R&D will flourish, empowered by remote work and global connectivity, resulting in diverse and innovative solutions. As R&D activities surge, businesses must navigate evolving regulatory landscapes, particularly in the finance, healthcare, and biotechnology industries. Adhering to changing regulations will be a pivotal aspect of R&D strategies. Additionally, ethical considerations will gain prominence, especially in AI and data-driven research. Companies will need to address data privacy and ethics in their R&D efforts, ensuring they align with evolving societal and regulatory expectations.

The future of R&D in business is a dynamic landscape marked by technology-driven innovations, sustainability, collaborative endeavors, and ethical consciousness. Businesses that adapt to these transformative trends will be better positioned to maintain their competitiveness and relevance in the ever-evolving global marketplace.

Emerging trends in research and development (R&D) are reshaping how organizations innovate and adapt to an evolving business landscape. Artificial intelligence (AI) and machine learning are revolutionizing R&D across various industries, accelerating the development of new products and services. Open innovation is gaining momentum, with companies collaborating with external partners, start-ups, and research institutions to access new ideas and expertise. Sustainability is an increasing focus in R&D, with efforts centered on developing eco-friendly products, renewable energy solutions, and efficient waste reduction processes. Digital transformation is at the core of R&D efforts, as companies leverage it to meet the demands of the digital age and enhance customer experiences.

R&D in biotechnology and healthcare is yielding breakthroughs in personalized medicine, gene therapy, and digital health solutions, improving patient outcomes and healthcare delivery. Automation and robotics are transforming R&D in manufacturing and logistics, improving efficiency and reducing human error. Cybersecurity R&D is crucial as companies invest in research to develop innovative security solutions to protect their data and systems. Agile methodologies are being applied to R&D processes, allowing for faster development cycles and adaptability to changing market conditions.

Cross-disciplinary collaboration is breaking down the barriers between traditional fields of science and technology, driving new discoveries and products. Global collaboration is facilitated by remote work and global connectivity, enabling cross-border R&D partnerships with experts and organizations from around the world. Data privacy and ethics are increasingly important in R&D, particularly in AI and data-driven research. These emerging trends in R&D are shaping the future of innovation, product development, and problem-solving in the business world. Organizations that embrace these trends will remain competitive and relevant in their respective industries.

Preparing for disruptive technologies in R&D is essential for organizations looking to stay competitive in a rapidly evolving technological landscape. This entails fostering a culture of innovation and continuous monitoring of emerging technologies. Organizations can identify disruptive innovations early by maintaining a dedicated team to scan the technology landscape.

Additionally, establishing strategic partnerships with start-ups, research institutions, and technology experts allows access to cutting-edge technologies and external expertise that may only be available in some places. Investing in R&D resources and talent, such as hiring or upskilling employees with expertise in relevant areas, is also crucial. Flexible R&D frameworks that accommodate rapid adaptation to new technologies, scenario planning, and risk assessments provide a structured approach to dealing with disruptive innovations. Furthermore, a customer-centric focus ensures that emerging technologies align with addressing customer needs and improving their experiences.

Staying informed about regulatory compliance, data security, and privacy measures is paramount, especially in highly regulated industries. Practical steps include encouraging cross-functional collaboration and conducting small-scale pilot projects to assess feasibility and impact. Safeguarding intellectual property and developing a long-term strategic vision for integrating disruptive technologies into the overall business strategy round out the strategies for effective preparation. Ultimately, agility and adaptability are critical traits for organizations seeking to harness the potential of disruptive technologies while effectively managing associated risks.

References

1. J. Mitra, *Entrepreneurship, Innovation and Regional Development: An Introduction* (Routledge, 2019)
2. M.M. Keupp, M. Palmié, O. Gassmann, The strategic management of innovation: A systematic review and paths for future research. Int. J. Manag. Rev. **14**(4), 367–390 (2012)
3. L.P. Santiago, V.M.O. Soares, Strategic alignment of an R&D portfolio by crafting the set of buckets. IEEE Trans. Eng. Manag. **67**(2), 309–321 (2018)
4. X. Zhu, A. Zhu, *Emerging Champions in the Digital Economy* (Springer, 2019)
5. I.C. Kerssens-van Drongelen, A. Cooke, Design principles for the development of measurement systems for research and development processes. R&D Manag. **27**(4), 345–357 (1997)

Chapter 3
R&D Strategy and Planning

It has become clear that research and development (R&D) is essential to innovation and competitiveness. Organizations seeking to flourish and achieve in their respective industries no longer have the option to forgo the persistent search for fresh knowledge, technological innovations, and creative solutions. The strategic alignment of R&D efforts with more general corporate goals is emphasized as this chapter digs into the foundational ideas of R&D management. Setting SMART goals, utilizing market information, making the most of resources, and managing risks are all topics covered in this chapter. The significance of technological roadmaps, project selection, and teamwork are also covered. Successful R&D relies on monitoring and adaptation. Therefore, this chapter serves as a useful manual for businesses looking to lead and innovate in a changing market.

> **Learning Objectives**
>
> - Understand the concept of R&D strategy and its role in achieving business objectives.
> - Develop specific, measurable, achievable, relevant, and time-bound (SMART) R&D goals.
> - Learn how to gather and analyze market data for informed R&D decisions.
> - Gain skills to allocate resources effectively for R&D projects.
> - Develop technology roadmaps to guide R&D efforts.
> - Evaluate and prioritize R&D projects using strategic criteria and decision-making tools.

3.1 Significance of R&D in Modern Businesses

R&D is essential because it fuels innovation, economic competitiveness, and long-term prosperity. R&D is the process of systematically investigating and developing novel concepts, goods, and services and creating and developing new technologies. It is important to the strategies of many thriving businesses in today's dynamic environment, and its importance spans many domains. R&D plays a crucial role in driving innovation. It is the melting pot where concepts are developed and given form. R&D encourages a mindset open to new ideas and risks, which helps to expand the horizons of what is possible. Businesses nowadays can only survive with this kind of innovation, which helps them adapt to the changing needs of their customers.

R&D also gives businesses a long-term edge in the market. Companies that put money into R&D typically outperform their rivals because they provide customers with better products and services. These breakthroughs can garner premium pricing and snag a sizable chunk of the market, establishing a firm foothold in the business. R&D also gives companies the flexibility to respond quickly to shifts in the market. Companies may stay ahead of the curve and adapt quickly thanks to R&D in today's fast-paced environment of shifting customer tastes and disruptive technologies.

To reduce danger, R&D is essential. Thorough investigation and testing help companies find and fix problems before they spiral out of control and cost them money. This protects the company's credibility and, by extension, its bottom line. R&D also helps lower costs by improving manufacturing efficiencies, cutting down on waste, and maximizing the use of available resources. It can potentially spur the creation of ecologically friendly and energy-saving solutions, which is good for both businesses and the world [1].

3.2 Defining R&D Strategy

A company's innovation activities are guided by its R&D strategy, which systematically directs how it conducts research, creates new products, and competes in a competitive market. This strategic framework is essential for coordinating long-term organizational goals with innovation, allocating resources efficiently, and ensuring a business stays at the forefront of its sector. R&D strategy includes several important components, such as goal alignment, priority areas, resource allocation, risk tolerance, developing an innovative culture, setting milestones, and ongoing review. It is a dynamic plan that changes as the business environment does, not a static document.

An R&D strategy should align with the organization's overarching mission and vision, establishing a link between innovation initiatives and the company's bigger objectives. It determines where the company will allocate its R&D funds, whether for technological advancement, product creation, process improvement, or a combination of these areas. Another crucial component is resource allocation, which describes how money, people, and infrastructure are allocated to various projects and prioritized by strategic goals. It also discusses risk tolerance, acknowledging the necessity to balance incremental and disruptive innovations to promote sustainable growth.

Fostering an innovation culture is integrated into the R&D strategy, highlighting how crucial it is to promote experimentation, cooperation, and creativity. The plan provides a clear roadmap for accomplishing particular objectives and innovations by setting timetables and milestones. The dedication to ongoing examination and adaptation, in recognition of the always-shifting business environment, is equally important. Organizations create a road plan to navigate the complex landscape of innovation by defining the concept of R&D strategy. Decision-makers are given the tools to make wise decisions, allocate resources effectively, and control risks, eventually leading to sustainable development, competitive advantage, and game-changing innovations.

R&D is a crucial engine for innovation and business expansion, not merely a minor component of a firm. A well-organized and properly coordinated R&D strategy is crucial for firms to succeed, and its significance must be recognized. Ensuring that a good R&D strategy is well aligned with the larger corporate goals and objectives is one of the essential parts of doing so. The success of R&D initiatives is based on this alignment, which acts as their foundation.

When the overall company plan and the R&D strategy are in sync, it guarantees that R&D efforts are concentrated on the most important areas of the organization's success. This strategic alignment enables R&D teams to prioritize their work, allocate resources, and focus their creative energies on projects that directly advance corporate goals. R&D initiatives frequently require a lot of resources, including time, money, and qualified staff. An integrated R&D strategy ensures the effective and efficient utilization of these resources. Without this coordination, resources may be dispersed among several programs, weakening their impact and possibly resulting in financial waste.

Business objectives are frequently strongly correlated with consumer and market demands. A coordinated R&D strategy ensures that the innovations, services, or products produced align with market demands. This increases the likelihood of a successful launch and widespread market acceptability, thereby affecting the revenue and profitability of the business. An R&D strategy that aligns with company objectives can also result in the development of distinctive and competitive features, goods, or services. Whether through cutting-edge technology, greater quality, or creative solutions, such distinction can give the business a competitive advantage.

3.3 Setting R&D Goals and Objectives

Setting precise, well-defined goals and objectives is the basis of any successful R&D plan. These goals act as a guide for the R&D efforts, ensuring that they are in line with the organization's larger mission and contribute to its success. In this part, we will go into establishing R&D goals and objectives and the fundamental ideas that guide this important facet of strategic planning.

3.3.1 The Role of SMART Objectives in R&D

One essential component of strategic planning is the creation of clearly defined objectives. These goals are the cornerstone of the R&D strategy, guiding the work and laying out a precise course for advancement. It is essential to follow the SMART criteria to make sure the R&D goals are practical and executable (Fig. 3.1).

Specific: The "S" in SMART objectives emphasizes the need for specificity. Specific objectives provide an unambiguous description of what aims to be achieved. Vague or general goals can lead to clarity and alignment within the R&D team. In R & D, specific objectives involve technical achievements, product enhancements, or process improvements. For instance, "Enhancing the energy efficiency of the manufacturing process by 15%" is more specific than simply stating "Improving energy efficiency."

Measurable: "M" highlights the importance of measurability in the objectives. Measurable objectives are quantifiable, enabling tracking of progress and evaluation of success. Measuring outcomes against defined metrics and key performance indicators (KPIs) is essential in R&D. For example, "Increasing the tensile strength of the new material by 20% in laboratory tests" provides a measurable goal that can be objectively assessed.

Achievable: Setting achievable objectives means ensuring the R&D goals are realistic and within reach. More ambitious or attainable goals can lead to satisfaction and demotivation. Assess the organization's capabilities, available resources, and

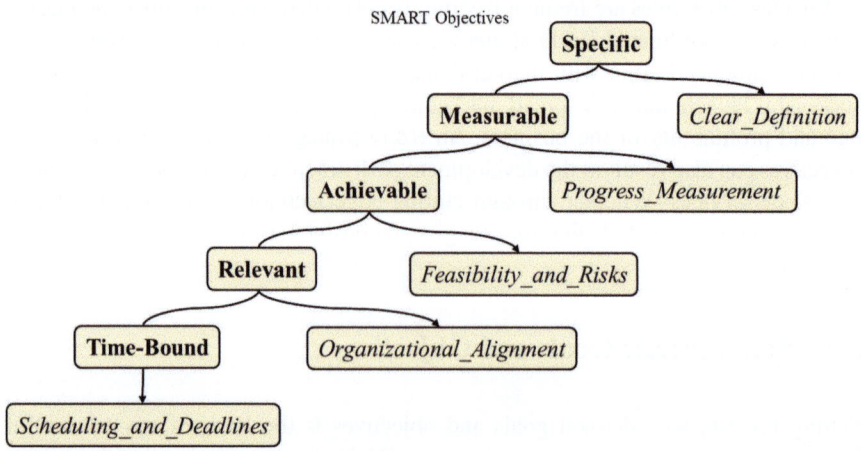

Fig. 3.1 SMART objectives in R&D strategy and planning

technical expertise to determine whether the objectives can be realistically accomplished. For R&D, an achievable objective might involve developing a prototype, given the available budget and expertise.

Relevant: The "R" underscores the need for relevance in the R&D objectives. Relevant objectives are directly aligned with the organization's strategic goals and the broader market and industry context. In R&D, relevance might entail objectives that address specific market needs, technological trends, or competitive pressures. For instance, "Integrate AI-driven customer support features to meet evolving customer service expectations" is relevant in a technology-focused industry.

Time-Bound: The "T" in SMART objectives emphasizes setting a timeframe for achieving the goals. Time-bound objectives create a sense of urgency, helping with planning and resource allocation. In R&D, a time-bound objective might involve completing a research project within a specified timeframe, such as "Launch a proof-of-concept prototype within 6 months."

3.3.2 Developing R&D Goals That Reflect Market Needs and Competitive Dynamics

The creation of specific goals and objectives is the basis for success. These goals provide a clear road map for R&D initiatives because they are articulated inside the SMART framework. Measurable objectives make it easier to track progress using key performance indicators (KPIs), while specific targets eliminate any potential for ambiguity. These goals must be achieved within reasonable limitations because exceedingly ambitious goals might demotivate people. These goals, which address the market's needs and organizational mission, should always be current. They also need to follow a specified timeframe, which gives the plan a sense of urgency.

Beyond SMART criteria, creating effective R&D objectives depends on a deep comprehension of market dynamics and competition. Market research reveals crucial insights through the investigation of client demands, trend forecasts, and competition intelligence. Organizations can create goals that resonate with the surrounding environment by matching R&D objectives with this market information. This could result in creating new products in response to consumer demands, improvements to the customer experience, or modifications to comply with legislative changes.

Sustainability in R&D objectives necessitates ongoing evaluation and modification, reflecting the constantly shifting market environment [2]. To make sure that R&D goals are in line with altering market and competitive situations, periodic evaluations are essential. With these tactics, businesses can better anticipate market trends, stand out from rivals, and give customers long-lasting value.

3.4 Environmental Scanning and Market Research

A strong R&D plan needs to include both market research and environmental scanning. These processes are essential for gathering information about the shifting commercial environment, market dynamics, and upcoming technology. Effective R&D planning relies on environmental scanning, offering strategic advantages by revealing external elements significantly influencing the course and success of R&D activities. The primary goal of environmental scanning is to facilitate adjustment to shifting circumstances. Given the continuous changes in the business environment, such as the development of new technology, evolving consumer tastes, and changing market dynamics, R&D teams can adapt their strategy and align with these changes, thanks to the data provided by environmental scanning.

Additionally, environmental scanning is a priceless tool for spotting new prospects. R&D personnel can proactively identify creative opportunities by analyzing market trends and technology developments attentively. This strategy frequently results in the creating of innovative goods and services that provide a clear competitive advantage. Environmental scanning is essential for risk mitigation along with capturing opportunities. R&D teams can proactively address threats and problems by anticipating them early. For instance, keeping up with regulatory, economic, and competitive changes allows for good risk management and reduces interruptions to ongoing initiatives.

Environmental scanning also encourages customer-centric growth. It offers insightful information on client wants and preferences, enabling R&D teams to customize their efforts to suit market demands. This leads to developing goods that appeal to customers, increasing their acceptance and marketability. Furthermore, to stay competitive in any field, one needs always to keep a sharp eye on the competition. Environmental scanning provides R&D teams a thorough awareness of rivals' R&D activities, product launches, and business strategies. With this knowledge, R&D planning can be more informed, allowing businesses to outsmart rivals and make tactical choices that provide them a competitive advantage.

3.4.1 Techniques for Gathering and Analyzing Market Data

Effective market data gathering and analysis skills are essential. This method lays the groundwork for well-informed choices, enabling R&D teams to coordinate their work with market demands and trends.

Surveys and Questionnaires

Surveys and questionnaires are reliable tools for acquiring crucial market information that helps to build R&D plans. They provide a methodical way to gather honest feedback from the target market, allowing for making well-informed choices. When designing effective surveys, it is critical to start with well-defined objectives and comprehension of the data sought. Key components of survey design include choosing the appropriate sample, creating impartial and clear questions, and arranging them rationally. Open-ended questions can produce rich qualitative insights, whereas closed-ended questions with predefined response alternatives are frequently utilized to simplify analysis. Pilot testing assists in resolving any flaws before the survey is released.

After the survey, the data is analyzed, which may use quantitative and qualitative techniques. For closed-ended questions, quantitative analysis is used, which involves statistical methods and software to tabulate and summarize answers. To find themes and patterns in text responses to open-ended questions, qualitative analysis organizes and summarizes the material. Data segmentation can also show response variations depending on customer or demographic groups. Data visualization tools produce charts, graphs, and reports that make complex data more understandable to decision-makers to communicate findings effectively.

The firm will enhance its ability to stay in tune with changing market demands by incorporating surveys and questionnaires into the R&D plan. This methodical approach to data collection provides insights into client preferences and expectations. It promotes flexibility and competitiveness, ultimately assisting in creating goods and services that appeal to the target market. Regular, well-conducted surveys and questionnaires give firms a window into the changing environment and aid in helping them make decisions that will spur innovation and growth.

Competitor Analysis

Effective R&D strategy and planning include competitor analysis, which provides important insights into the competitive landscape. To start, it is critical to comprehend the rivals. To do this, it is necessary to classify them into different groups, such as direct, indirect, or potential disruptors, and to carefully consider important factors, including their product and service offerings, market positioning, target market, geographic reach, market share, and growth rates. A SWOT analysis thoroughly evaluates a company's market position by delving further into its strengths, weaknesses, opportunities, and threats.

The examination also includes price and product strategy. Finding possible areas for innovation and difference can be aided by analyzing the competitors' products' qualities, features, and pricing strategies. The strategy decisions are influenced by assessing value propositions and investing in R&D and innovation. Additionally, comments and reviews from customers offer insightful information. Customer satisfaction levels, pain issues, and chances for improvement can be discovered via monitoring social media, review sites, and surveys.

The R&D strategy gains a competitive edge through a detailed competitor study, allowing prediction of market changes and strategic placement of inventions. This information guides decisions regarding where and how to concentrate R&D efforts, aligns pricing strategies, and impacts product development, ensuring that R&D projects align with market demands and stand out in the competitive environment.

Market Segmentation

A key strategy in R&D planning is market segmentation, which enables organizations to divide a large target market into various categories depending on various characteristics. Demographics, psychographics, behavioral tendencies, and location are frequently included in these criteria. Market segmentation's main goal is to give companies a better knowledge of the varied demands and preferences of various client segments.

For instance, demographic segmentation divides the market according to factors like age, gender, income, and education, enabling businesses to create products specifically for particular age groups. By examining customers' psychological and lifestyle traits, psychographic segmentation enables firms to develop products that appeal to particular beliefs and attitudes. On the other hand, behavioral segmentation enables businesses to create goods that align with various customer behaviors by classifying consumers based on usage patterns and brand loyalty. Geographic segmentation is crucial for firms with regional or global operations since it considers variables like climate, culture, and legal variances that call for R&D efforts to be adjusted.

Market segmentation offers numerous benefits for R&D. It makes customization easier since R&D teams may create goods that perfectly meet the particular requirements of each market segment. This customized strategy boosts market success and consumer delight. Additionally, segmentation encourages resource allocation efficiency, lowering the risk of creating products that do not appeal to the target market. Segmentation also helps businesses create marketing and advertising efforts that have a strong emotional connection with their target audiences. This tactic may create a competitive advantage, allowing companies to rule niche markets.

Market segmentation presents several difficulties. It depends on the veracity of the data; good data might result in wasteful targeting and efficient use of resources. Over-segmentation is a common mistake that can result in a complicated approach and reduce the effectiveness of R&D efforts. Businesses need to strike a balance between practicality and granularity. Furthermore, market dynamics might alter over time due to changes in consumer behavior and technological improvements, demanding a continuous reevaluation of segmentation strategies to remain competitive and responsive to changing customer needs.

3.4.2 Using Competitive Intelligence to Inform R&D Decisions

The systematic process of obtaining, examining, and interpreting data about the rivals and the competitive environment is known as competitive intelligence (CI). When correctly incorporated into R&D, CI can offer priceless insights that guide and enhance the R&D activities. In this chapter, we examine the crucial part that competitive information plays in directing R&D decisions.

Technology Monitoring

Monitoring technological advancements in the industry is one of the main goals of competitive intelligence in R&D. Spotting new trends, exciting technology, and potential disruptions can be achieved by taking a proactive approach. Patent analysis, where an analyst examines competitor patents to comprehend their technological advancements, white spaces with little patent activity, and informing the product development schedules based on patent expiry dates are key elements of technology watch. Monitoring peer-reviewed research papers from rival companies, subject matter experts, and academic institutions can also assist in finding partnership opportunities and gaining useful insights from peer-reviewed studies. Additionally, by examining the product development roadmaps of the rivals, it becomes possible to predict their upcoming releases, align the R&D initiatives accordingly, and take advantage of market opportunities.

Benchmarking

To define performance standards and strengthen the competitive position, benchmarking is a potent tool for assessing R&D processes and performance relative to peers and industry benchmarks. Process benchmarking, involving examining R&D processes, workflows, and techniques to identify areas where efficiency can be improved, is one of the key components of benchmarking. Streamlining operations and reducing time-to-market can be achieved by comparing these procedures to industry best practices. Performance metrics play a crucial role in benchmarking. Identifying key performance indicators (KPIs) for R&D and comparing them to industry norms and competition performance can help identify areas where the R&D strategy needs improvement and potential weaknesses.

Another crucial component of benchmarking is product benchmarking. It comprises comparing the products' quality, features, and performance against those of the rivals' products. This kind of benchmarking can direct decisions regarding product development and help preserve or obtain a competitive advantage in the market.

Intellectual Property Analysis

Making wise choices regarding R&D requires a thorough awareness of the sector's intellectual property (IP) landscape. Analyzing intellectual property (IP) involves evaluating patents, trademarks, and copyrights to learn more about competitive positioning and prospective collaborations. Examining rivals' patent portfolios to find possible threats to the R&D initiatives is one of the key components of IP analysis. Consider other strategies or licensing alternatives if a rival has a dominant patent position in a key technological field. It also entails examining the IP environment to find chances for partnership, licensing, or cross-licensing with other businesses or academic institutions. Collaboration can be a smart approach to access technologies and knowledge supporting R&D efforts. It can also assist in undertaking a Freedom to Operate (FTO) analysis on the R&D projects to determine whether they might violate patents or other existing IP rights. Conducting an FTO analysis is crucial to prevent legal issues and establish a clear path for commercializing the discoveries.

3.5 Resource Allocation and Budgeting

An effective R&D strategy needs to include budgeting and efficient resource allocation. They are crucial in determining if R&D efforts are successful. The core of this strategy is the allocation of financial resources. It begins with thorough budget preparation, ensuring that the funds allotted align with major R&D objectives. Since not all projects are created equal, prioritization is essential. A larger financial budget should be allotted to high-priority projects with strategic relevance, commercial potential, or technological viability. Additionally, the budget needs to be flexible to allow for the reallocation of cash as priorities change or new possibilities present themselves.

Human resource allocation and financial resource allocation are the lifeblood of any R&D project. It is critical to pinpoint the precise knowledge and abilities needed for each project. To ensure a smooth project progression, it is essential to assign team members with the appropriate skills (Fig. 3.2). Staffing, training, and succession planning are all included in workforce planning, guaranteeing a steady flow of

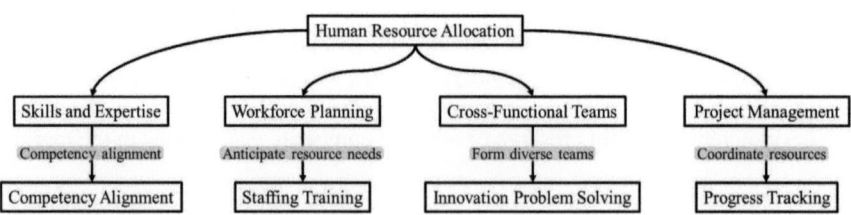

Fig. 3.2 Human resource allocation

qualified personnel. Cross-functional teams unite people with various talents and encourage creativity and comprehensive problem-solving. Coordinating human resources, monitoring progress, and preventing possible bottlenecks require effective project management techniques.

In R&D, a portfolio approach to resource allocation balances risk and reward. This strategy entails allocating resources among many projects. Low-risk, incremental innovation projects might get a smaller part of the resources, but high-risk, high-reward projects might need a larger allocation. The danger of investing all R&D resources in a single project that could not produce the desired outcomes is reduced by a well-managed portfolio. Resource pools offer flexibility by enabling resource allocation based on project needs and priorities. Resource pools are tapped into as projects need more assistance instead of designating specific people or money to one project, improving adaptability and responsiveness.

Stage-gate models can be used in resource allocation to offer an organized method. Projects are broken up into phases, each with set resource allocation checkpoints. Resources are released in phases depending on how well the project is coming along and how well it fits into the R&D strategy. This strategy ensures that resources are used effectively and enables the early detection and rerouting of underperforming initiatives. A thorough awareness of the organization's strategic goals, the competitive landscape, and the market's changing needs is necessary for maintaining a balance between the allocation of financial and human resources. It is a crucial component of R&D strategy and planning, and its successful implementation can greatly increase the likelihood of R&D success.

A combination of responsible financial management and strategic decision-making is essential to maximize the utilization of the R&D budget. Implementing strict cost management measures is crucial for achieving the most from the R&D spending. This involves closely monitoring expenditures, conducting thorough cost analyses, and periodically performing internal audits. Additionally, obtaining cost-effective resources and services through negotiations for advantageous terms with suppliers and contractors can extend the budget.

Resource efficiency plays a vital role in optimizing the R&D budget. Wisely using workers, tools, and supplies is key to maximizing resource efficiency and minimizing waste. To achieve this, R&D teams should be motivated and well organized. Lean approaches should also streamline procedures and promote resource sharing among projects. These procedures enhance effectiveness and prevent resource underutilization, reducing budgetary allocation costs.

The value of R&D investments needs to be evaluated to inform future budget allocations [3]. Organizations can evaluate the effectiveness of their R&D efforts by establishing and tracking innovation indicators like ROI, time-to-market, output quality, customer feedback, and the health of the innovation pipeline. Planning for different outcomes and creating a contingency budget help firms be ready to deal with unforeseen difficulties without jeopardizing their R&D strategy. A continuous improvement culture also ensures that R&D budget optimization is a continual activity, responding to new situations and looking for creative ways to enhance efficiency while providing benefits to the business.

3.6 Technology Roadmaps

Technology roadmaps are essential instruments for directing R&D activities. These visual representations offer a strategic framework for comprehending technology development, coordinating R&D efforts with long-term corporate objectives, and guaranteeing that businesses maintain competitiveness in fast-moving markets. A well-designed technological roadmap has to provide a clear path for advancing goods, services, and technologies over a specific timeframe, usually many years.

3.6.1 Creating Technology Roadmaps to Guide R&D Efforts

Strategic R&D planning needs to include the development of technology roadmaps since they offer a well-structured way to accomplish long-term technological goals. A thorough evaluation of the organization's technological landscape is the first important step in creating a technology roadmap. This first step is thoroughly examining the technology infrastructure, products, and services already in place. To establish the beginning point for the plan, it is essential to understand where the firm stands.

A solid technology plan needs to be built on a foundation of imaginative thinking. Aligning the long-term technical vision with the overarching goals and aspirations of the firm is important. What improvements and advancements in technology do you hope to see? The plan's purpose is framed by these lofty objectives, which act as the compass and North Star.

The long-term vision needs to be divided into short-term objectives or milestones that each fit the SMART criteria. These achievements act as significant first steps in achieving the long-term goals. A well-organized timeline is required to make it apparent when each milestone should be reached. This facilitates efficient resource allocation and scheduling in addition to helping to establish reasonable expectations. The R&D initiatives will be effectively supported in pursuing these milestones provided the relevant resources, including financing, human capital, infrastructure, and equipment, are identified and allocated.

The process of developing a technological roadmap is continual and iterative rather than being a one-time activity. The roadmap should change concurrently with how the organization changes and adapts to new situations. Regular evaluations and updates are necessary to keep it current and trustworthy as a guide for the R&D efforts, connecting them with the organization's strategic vision and adjusting to market circumstances.

3.7 Project Selection and Prioritization

Table 3.1 Striking the balance: short-term versus long-term technology development

Aspect	Short-term technology	Long-term technology
Goals	Immediate improvements	Transformative innovations
Importance	Customer needs, revenue	Innovation, future-proofing, sustainability
Customer focus	Feedback, trends	Anticipating future needs
Revenue generation	Essential for revenue	Potential for future income
Agility	Quick responses	Long-term perspective
Innovation leadership	Opportunities for breakthroughs	Innovation leader status
Future-proofing	Short-term relevance	Long-term sustainability
Sustainability & impact	Current needs	Lasting societal impact
Examples	UI enhancement, new features	Renewable energy research, disruptive technology

3.6.2 Balancing Short-Term and Long-Term Technology Development

Balancing short-term and long-term technological progress is key to developing a thorough technology roadmap. Short-term objectives are focused on making quick changes, little improvements, and satisfying market demands. These objectives help businesses remain responsive to changing market dynamics and client preferences while generating income and preserving corporate agility. On the other hand, long-term goals take a strategic, forward-looking approach and concentrate on game-changing, potentially disruptive technologies that can firmly establish a company's position as an industry leader. By developing solutions that have long-lasting effects on society and the environment, long-term goals are essential for encouraging innovative leadership, future-proofing corporate operations, and contributing to sustainability and long-term success.

A technology roadmap's success ultimately depends on finding a balance between short-term and long-term goals, considering the need for quick wins while looking for developments that could reshape industries and guarantee the company's long-term success. Table 3.1 compares the objectives, significance, customer focus, revenue, agility, innovation, sustainability, and examples of short-term versus long-term technology development.

3.7 Project Selection and Prioritization

Only some initiatives or concepts are created equally. R&D strategy and planning need to consider project selection and prioritization carefully. The choices taken during this era may strongly impact a business's success and expansion. The key to effective innovation is choosing the correct R&D projects. Making sure that R&D

initiatives are in line with the organization's overarching strategic strategy is the first step. A project's strategic alignment with the long-term goals and objectives of the organization is assessed. Organizations can systematically evaluate the alignment of each project by developing a set of criteria based on the company strategy. Projects strongly connected with the organization's strategic direction are more likely to contribute to its long-term success substantially.

A crucial component of the appraisal of R&D projects is figuring out a project's technical viability. This step determines whether the required technology can be created or bought within the specified timescales and financial constraints. A thorough technical investigation is necessary to find any obstacles and problems that can hamper the project's progress. Organizations can rank or score projects based on how probable they are to be completed successfully, giving them higher priority.

To make sure that R&D efforts are focused on addressing true market needs, it is essential to have a thorough understanding of the market potential. Analyzing market size, growth potential, and competition dynamics is necessary. Organizations should assess the resources needed for the project's implementation simultaneously, including financial and human resources. Resource distribution needs to be carefully considered to avoid jeopardizing other crucial projects. To prioritize initiatives that are technically feasible, market-responsive, and resource-efficient in addition to being strategically aligned, it is imperative to conduct these thorough evaluations. These factors work together to ensure that R&D investments are made intelligently, boosting project success rates and fostering long-term business growth.

3.7.1 Criteria for Project Selection

Any organization's choice of R&D projects is crucial since it can greatly impact the organization's development and success. It is critical to set precise and well-defined criteria for project selection to make well-informed decisions. These standards will be a foundation for assessing possible projects and ensuring they complement the strategic goals. The important factors to consider cover various aspects (Table 3.2). For initiatives to remain relevant to the business goals, contribute to the mission, and establish a competitive edge, a strategic alignment needs to be established initially. They ought to make use of the primary skills of the company. Another important consideration is the project's market potential, which considers the target market's size, growth potential, and the particular consumer segments it intends to target. The technical difficulty of the project, the availability of necessary resources, and any ramifications for intellectual property are all considered when determining technical feasibility. Financial viability, which includes ROI, cost-effectiveness, and budget alignment, is crucial. Last but not least, to ensure successful project execution, resource availability, including human and financial resources and necessary infrastructure and equipment, should be thoroughly examined.

3.7 Project Selection and Prioritization

Table 3.2 R&D project selection criteria

Criteria	Sub-criteria	Description
Strategic alignment	Relevance to business goals	Alignment with long-term organizational goals and mission
	Competitive advantage	Potential to provide a competitive advantage or market strengthening
	Fit with core competencies	Utilizing core strengths and existing expertise
Market potential	Market need	Addressing a clear and pressing market need
	Market size and growth	Evaluating the target market's size and growth prospects
	Customer segmentation	Identifying specific customer segments being targeted
Technical feasibility	Technical complexity	Assessment of technical complexity and risks involved
	Availability of resources	Ensuring the availability of required technical resources
	Intellectual property considerations	Evaluating IP challenges and patenting opportunities
Financial viability	Return on investment (ROI)	Calculating expected ROI, payback period, and net present value
	Cost-effectiveness	Evaluating cost-effectiveness without compromising quality
	Budget availability	Ensuring the project aligns with the available R&D budget
Resource availability	Human resources	Availability of qualified personnel with necessary expertise
	Financial resources	Matching the project's financial requirements with the budget
	Infrastructure and equipment	Identifying specific infrastructure or equipment needs

3.7.2 Decision-Making Processes and Tools for Project Prioritization

R&D project prioritization is a challenging task that necessitates a methodical strategy. Organizations use a variety of techniques and technologies to encourage informed decision-making. The first strategy uses scoring models, which rank projects by comparing them to predetermined criteria and assigning weights to each. This methodical approach requires careful criterion selection and weighting to maximize transparency. Decision matrices, the second method, provide a visual tool for evaluating projects against several criteria. Understanding the strengths and limitations of a project is made easier by scores, weights, and visual representation. Decision matrices are especially beneficial when working with a moderately large number of projects and criteria.

Additionally, portfolio analysis, which entails categorizing projects, allocating resources, controlling risk, and establishing a balanced portfolio, is included in

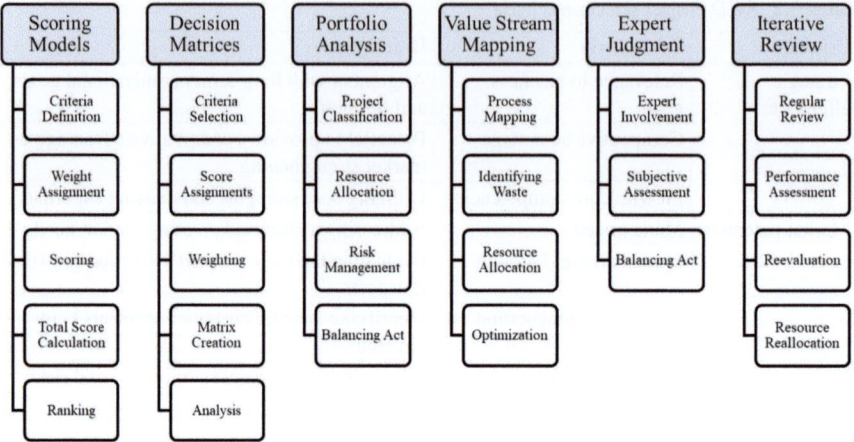

Fig. 3.3 Project prioritization process and methods

project prioritization beyond just individual projects. The entire project portfolio's resource allocation and risk mitigation are optimized. Value stream mapping is a different method that distinguishes between activities that provide value and those that do not, improving efficiency and resource allocation. The fifth method, expert judgment, emphasizes that qualitative views from knowledgeable people can supplement quantitative approaches, offering a more thorough perspective on project selection. The final strategy, an iterative review process, guarantees ongoing project priority adaption as market conditions and strategic goals change. Regular performance reviews and resource reallocation aid in keeping organizational goals in line with change.Making decisions about R&D projects can be done using the method shown in Fig. 3.3. The process includes several techniques and phases, starting with scoring models, where criteria are specified and weighted, and projects are ordered according to total scores. With an emphasis on resource allocation and risk avoidance, portfolio analysis and decision matrices help manage the full portfolio of R&D initiatives. By mapping value streams, project development processes can be made more efficient and less wasteful. Expert judgment adds qualitative insights to quantitative procedures. An iterative review process assures continual alignment with business goals as market conditions change.

3.8 Collaboration and Partnerships

An effective R&D plan must include collaboration and partnerships because they help firms use outside expertise, pool resources, and speed up innovation; leveraging external collaborations and partnerships is essential for firms looking to

3.8 Risk Management and Contingency Planning

maintain their competitiveness and innovation in their R&D initiatives. A key tactic in this area is the idea of "open innovation," which emphasizes the necessity to search outside the confines of an organization for new concepts, know-how, and resources. Open innovation promotes proactive interaction with outside partners, such as vendors, clients, institutions, start-ups, and rival businesses. It includes various elements, including idea sharing, technology reconnaissance, and cooperative research. Companies may increase their knowledge base, speed up product development, and save R&D expenditures by embracing open innovation. Notably, it frequently results in developing groundbreaking goods and services that might have been impossible with internal R&D efforts.Academic partnerships provide firms with a useful way to access specialized resources, top-tier personnel, and cutting-edge research. The centers of groundbreaking research in many subjects are found in universities and other research facilities. Access to the most up-to-date innovations and technology is made possible via collaborations with these institutions. Partnerships with academic institutions can also make hiring highly qualified researchers and students who can contribute their knowledge to actual R&D initiatives easier. These partnerships can also result in resource sharing, giving businesses access to cutting-edge resources that might not otherwise be available. For instance, a technology company might partner with an engineering school to create cutting-edge hardware, or a pharmaceutical company would work with a university's chemistry department to investigate novel medication compounds.

Start-up: Another opportunity for foreign cooperation is through incubation. An organization can foster innovation by funding or investing in start-ups. Start-ups frequently work on technology's bleeding edge and offer novel, disruptive concepts. This creative energy can help established businesses explore new markets and technology. Additionally, companies usually focus on particular specialized industries or recent developments. Working with start-ups can give existing companies access to new markets and a foothold in industries that are undergoing fast change. Notably, collaborations with start-ups can result in chances for acquisition as well. Businesses can find promising initiatives to invest in or incubate, and when those enterprises are ready to be acquired, they can take advantage of their unique technologies and market potential. Tech behemoths like Facebook and Google have made it a habit to buy companies with groundbreaking innovations to expand their product lines. Clear objectives, clearly defined roles and duties, and effective communication channels are necessary for successful external cooperation. To create a fruitful and mutually beneficial partnership, it is essential to establish trust and shared ideals. Managing any IP-related issues that can develop during the partnership also depends on having transparent IP agreements and an awareness of the implications of intellectual property (IP). In a corporate environment that is rapidly changing, productive external collaborations can boost innovation, lower risks, and give a company a competitive advantage.

3.9 Risk Management and Contingency Planning

A comprehensive R&D strategy needs to include effective risk management and emergency preparation. Organizations need to be ready to deal with unforeseen obstacles and setbacks since R&D activities are inherently uncertain. Methodically identifying, evaluating, and minimizing risks are essential to increase the likelihood of a project's success and reduce setbacks. This section covers the crucial steps in identifying and reducing risks in R&D activities.

Risk identification is the first and most important phase in the risk management process. Various possible risks can be found by involving specialists and stakeholders from various backgrounds within the firm, holding brainstorming sessions, and reviewing historical data. Once these hazards have been discovered, classify them according to their traits, consequences, and likelihood of happening. Effective resource allocation and effort prioritization are made possible by this category. Technical, market, and regulatory risks are prominent categories for risks, among others. It is crucial to evaluate the potential repercussions of each risk and ascertain the likelihood of its occurrence with a categorized list of risks. Effect analysis and likelihood assessment make the ability to prioritize risks and concentrate efforts on the most pressing ones possible.

A critical component of managing R&D projects is contingency planning. It entails the creation of intricate plans to deal successfully with unforeseen difficulties that can conceivably impede the project's advancement. Establishing risk mitigation techniques relevant to recognized threats and including technical and operational solutions is essential before creating contingency plans. These mitigating measures are intended to stop the dangers from happening or decrease their effects. Contingency plans are implemented if high-impact and high-likelihood risks materialize, specify the steps, liabilities, and deadlines necessary to manage and remedy the problem, and ensure the project stays on schedule. To ensure that the contingency plan can be performed without delays due to resource constraints, adequate resource allocation, including budget, trained individuals, equipment, or alternative suppliers, is crucial. To evaluate the effectiveness of the contingency plans, simulations or "war games" may be held. This allows the team to practice handling emergencies and make any necessary improvements.

Communication protocols need to be created to specify roles, duties, and the chain of command to ensure a seamless flow of information and rapid decision-making during crisis events. The contingency plans need to be reviewed and updated frequently to consider changes in the project's scope, risk assessments, or resource availability. A thorough record of all contingency strategies, resource allocation choices, and actions during risk occurrences is used as a historical record and resource for ongoing R&D projects. Effective contingency planning increases the likelihood that the project will succeed by ensuring flexibility and adaptation when issues arise and preparing the team for unforeseen obstacles.

3.10 Monitoring and Evaluation

Projects must be continuously monitored and evaluated to ensure they stay on track, achieve their goals, and produce noticeable outcomes. The R&D strategy is guided by this phase, which also provides the tools for tracking progress, evaluating the success of efforts, and making wise adjustments. The process of R&D is dynamic and frequently characterized by unknowns and changing conditions. Successful R&D initiatives differ from those that stall by having the flexibility to modify and improve their strategies in response to new knowledge and shifting market circumstances. The real-time insight and intelligence instruments required for evidence-based decision-making include monitoring and assessment.

3.10.1 Implementing Key Performance Indicators (KPIs) for R&D Projects

Setting the compass for the innovation journey is similar to implementing key performance indicators (KPIs). KPIs are quantitative measurements chosen with care to evaluate the development, efficacy, and impact of the R&D projects. These metrics are crucial for tracking current performance and coordinating the R&D activities with overarching objectives.

The process of choosing appropriate KPIs for an R&D project is complex. It requires a thorough understanding of the project's goals and the specific elements to assess. Typical categories of R&D KPIs include technical benchmarks, market performance, financial indicators, innovation output, risk management, and cooperation efficiency. The most suitable KPIs depend on the nature, goals, and key performance areas of the R&D project. It is essential to strike a balance to assess progress without overwhelming measurements. Periodic examination and modification of KPIs are advisable as the project evolves to ensure alignment with project objectives. Monitoring the R&D project's status, making data-driven decisions, and ensuring efforts align with producing valuable results can be achieved with these KPIs [4].

3.10.2 Regular Monitoring and Evaluation of R&D Progress

Successful R&D initiatives require rigorous monitoring and assessment that is conducted regularly. During this crucial stage, pertinent data must be gathered, real-time feedback must be given, and periodic performance reviews must be conducted. Tracking technical data, financial metrics, results of market research, and project management data that align with recognized key performance indicators (KPIs) is crucial. These KPIs are quantitative measurements that allow real-time evaluation of success and growth. Quality control measures should be implemented to verify further that project components satisfy predetermined standards and objectives.

To make sure that the project stays within the budget allotted and that resource allocation is optimized, tracking the project's budget and resources is an essential part of monitoring and assessment. Additionally, risk assessment is a continuous process that helps to identify and resolve potential problems and facilitates the creation of backup plans. Effective internal and external stakeholder interaction is essential, and progress reports, milestone updates, and strategic alignment are important communication points. Iterative monitoring and evaluation facilitate adaptive management by enabling modifications to the R&D strategy and project plan as new information and insights become available.

3.10.3 Adjusting Strategies Based on Performance Data

Performance information gleaned via thorough monitoring and assessment acts as a strategic compass, allowing R&D teams to negotiate the always-changing terrain of R&D. Swift's successful and well-informed response becomes necessary when performance data reveals departures from the original plan or unanticipated difficulties. The first stage is finding variations and differences between actual findings and anticipated outcomes. This critical examination highlights potential weak points in projects and identifies areas where they might be strengthened. The next critical stage is root cause analysis, which helps identify the underlying causes of these deviations, including technical problems, resource limitations, or changes in market dynamics. With these insights, R&D teams can set out on a road of innovation and adaptability, modifying their plans as necessary [5].

Resource reallocation, changed project timetables, or a strategy pivot in reaction to customer feedback and technical developments may be part of adaptation and innovation in response to performance data. In this stage, flexibility and agility are crucial for keeping R&D initiatives aligned with organizational objectives. A guiding idea is continuous improvement; R&D is an iterative process, and constant evaluation of the effects of strategy changes is crucial. Effective communication is crucial to establish a sense of purpose and shared responsibility among the team members and ensure that everyone knows and accepts these changes.

R&D leaders enable their teams to proactively solve problems and seize opportunities by integrating performance data into decision-making. As a result, projects not only meet but significantly surpass expectations. Organizations may thrive in uncertainty and shifting market conditions thanks to this dynamic approach to R&D strategy change, which fosters continued innovation and success.

References

1. W.L. Lin et al., Influence of green innovation strategy on brand value: The role of marketing capability and R&D intensity. Technol. Forecast. Soc. Chang. **171**, 120946 (2021)
2. D. Zhang, M. Mohsin, F. Taghizadeh-Hesary, Does green finance counteract the climate change mitigation: Asymmetric effect of renewable energy investment and R&D. Energy Econ. **113**, 106183 (2022)
3. H. Jang, A decision support framework for robust R&D budget allocation using machine learning and optimization. Decis. Support. Syst. **121**, 1–12 (2019)
4. K. Brock et al., Front end transfers of digital innovations in a hybrid agile-stage-gate setting. J. Prod. Innov. Manag. **37**(6), 506–527 (2020)
5. C. Li, S. Wang, Digital optimization, green R&D collaboration, and green technological innovation in manufacturing enterprises. Sustain. For. **14**(19), 12106 (2022)

Chapter 4
R&D Funding and Budgeting

Research and development (R&D) is at the vanguard of innovation, advancing a wide range of industries, from manufacturing to technology and healthcare. It is the driving force behind industry innovation, the source of breakthroughs, and the lifeblood of growth. However, for R&D to succeed, it needs funding, careful planning, and efficient budgeting. This chapter examines various funding options, such as government grants and investments from the private sector, and budgeting techniques for allocating resources such as personnel, equipment, and materials. We also look at budget management for R&D, maximizing return on investment (ROI), and the functions of public and private investors. This chapter offers a thorough guide to navigating this difficult area and insights into new trends reshaping the R&D financing market.

> **Learning Objectives**
> - Understand the sources of R&D funding and evaluate their pros and cons.
> - Learn how to set and manage R&D budgets effectively.
> - Calculate and improve ROI for R&D projects.
> - Grasp the basics of government grants and incentives for R&D.
> - Explore private sector investment options for research and development.
> - Discover emerging trends in R&D funding.

4.1 The Basics of R&D Funding

R&D is the backbone of innovation, advancing industries, and assisting businesses in remaining competitive in a world that is constantly changing. R&D includes a broad range of activities, such as scientific research and product development, all of which are geared toward generating novel insights, technologies, and remedies. Organizations need sufficient cash to support their R&D projects to fuel their endeavors.

To secure the proper funding, one needs to not only secure the necessary financial resources but also make sure that they are used wisely to achieve the company's strategic goals. Funding for R&D is a vital enabler that makes it possible to pursue innovative concepts and create goods and services that have the potential to influence the future. Effective R&D funding is a difficult and dynamic process driven by economic conditions, industry developments, and the organization's particular goals. Understanding the fundamentals of R&D finance can help organizations better understand how to support their creative ideas and advance success in the dynamic field of research and development.

4.1.1 Sources of R&D Funding

Before we delve into the intricacies of R&D financing sources, it is critical to realize that there are many different funding sources for this field. These resources are essential for expanding scientific understanding, encouraging innovation, and stimulating economic progress. R&D financing supports the creation of cutting-edge goods and services that have the potential to reshape industries and enhance our quality of life.

The distribution of R&D funds is influenced by several variables, including the research type, the organization's objective, and the sector or industry in which it works. Research is a multifaceted and dynamic activity that can be carried out in various subjects, such as the natural sciences, engineering, social sciences, and humanities.

Government Grants and Contracts

Grants and contracts from the government are important sources of funding for R&D projects. Governmental organizations at different levels are in charge of managing these financing channels, which support research projects that are in line with their particular priorities and goals. Government support for R&D covers various scientific and technological pursuits thanks to various agencies, each specializing in a different area. To be considered for this funding, researchers and organizations need to submit thorough proposals that fully describe the aims, methods, anticipated results, and alignment of the research with the agency's goals.

Long-term, risky, and fundamental research are well suited to government R&D funding. This method of funding encourages scientific innovation and discovery that may have little economic applications but have the potential to influence society. It also offers the benefit of large financial resources and gives the projects it supports respect and legitimacy. However, it comes with strict compliance and reporting standards that necessitate strict adherence. Furthermore, receivers of government assistance may need to change due to shifting political goals.

Government contracts and grants are essential to R&D funding since they enable various research projects. Their value comes from their ability to provide significant financial resources for developing knowledge and innovation. Government financing has many benefits, such as legitimacy and support for fundamental research. However, it is also fiercely competitive, has complicated reporting requirements, and is subject to shifting political objectives. To make the most of these rich funding sources, organizations and researchers looking for government R&D financing should carefully match their proposals with the priorities of the appropriate agencies and be ready to comply with all requirements.

Private Sector Investment

R&D projects are crucially financed by private sector investment, especially for start-ups and fast-growing businesses. Private investors come in many forms, such as venture capitalists, angel investors, corporations, and private equity firms. For instance, venture capitalists (VCs) are experienced investors who oversee pooled funds and contribute not only finance but also crucial business knowledge. On the other hand, angel investors invest their own money and frequently take a more flexible approach to investing. Corporate investors may offer extra resources and experience while frequently trying to match R&D initiatives with their corporate objectives [1].

To receive money from the private sector, equity in the organization is often exchanged. This implies that investors—whether venture capitalists, angel investors, or corporate entities—become part owners of the business. They want a return on their investment, which increases the pressure on R&D initiatives to provide outcomes and become commercially viable. Returns can come in various forms, including dividend payments, capital gains from the future sale of shares, or firm earnings.

R&D programs also benefit from the knowledge and advice of private sector investors. Beyond providing financial assistance, their mentoring can assist projects in overcoming obstacles, improving their business plans, and establishing vital relationships within the sector. However, to attract private sector funding, investors need to conduct thorough due diligence in assessing the viability and potential of R&D initiatives. To increase their chances of receiving investment, organizations looking for this kind of finance should work to connect their goals with those of the investors.

Private sector investment has chances for networking, access to finance, and knowledge but also drawbacks and trade-offs. These include potential ownership dilution, decentralized decision-making, and meeting investors' expectations. A key component of managing private sector investment in R&D initiatives is striking a balance between the organization's and its investors' interests. Private sector investors are critical allies in pursuing creative projects and commercial success because they promote R&D initiative growth and drive innovation.

Internal Company Budgets

Many firms set aside internal resources to pay for their R&D efforts in addition to looking for money from outside sources. This self-funding strategy might have several benefits but has its share of difficulties. Unlike external sources, which are prone to volatility and uncertainty, internal budgets offer a constant and reliable source of revenue. This consistency guarantees that R&D projects can move forward without being hampered by a lack of funds. Additionally, it gives businesses a great deal of autonomy and control over their R&D endeavors, enabling them to precisely match research programs with their broader strategic goals and long-term objectives.

The capacity to strategically connect internal budgets with the organization's overarching objectives is one of the main advantages of doing so. This enables businesses to allocate funds to R&D initiatives that support their core operations, strengthen their competitive edge, or spur innovation in particular fields. However, there are limitations to this strategy as well. Since internal budgets are typically small and finite, careful resource allocation is required. This balancing is figuring out how much money should be set aside for R&D for other financial commitments including capital expenditures, operating expenses, and marketing.

There is a risk of underinvestment in R&D for firms that rely primarily on internal budgets, especially for smaller or financially limited entities. This may hurt an organization's competitiveness over the long run and limit the size and scope of R&D activities. Consequently, careful money management and project prioritization are crucial. To ensure that resources are used efficiently and that R&D efforts contribute to the organization's success, organizations need to set up systems for calculating the ROI for those activities. Based on their strategic value and possible impact, decisions need to be taken regarding which R&D projects to fund.

Crowdfunding and Other Alternative Sources

Alternative funding methods and crowdsourcing websites have been increasingly popular in recent years as creative ways to support R&D projects. These techniques allow businesses, especially start-ups and creative initiatives, to acquire money from various sources or use innovative ways.

In recent years, crowdfunding has become a viable alternative financing source for R&D projects. By obtaining small donations from a large number of people or organizations through online platforms, crowdfunding taps into the power of the public to promote creative ideas and projects. Crowdfunding is a desirable choice for R&D finance since it provides a variety of backers, the approval of ideas, and the protection of intellectual property. However, it also presents difficulties like fierce competition and the requirement for effective project completion [2].

Innovative people and organizations might look into additional nontraditional forms of R&D funding besides crowdfunding. Support from the general public, participation in contests and awards, teamwork with open source communities, and sponsored research agreements with influential companies or larger organizations are all innovative ways to raise money and resources for R&D activities. Each of these alternate sources has particular benefits and things to keep in mind.

While alternative funding options like crowdfunding offer adaptability and innovative methods to finance R&D, they also need strong marketing, communication, and occasionally a standout value proposition to attract supporters. These techniques suit R&D initiatives with broad appeal or strong affinities with particular interest groups.

4.1.2 Advantages and Disadvantages of Different Funding Sources

It is critical to weigh the benefits and drawbacks of various financing sources when looking for funding for R&D efforts (Table 4.1). Especially in academic and scientific research, government grants and contracts offer significant financial support and raise an organization's credibility. However, they are subject to shifting political agendas, loaded with onerous reporting and compliance obligations, and extremely competitive. On the other hand, private sector investment, frequently from angel and venture capitalists, can offer sizeable funds and priceless knowledge and advice. Usually, this investment strongly emphasizes market viability and rapid expansion.

Table 4.1 Pros and cons of R&D funding sources

Funding source	Advantages	Disadvantages
Government grants and contracts	Significant financial resources	Highly competitive
	Prestige and credibility	Stringent reporting and compliance
	Support for fundamental research	Subject to changing priorities
Private sector investment	Substantial capital injection	Equity dilution and loss of control
	Expertise and guidance	Pressure for quick returns
	Focus on rapid growth	Alignment with investors' objectives
Internal company budgets	Autonomy and control	Limited budget size
	Long-term strategic alignment	Constrained by other financial commitments
	Stable funding source	Potential to overlook external opportunities
Crowdfunding and alternative sources	Access to diverse funding	Not suitable for all projects
	Market validation	Effective marketing and outreach required
	Creative and unconventional funding options	No guaranteed success

However, it comes at the expense of share dilution, a possible loss of control, and the need to produce speedy returns, necessitating alignment with investors' goals.

Internal corporate budgets give R&D projects autonomy and authority and can be set up for long-term strategic alignment. Additionally, they provide a reliable source of funding, lowering uncertainty. However, they can work with tight budgets, be bound by other financial obligations, and run the danger of missing out on possibilities for outside support. Crowdfunding and other unconventional financing methods can confirm market interest while diversifying funds from a large audience. They do not, however, guarantee success and are only appropriate for some initiatives. They also call for efficient marketing and communication. When choosing the best financing source for R&D, weighing these benefits and drawbacks is critical. In many cases, combining funding sources is the most realistic strategy for ensuring consistent financial support for initiatives in innovation and research.

For R&D initiatives, choosing a suitable financing source is crucial, with substantial ramifications for the project's success. Several crucial elements must be carefully considered to navigate this decision properly. The overall project goals should be clarified because different financing sources work best when paired with focused objectives. Furthermore, the choice of source is heavily influenced by the organization's resource capacity, with government grants being fiercely competitive and private sector investments demanding equitable considerations. When considering private sector investment, it is important to assess the project's financial requirements, deadlines, risk tolerance, adaptability, and compatibility with investors' goals. Understanding how flexible and adaptable financing sources are is critical because some may permit strategic alterations while others do not. A comprehensive approach to diversifying and reducing risk in R&D endeavors is to use various funding sources.

The choice of an acceptable financing source in the context of R&D funding is far from a one-size-fits-all activity. Making educated selections requires careful planning, strategic thinking, and a thorough analysis of the benefits and drawbacks of each source. Making decisions that better position the organization for fruitful R&D activities requires understanding the complexities of various funding possibilities, taking into account the organization's distinctive qualities and the research project's specific goals. This strategy guarantees that financial resources are used as effectively as possible in pursuing innovation and excellence and that the chosen source is seamlessly in line with the main goals and requirements of the R&D efforts.

4.2 Budgeting for R&D

R&D budgeting is a crucial activity that necessitates careful management and planning. A strong R&D budget supports the organization's innovation initiatives and aligns with its strategic objectives. In this part, we will go into the essential elements of budgeting for R&D, from establishing goals to allocating funds. Research and development are the driving forces behind innovation in both the public and private sectors. It is where innovations are developed, technology is improved, and groundbreaking discoveries are turned into workable solutions. However, this is only possible with a wise use of financial resources. R&D budgeting is a strategic exercise that calls for a thorough comprehension of the organization's goals and the ability to translate those goals into concrete projects.

4.2.1 Setting R&D Budget Goals and Objectives

Setting well-defined goals and objectives is the cornerstone of good R&D budgeting. Budgets for R&D should be perfectly aligned with an organization's ultimate strategic goal. These goals are crucial in determining how R&D projects will proceed. They determine whether the emphasis is on the creation of novel products, the improvement of current products, the investigation of groundbreaking technologies, or the expansion of scientific knowledge. These goals need to be carefully chosen, aligned with long-term strategic plans, and guided by the organization's mission.

Prioritizing R&D projects according to their potential impact, viability, and alignment with the established objectives is essential to ensuring that the budget aligns with them. Due to their immediate business relevance, capacity for disruptive innovation, or strong resonance with client expectations, certain initiatives may be given priority. The setting of measurable objectives is also crucial. These objectives should be measurable so that progress can be monitored and R&D performance can be evaluated. They offer a practical way to assess achievements, whether they are measured in terms of developed prototypes, time-to-market, revenue estimates, or indicators for customer happiness.

R&D requires special consideration due to the inherent risk and uncertainty. These elements should be considered while setting targets, together with potential difficulties, technical uncertainty, and market volatility. Budgeting that considers unforeseen events and exploratory innovation is made possible by a pragmatic approach that considers these factors. As R&D goals should change in parallel with shifting strategic priorities, market realities, and technical breakthroughs, regular assessment and adaption of the objectives is essential. A dynamic environment demands dynamic goals that can adapt and be refined as new information is discovered.

4.2.2 Creating a Budgeting Process

To allocate resources and manage projects effectively, R&D must have a budgeting procedure. The procedure consists of several organized procedures that guarantee financial resources align with a business's strategic objectives (Fig. 4.1). Clear objectives and targets for R&D projects are first established, along with a list of the

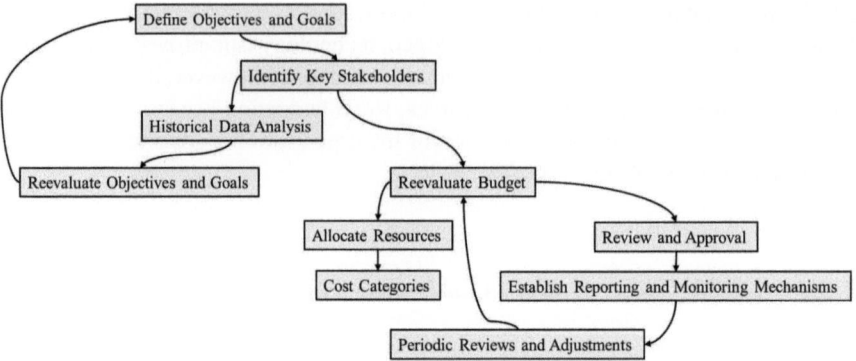

Fig. 4.1 R&D budgeting process with feedback mechanisms

major players in the budgeting process. Making informed budgeting decisions is made easier with the support of cost structure insights gained from the analysis of past expenditure data. Budgets for R&D typically include sections for human costs, direct project expenditures, overhead costs, and contingency funds.

A crucial step in the process is allocating resources to particular projects or initiatives while considering project priority, possible ROI, and project timescales. A rolling budget strategy is frequently preferred because it allows flexibility and adaptability as conditions change. Following the creation of the initial budget, it is reviewed and approved by important decision-makers and stakeholders to ensure it is in line with the organization's overall financial strategy. To consistently track budget vs. actual spending, it is essential to establish reliable reporting and monitoring procedures. The R&D budget is kept on track and flexible by regular assessments and modifications that consider project milestones, market conditions, and unforeseen developments.

4.2.3 Factors to Consider When Budgeting

It is important to comprehend the many aspects that can significantly affect the R&D budget before getting into the intricacies of budgeting for R&D. A well-planned budget considers various factors, allowing for efficient resource allocation and raising the likelihood that the project will be successful.

Personnel and Talent Acquisition

The hiring of personnel and the development of talent are crucial. The team's competence directly influences the quality and results of the research initiatives. Budgeting for personnel needs to include money set aside for attractive pay and benefits plans that can draw and keep top talent. Benefits and incentives, such as health insurance, retirement plans, and bonuses, not only help to recruit qualified workers but also increase their loyalty and job happiness. The staff must also continually expand their skills through workshops, courses, and conferences to stay on the cutting edge of their respective industries.

In the budgeting of personnel, balancing staffing levels is equally important. It is crucial to tailor the team size to each project; more complex or time-sensitive tasks may require more team members. Keep a tight watch on workload management to prevent overworked workers from being burned out and producing less. Budgets for personnel need to be flexible to account for project changes, with contractors or temporary workers offering the most cost-effective options for specialized, short-term tasks.

Create mentorship programs and job promotion paths to improve retention and career growth. Outstanding work should be acknowledged and rewarded; this is an effective retention tactic. Employee happiness and loyalty are further increased by fostering a work environment that balances work and personal life. A competitive environment, the unique requirements of R&D efforts, and the promotion of a supportive and growth-oriented workplace are all considered when allocating resources to personnel and talent acquisition. Such a strategy aids in recruiting, developing, and maintaining a high-performing team, which is essential to accomplishing the R&D projects.

Equipment and Technology

The purchase and upkeep of sophisticated machinery and cutting-edge technologies are essential to the projects' success. For the R&D activities to be efficient and effective, carefully budgeting these resources is essential. It is wise to consider several things regarding technology and equipment. Start by allocating funds for the first purchase of machinery and technology, including software licensing to sophisticated laboratory gear. Be sure to account for the price of equipment warranties and service agreements, as these contracts can be crucial in maintaining the equipment's peak performance and minimizing downtime.

Additionally, keep in mind to set aside money for routine preventive maintenance to increase the equipment's lifespan and performance. Additionally, it is important to remember that equipment and technology can become obsolete or less effective as they age. This may require budget allocation for essential upgrades or replacements to uphold research quality.

Additionally, it can be taken into account the adaptability that renting or leasing choices provide in some circumstances. This strategy may be particularly useful for initiatives requiring temporary equipment or using quickly changing technology. When choosing rental or leasing agreements, it is crucial to consider the long-term financial effects and bargain advantageous terms with vendors, including rental duration, maintenance agreements, and return conditions.

Materials and Supplies

Whether the goal is to produce new products, conduct scientific studies, or develop new engineering techniques, materials, and supplies, these are fundamental components of any R&D project. Maintaining the smooth and efficient progression of R&D initiatives necessitates smartly budgeting these resources. Setting a budget for materials and supplies requires considering several important factors. In this aspect, inventory management is essential. A thorough list of the materials and supplies required for the projects, identifying quantities, quality requirements, and estimated consumption rates must first be compiled as part of an assessment of the project-specific needs. To prevent project delays brought on by shortages, inventory control methods should be implemented to monitor utilization, with systems that warn stakeholders when stock levels are getting low. Whenever possible, Just-In-Time (JIT) procurement techniques should be used to reduce the requirement for lengthy storage and related carrying expenses and the danger of products becoming outdated.

Another important aspect is quality control. Ensuring that these resources continually comply with these requirements and maintaining the integrity of R&D initiatives requires defining and documenting quality standards for the materials and supplies. To maintain constant quality, it is crucial to choose trustworthy and reputable suppliers, with the negotiation of advantageous terms and contracts being a key step in securing a steady supply. It is crucial to regularly check and test incoming materials and supplies since this proactive strategy can identify and fix quality problems early on, reducing the possibility of project setbacks. A proactive strategy for reducing waste should also be undertaken. This entails making efforts to optimize material usage through the creation of effective work procedures and processes, hence ensuring wise utilization. The investigation of material reuse and recycling options is equally crucial, as these actions can reduce costs and support more general environmental sustainability goals.

Proactive price negotiation with suppliers, particularly for large orders, becomes a key strategy for cost control. Investigating opportunities for discounts, volume-based pricing, or long-term contracts can be profitable. It is also advised to set aside a certain amount of the R&D budget for materials and supplies to ensure these resources get the required funding. Cost tracking should be a continuous process that calls for being ready to modify the budget if market conditions or supplier pricing dynamics change. Last but not least, safety and compliance issues should not be disregarded. Ensuring that all materials and supplies adhere to safety standards and legal requirements is crucial because failure to do so may result in expensive delays or legal problems. To ensure complete adherence to safety laws throughout the R&D initiatives, Material Safety Data Sheets (MSDS) for all materials, particularly hazardous compounds, need to be kept up to date and easily accessible.

Figure 4.2 illustrates the integrated R&D funding and budgeting process for managing materials and supplies. Prioritizing a budget and establishing the project's requirements come first, then vendor selection and cost projection. The flowchart then divides based on cost comparison, with one branch directing procurement if costs are within budget and the other triggering a review if expenses are above

4.2 Budgeting for R&D

budget. Additional processes include inventory control, quality assurance, and expense tracking, all rigorously compared to the given budget. Corrective measures are implemented if the budget is exceeded, yet successful projects keep moving forward with project implementation. Adjustments are made as necessary to complete the project and bring the process to an end.

Overhead Costs

Any R&D budget needs to include overhead costs since they are essential to the operations, infrastructure, and personnel of R&D. These indirect costs may not be directly related to particular R&D initiatives. However, they are nonetheless crucial to the overall operation of the research organization. For wise budgeting, it is essential to comprehend and manage overhead costs. The R&D budget's overhead costs should consider utilities, office space, administrative assistance, and risk management.

Utilities, including water, electricity, heating, and cooling costs power the R&D facilities. Utilities are also necessary for running laboratories. Using energy-efficient devices and practices can aid in the management of utility expenses and the reduction of costs. Budget concerns for office space include rent or lease costs and

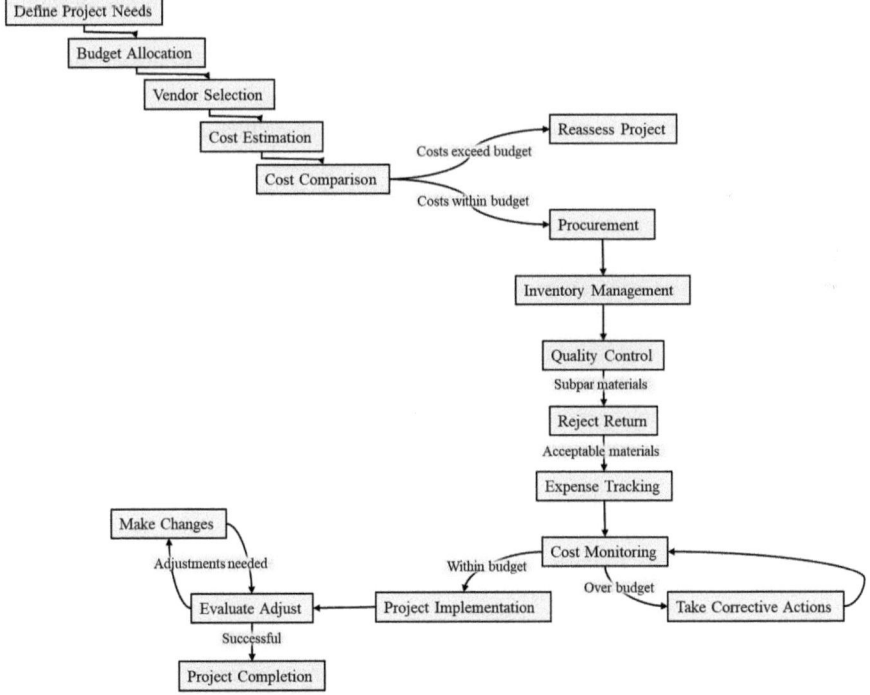

Fig. 4.2 Materials and supplies management in R&D budgeting

maintenance and repair charges, frequently used for administrative and project management tasks. Cost-effectiveness and the requirement for a positive work environment need to be balanced.

The efficient running of R&D operations depends on the administrative support staff, and budgetary considerations for this workforce include pay, benefits, and office supplies. Optimizing personnel levels and improving administrative procedures using technological solutions are essential for effective cost management. Additionally, even though it is not a direct overhead expense, the R&D budget needs to include money for risk management to handle unforeseen difficulties. To ensure the R&D organization is ready for any unforeseen setbacks; this involves setting aside contingency money for unforeseen expenses and looking into insurance solutions to reduce financial risks connected with disruptions.

Contingency Planning

Recognizing creative initiatives' fundamentally dynamic and unpredictable character, contingency planning is a crucial part of the budgeting process. The allocation of resources and the creation of methods to deal with unforeseen costs and potential risks that may surface throughout an R&D venture are all included in this planning. It begins by identifying potential sources of unplanned costs, such as equipment failures, the demand for further testing or analysis, changes in the market, or modifications to regulatory requirements. Consequently, a percentage of the R&D budget should be set aside for contingency funding, often 10–20%, depending on the project's particulars.

Effective contingency planning includes risk management, which includes a review of potential dangers in the technological, market-related, legal, and financial realms. Strategies are developed to promote early discovery and reaction to reduce these identified risks, and continuous monitoring of potential risks happens throughout the project's lifecycle. Contingency funds need to be flexible; they need to be able to respond to unforeseen problems without becoming a catch-all for unforeseen costs. The recording and distribution of contingency funds are conducted with transparency and accountability, ensuring that team members with relevant duties know the budget's flexibility for dealing with unforeseen difficulties. Implementing a well-structured contingency plan not only protects R&D investments but also shows a steadfast dedication to the accomplishment of creative undertakings.

4.2.4 Allocating Funds for Different R&D Projects

The cash distribution among various projects is a crucial phase. The first phase, prioritization, comprises matching projects with organizational goals, assessing their potential impact, and determining resource needs. Equally important is maintaining equilibrium in the distribution of resources. Diversifying risk among projects, considering both short- and long-term objectives, and realizing resource limitations are all part of this. Regular reviews and modifications are required to ensure that funds are distributed as efficiently as possible. Performance indicators, milestone evaluations, and the flexibility to reallocate resources as necessary all play critical roles in this process. Effective project team communication further assures openness, understanding, and flexibility within allotted budgets.

4.3 Managing R&D Budgets

For R&D activities to be successful, R&D budgets need to be managed effectively. It guarantees effective resource allocation, on-time project completion, and the organization's ability to meet its innovative objectives. For R&D projects to be completed successfully, budget management for R&D needs to be monitored and track expenditures. The cornerstone of efficient budget management is regular monitoring, which calls for creating a reporting schedule, defining defined roles and duties, and using cutting-edge budgeting and spending tracking software. The R&D teams ensure that budget tracking becomes a regular part of their operations by maintaining a defined timetable, enabling them to identify and correct violations quickly. Accountability is encouraged by designating specific people or groups to handle these activities, and the organization of budget-related data is aided by using distinct spending categories. Additionally, budgeting and cost-monitoring software makes it easier to track budget changes, improves transparency, and offers real-time information.

The classification and itemization of expenses are essential steps in the spending monitoring process. This level of detail offers a more thorough understanding of resource allocation and budget management. For instance, spending for different positions, salaries, and perks should be separated within the people category to evaluate labor costs and optimize teams. Similarly, tracking individual purchases, maintenance expenses, and depreciation within the equipment and technology category provides data crucial for asset management and replacement planning. It is possible to identify potential cost-saving possibilities and better effectively use resources by itemizing and categorizing costs associated with goods and supplies.

Effective R&D budget management necessitates the development of a spending review and approval process in addition to ongoing monitoring and thorough classification. Representatives from project management, finance, and other pertinent departments participate in this process, typically overseen by a budget monitoring committee. To improve control over budget expenditures, the committee creates criteria requiring further approvals for expenses, particularly for high or unusual costs. Additionally, keeping thorough records of all expenditures, such as receipts, invoices, and other pertinent documents, is essential for transparency and auditing. This paperwork guarantees that spending aligns with project objectives and financial restrictions, encouraging prudent budget management techniques.

4.3.1 The Role of Financial Reporting in R&D Budget Management

Financial reporting provides a systematic framework for monitoring, analyzing, and optimizing resource allocation within R&D activities, making it a crucial pillar in the field of R&D budget management. Timely and accurate financial reporting is the cornerstone of efficient R&D budget management. All financial transactions need to be meticulously documented, data need to be current, and a thorough breakdown of R&D expenses needs to be provided. It is impossible to emphasize the significance of accuracy since it guarantees that decision-makers have access to reliable data. In contrast, inaccuracies may result in poor budget choices that could jeopardize the success of R&D projects. Timely reporting holds equal significance, as delays could hinder the ability to rectify budget inconsistencies and make wise resource allocations. Financial reporting also includes variance analysis, contrasting planned amounts with actual spending. It continues beyond data entry. This study offers priceless insights into how well the budget is being used, empowering decision-makers to make well-informed decisions about resource reallocation, cost containment, and project scope changes. Additionally, the automation, data visualization, and customization tools used to modernize financial reporting have changed its effectiveness and efficiency in managing R&D budgets, improving communication and alignment with budget goals.

Another crucial element is the creation of a reporting schedule, which promotes collaboration and guarantees consistency among stakeholders like project managers, financial analysts, and decision-makers. Reporting intervals can range from monthly to quarterly depending on the project's complexity, but the goal is to maintain regularity for ongoing budget performance monitoring. This methodical approach also applies to variance analysis, which involves looking at budget variances, whether positive (under budget) or negative (over budget), and then determining their main causes. Adjustments in project priorities, scope adjustments, or unforeseen costs are common root causes that offer crucial information for course correction. Modern software platforms and solutions are essential for maximizing the effectiveness of financial reporting. Data visualization tools convert complex financial data into clearly understood charts and graphs, improving communication with various stakeholders. Automation streamlines the reporting process, increasing its efficiency and decreasing its propensity for errors. Additionally, customization enables businesses to create reports tailored to their requirements and preferences and aligned with their budget management goals. The interrelated elements of financial reporting are shown in Fig. 4.3 for R&D budget management, emphasizing the value of accuracy, timeliness, regular reporting, variance analysis, and reporting tools.

Fig. 4.3 Financial reporting in R&D budget management

4.3.2 Making Budget Adjustments When Necessary

Budget management for R&D calls for agility and flexibility. Budget modifications are necessary when circumstances change, or unforeseen difficulties or opportunities appear. Flexibility is fundamental, and emergency funds are set aside to cover unanticipated costs without affecting the total budget. In these circumstances, prioritization is essential, with an emphasis on initiatives that closely match strategic goals and have the potential to generate a better ROI. When choosing which projects to prioritize, timely completion, effective resource use, and strategic alignment are important factors. Understanding potential budget risks, such as technical difficulties, market upheavals, or legislative changes, requires regular risk assessments. The allocation of resources for risk mitigation, such as budgetary allowances for unforeseen technical difficulties or delays, is informed by this knowledge. When adapting the R&D budget to changing conditions, effective communication and cooperative decision-making guarantee that budget adjustments are done with a thorough understanding of the organization's goals and priorities.

4.3.3 Avoiding Common Budgeting Pitfalls

In addition to making the necessary modifications, effective R&D budget management requires being proactive in avoiding typical budgeting errors that can result in budget overruns, delayed projects, and problems with resource allocation. Scope creep is a common problem where the project's scope grows beyond its original description, causing unexpected costs and delays. All R&D projects need to have clear objectives and scopes, be documented, and have a change control procedure that requires official approval for any scope changes to minimize scope creep.

Overly optimistic expectations can create actual expectations and budget overruns, which is another typical mistake. To prevent this, the budget should be based on a reasonable estimation of costs, taking historical data and industry benchmarks into account, and it should include contingency funds to cover unforeseen costs. Another area that might result in misconceptions and financial problems is better communication between project teams, budget managers, and decision-makers. Set up regular meetings to enhance transparency and facilitate communication by disclosing budget information and project status to all pertinent parties. Additionally, a lack of accurate financial and spending records can make managing expenses clearer and more manageable. Maintain thorough and detailed documentation of the budget management and approval procedures to overcome this problem.

Finally, lacking contingency planning may lead to resource shortages and budget issues. To avoid this, set aside some of the budget for a contingency fund, think about probable risks and difficulties, and create scenarios that show how the budget would change in response to various situations. Table 4.2 summarizes frequent budgeting errors in R&D, preventative actions, and probable effects on R&D initiatives.

It is wise to avoid typical budgeting problems in R&D budget management by being watchful for scope creep, avoiding excessively optimistic assumptions, encouraging communication, maintaining thorough documentation, and considering contingency planning. These steps will assist in preserving financial restraint and guarantee that the R&D programs remain on course.

Table 4.2 R&D budget management pitfalls and preventative measures

Pitfall	Description	Preventative measures	Impact on R&D
Scope creep	Expanding project scope beyond initial definition	Define clear objectives Implement change control	Unanticipated expenses and delays
Overly optimistic assumptions	Unrealistic cost and timeline estimates	Base budget on reality Include contingency funds	Budget overruns and unrealistic expectations
Lack of communication	Inadequate stakeholder communication	Schedule regular updates Promote transparency	Misunderstandings, missed opportunities, budget challenges
Inadequate documentation	Insufficient financial record-keeping	Maintain clear records Document processes	Confusion and difficulties in tracking expenses
Lack of contingency planning	Failure to prepare for challenges	Allocate contingency funds Develop scenario plans	Budget crises and resource shortages

4.4 Ensuring ROI

Realizing a positive ROI is one of the core objectives of every firm involved in R&D. Even though the intrinsic nature of R&D frequently entails risks and uncertainties, building processes for gauging and improving ROI to support long-term innovation and justify resource allocation is crucial.

4.4.1 Measuring the Success of R&D Projects

The success of R&D initiatives depends on carefully evaluating many key performance indicators (KPIs), which are crucial for determining the effectiveness and impact of investments. Accurately measuring key indicators is essential to ensure that the R&D projects align with the company objectives. Time-to-market, cost-effectiveness, market acceptance, creation of intellectual property, and customer feedback are important criteria for judging the success of R&D projects. Critical KPIs for gauging the success of R&D projects are provided in Table 4.3.

Time-to-market is a crucial indicator measuring how quickly an R&D project moves from a concept to a product or service ready for the market. Getting to market faster results in competitive advantages, increased revenue, and quicker returns on investment. It entails computing the typical time from project inception to launch, considering regulatory processes and development phases, and comparing this timeframe to industry norms. Cost efficiency, or the capacity to achieve R&D objectives within the budget allotted, is equally important. To maximize ROI and prevent budget overruns, effective cost management is essential. It entails comparing actual expenses to anticipated prices and assessing how substantial variances affect the project's success. Market adoption gauges how well-received and in-demand R&D is in the intended market. Increased sales, market share, and revenue result from high market adoption. It is crucial to assess consumer demand, interest, and adoption rates using market research, sales statistics, consumer feedback, and market penetration. The creation of patents, copyrights, trademarks, or proprietary information are all examples of intellectual property generation. These are valuable assets that can be used to generate cash through partnerships, licensing, or legal protection. The measurement entails calculating the amount of intellectual property produced and estimating its market value and potential revenue. Client feedback measures the level of feedback and client satisfaction for R&D-driven goods or services. While negative feedback directs changes, positive feedback indicates successful design and development. For this evaluation, gathering and examining consumer feedback through surveys, reviews, and direct conversations is essential. It is also important to track customer sentiment and satisfaction changes over time.

Table 4.3 KPIs for measuring R&D project success

KPI	Definition	Significance	Measurement
Time-to-market	Measures the speed from concept to market-ready product or solution	Shorter time-to-market leads to competitive advantages and quicker ROI	Calculate the average time from initiation to launch and compare it to industry benchmarks
Cost efficiency	Evaluates the ability to achieve R&D goals within the allocated budget	Efficient cost management maximizes ROI and prevents budget overruns	Compare actual costs to budgeted costs and analyze deviations' impact
Market adoption	Assesses the acceptance and demand for R&D output in the target market	High adoption leads to increased sales, market share, and revenue	Conduct market research and analyze sales data, customer feedback, and penetration
Intellectual property generation	Measures the creation of patents, copyrights, trademarks, or proprietary knowledge	Intellectual property can generate revenue and long-term value	Quantify the number of patents, copyrights, trademarks, or proprietary knowledge. Assess market value
Customer feedback	Evaluates customer satisfaction and feedback among end-users of R&D-driven products or services	Positive feedback indicates successful development; negative guides improvements	Gather and analyze feedback through surveys, reviews, and direct communication. Monitor sentiment changes

4.4.2 Calculating ROI for R&D Initiatives

The ROI needs to be calculated to assess the effectiveness and worth of R&D initiatives [3]. ROI analysis aids firms in assessing whether their R&D initiatives yield a profit, align with strategic objectives, and support resource allocation. There is a simple formula used to calculate ROI:

$$\text{ROI} = \frac{\text{Gain from Investment} - \text{Cost of Investment}}{\text{Cost of Investment}} \times 100\%$$

The "Gain from Investment" encompasses the financial benefits and returns generated by the R&D project, such as increased revenue, cost savings, or new revenue streams. In contrast, the "Cost of Investment" includes all expenses related to the R&D initiative, such as salaries, materials, equipment, and overhead costs. The resulting ROI value can fall into one of three general categories:

1. *Positive ROI (greater than 0%)*: A positive ROI indicates that the R&D initiative has generated more financial gains than the costs incurred, resulting in a net positive return. This is the desired outcome for R&D projects.

2. *Negative ROI (less than 0%)*: A negative ROI suggests that the R&D initiative has not recouped the costs and has resulted in financial losses. Organizations need to assess and possibly reconsider the project.
3. *Break-even ROI (equal to 0%)*: A break-even ROI signifies that the R&D initiative's gains are precisely equal to its costs. While this may not represent a loss, it does not provide a net financial gain, which may prompt a review of the project's value.

When calculating ROI for R&D initiatives, several considerations come into play. These include defining the timeframe for measurement, ensuring data accuracy, attributing gains and costs correctly, discounting for the time value of money, and acknowledging nonfinancial factors that may influence the project's value. While ROI is a valuable tool for assessing R&D projects, it has challenges, primarily due to the uncertainties and long timeframes inherent in many R&D endeavors. Accurate data collection and rigorous evaluation are crucial for reliable ROI calculations. Overall, ROI analysis is pivotal in guiding decision-making, resource allocation, and strategic planning for organizations engaged in R&D.

Consider a hypothetical case: a technology company invests $1 million in an R&D project to develop a new software product. Over 3 years, the project generated $1.5 million in additional revenue and saved the company $200,000 in operational costs. The ROI calculation is as follows:

$$\text{ROI} = \frac{\$1,700,000 - \$1,000,000}{\$1,000,000} \times 100\% = 70$$

This hypothetical project yields a positive ROI of 70%, demonstrating a substantial return on the initial investment. While ROI analysis is valuable, it should be part of a comprehensive evaluation considering financial and nonfinancial factors to provide a holistic view of an R&D project's value.

4.4.3 Factors Impacting ROI in R&D

A variety of dynamic elements affect R&D ROI. Market variables, such as competition, consumer preferences, and trends, influence the prospective returns on R&D investments. Increased R&D spending may be necessary in highly competitive markets to maintain an advantage. Another important consideration is technical feasibility, which establishes whether it will be feasible to accomplish R&D objectives on time and within budget. The R&D team's expertise is crucial since a knowledgeable and cooperative group is better equipped to complete tasks and control expenses quickly. Effective risk management, which includes risk assessment and mitigation techniques, helps avoid unanticipated delays and overruns, favorably affecting ROI. Additionally, by preserving the value developed during R&D, protecting intellectual property through patents and other forms of protection helps guarantee future revenue streams and boost ROI.

4.4.4 Strategies for Improving ROI

R&D project ROI needs to be maximized, necessitating a multidimensional strategy that includes project selection, efficient project management, and ongoing review. An essential first step in increasing ROI is project selection, which calls for thorough consideration of market relevance, technical viability, and alignment with strategic objectives. Organizations can increase their chances of attaining a good ROI by concentrating on projects that meet market demands, use organizational expertise, and support long-term objectives.

Effective project management techniques are essential for ROI optimization. Project management's key components are efficient cost control, prompt execution, and risk management. Cost management and ROI preservation depend on attentive monitoring of R&D budgets, adherence to project timeframes, and proactive risk management. Projects are kept on schedule, and adjustments are made as necessary by ongoing evaluation using milestone reviews, open feedback loops, and project-specific key performance indicators (KPIs). Throughout the R&D lifetime, this continual review offers the insights required to sustain and improve ROI.

To make sure that R&D investments are in line with market demands and take advantage of new opportunities, collaboration between R&D teams, marketing, and sales departments is essential. A positive ROI can be achieved by involving marketing and sales personnel in market validation and entry plans. This helps to improve R&D concepts, increase market acceptance, and generate early income. In addition, utilizing and effectively safeguarding intellectual property (IP) created during R&D is a significant ROI driver. This includes protecting intellectual property (IP) through patents, trademarks, and copyrights and looking into licensing possibilities to create new revenue sources. When these tactics are applied throughout the R&D process, a holistic strategy for raising ROI, maximizing innovation, and preserving competitiveness in a changing market environment is created.

4.5 Government Grants and Incentives

Government subsidies and incentives are critical in propelling R&D activities across various businesses. These initiatives aim to support innovation, promote economic expansion, and address pressing societal issues. This section goes into the world of government-sponsored R&D and provides information on obtaining this money and carrying out any necessary obligations.

Government support for R&D takes many forms, including contracts, tax incentives, grants for research, and grants. Federal, state, and local government organizations with distinct goals, qualifications, and application procedures primarily manage these initiatives. For instance, research funds involve competitive awards and comprehensive submissions that describe the study's goals, methods, and projected results. On the other hand, government contracts are agreements that charge businesses with specified R&D tasks. Tax benefits for innovative enterprises may include credits, deductions, or lower tax rates.

It can take time to obtain government incentives and grants. The initial steps entail choosing appropriate programs and carefully examining eligibility requirements to ensure alignment with the R&D objectives. Being able to effectively describe the study plan, the problem being treated, and alignment with program goals in a submission is crucial, especially for research funds. Gaining guidance and insights requires building relationships with the right government entities. To guarantee continuous financing and avoid legal issues, it is crucial to maintain compliance with reporting standards. Awareness of program modifications, opportunities, and deadlines is especially essential because the government funding landscape is subject to change. A methodical approach is needed during the application process, marked by diligence, transparency, and a clear alignment with program objectives to increase the likelihood of success. Government funding for R&D projects can significantly bolster research capabilities and financial resources.

4.6 Private Sector Investment

An essential component of funding for R&D activities is private sector investment. Whether one is a biotech company, a tech start-up, or an existing company seeking innovation, private sector investment provides the resources and expertise necessary to bring groundbreaking ideas to fruition. Venture capitalists (VCs) and angel investors from the private sector frequently look for promising businesses with the potential for disruptive innovation. They are drawn to R&D initiatives that show promise for the market and have the potential to fill unmet demands. A strategic approach that includes effective communication, an inspiring goal, and a clearly defined road to success is necessary to draw in these investors.

4.6.1 Attracting Venture Capital and Angel Investors

Early-stage and high-potential R&D projects depend heavily on VC and angel investors for financial support. It is crucial to create a persuasive pitch that highlights the unique R&D endeavor and conveys its market potential and competitive edge to grab their attention and win their investments. This pitch should outline the project's distinct value proposition and be supported by a thorough market analysis, a viable business plan, and tangible results. Indicators of market demand that are observable, such as customer interest, pre-orders, collaborations, or user interaction, are of special importance to investors. Investors examine the team in addition to the idea as a key component. Building a solid, varied staff with reputable advisers and mentors on board boosts investor trust. The pitch can be strengthened further by showcasing prior accomplishments, whether from prior business operations, fruitful R&D initiatives, or team members with relevant industry experience. To ensure that each team member's contribution effortlessly links in with the project's goals, it is equally crucial to identify roles and responsibilities within the group.

A crucial stage in luring venture capital and angel investors is market validation. Beyond theories and presumptions, verifiable evidence of market demand is persuasive. Customer interviews, pilot program implementation, getting pre-orders or early commitments, and competitor analysis are just a few techniques for validating the market. Investors want confirmation that the R&D innovation meets a genuine need and stands out from competing products. Furthermore, showing that the intellectual property is appropriately protected is crucial. Investors desire reassurance that patents, copyrights, and trade secrets protect intellectual property. Talking about the project with possible investors can call for the assistance of intellectual property lawyers and the use of nondisclosure agreements (NDAs). A consistent effort is needed to obtain money because there is fierce competition for venture capital and angel investments. Making constant improvements to the pitch, connecting with investors, and remaining flexible in the face of the industry's constant change are all crucial.

4.6.2 Corporate Partnerships and Strategic Investors

Thanks to corporate partnerships and strategic investments, R&D projects have a unique opportunity to secure funds and resources. These partnerships entail tight cooperation with reputable businesses with a strategic stake in the inventions or technology. Finding synergy between potential partners' R&D projects and strategic goals is the first step in creating a successful corporate cooperation. Finding businesses whose capabilities and objectives align with the R&D project is essential, as is performing in-depth market research and creating an effective value proposition that illustrates the advantages of the cooperation.

The next step involves drafting a comprehensive proposal that outlines the partnership's parameters once a suitable partner has been identified. This proposal should include the precise goals, objectives, anticipated results, project phases, milestone timeframes, the resources each partner will contribute, and the partnership's governance structure. Both sides must perform due diligence, which entails carefully evaluating each other's financial standing, reputations, and potential dangers, before entering into a corporate partnership. Reviewing financial statements, legal and compliance checks, background checks, and risk assessments are all part of the standard due diligence process.

Legally binding agreements that safeguard the interests of both parties form the basis of a fruitful corporate collaboration. The terms and conditions of the partnership, such as its duration, termination provisions, and the range of collaboration, as well as the handling of intellectual property, confidentiality and nondisclosure, and the procedure for resolving disputes, should be covered in these agreements, which legal experts should draft. Strategic investments and corporate alliances may be able to offer the resources and know-how required for the success of R&D projects. To ensure that the collaboration aligns with the long-term goals and continues to be advantageous to both parties, it is essential to approach these partnerships with openness, clear communication, and a well-defined legal framework.

4.6.3 Equity Financing and Its Impact on Ownership

A key element of private sector investment in R&D initiatives is equity financing, which involves issuing ownership shares in exchange for funding. Organizations frequently issue equity, which represents ownership shares in the company, when they seek outside finance from venture capitalists, angel investors, or strategic partners. Early-stage R&D ventures frequently use this type of funding, in which equity investors contribute money in exchange for a portion of ownership. The level of ownership interests is based on the company's worth. If new shares are issued, current ownership may be diluted, affecting control and decision-making inside the business.

Equity funding has a considerable effect on ownership and control. In exchange for investment, entrepreneurs and management need to give up ownership, which results in a loss of ownership even while it offers crucial funds for R&D projects. It is critical to determine how much ownership is willing to be given up because this transfer of ownership can impact decision-making authority. A healthy working relationship requires that the objectives of equity investors match those of the company's mission and vision, and discussions concerning exit alternatives should be included in the equity financing agreement.

4.7 Future Trends in R&D Funding and Budgeting

The methods for funding and budgeting R&D ventures change along with the research and development landscape. Organizations need to stay ahead of these new trends to stay competitive, inventive, and financially viable.

4.7.1 Emerging Funding Sources and Methods

Organizations are continuously looking for novel financing sources and techniques to support their projects. Through online platforms, crowdsourced financing has become a potent method for leveraging individual financial support. Using this strategy, one can diversify funding options, validate concepts, and gain valuable early feedback. Understanding the several types of crowdfunding, such as reward-based, equity-based, or donation-based, and choosing the best one for the R&D effort is vital. Public-private partnerships involving cooperation between governments, academic institutions, and private businesses are becoming more prevalent in the R&D scene. These collaborations pool resources, expertise, and infrastructure, which makes them especially useful for tackling difficult and expensive R&D problems. The benefits of these collaborations include risk sharing and regulatory support, but navigating them needs careful negotiation, IP management, and purpose alignment.

Another developing trend in fundraising is impact investment, which is motivated by a desire to provide advantageous social and environmental consequences and financial benefits. Impact investors aim for a "double bottom line," focusing on verifiable societal or environmental benefits and financial gains. It is crucial to show how the R&D project fits with impact goals and to be ready to track and report on these results. Advanced intellectual property techniques, like monetizing IP assets through licensing, developing IP collaboratively, and IP valuation, offer a special way to continue supporting and innovating inside R&D projects. The organization's ability to respond to the constantly changing R&D finance landscape can be improved by incorporating these new funding sources and techniques into the R&D plan. Selecting the best technique for the particular R&D project and objectives is crucial because each strategy has unique advantages and considerations.

4.7.2 Technological and Economic Trends Shaping R&D Financing

R&D financing dynamics are becoming increasingly entwined with economic and technical trends, making it crucial for companies to stay abreast of these developments to maximize their R&D efforts and successfully get funding. By enabling data-driven decision-making, predictive analytics for project selection and cost forecasting, and risk mitigation, big data and analytics have transformed R&D finance [4]. This has improved the effectiveness and efficiency of R&D financing. R&D financing is being revolutionized by blockchain and smart contracts [5], which increase collaboration transparency and confidence, automate financial transactions and royalty payments, provide secure intellectual property management, streamline operations, and increase trust among collaborators. Assembling virtual teams from across the world, cutting costs, and speeding up project timelines are all made possible by globalization and remote collaboration [6], which is made possible by new communication technology. These processes also increase access to talent, lower

Table 4.4 Key trends in R&D financing

Trend	Technologies	Impact on R&D financing
Public-private partnerships	Collaborative platforms, data sharing technologies	Diversifying funding sources enhances innovation
Venture capital and start-ups	AI for investment analysis, blockchain for fundraising	Attracts external capital and fosters tech innovation
Cross-functional teams	Project management software and collaborative tools	Optimizes resource allocation and accelerates projects
Data-driven decision-making	Big data analytics, business intelligence solutions	Informs budget allocation and reduces financial risks
Agile budgeting	Agile project management tools, real-time reporting	Enables adaptability and controls project cost overruns
Impact assessment	Key performance indicators (KPIs), impact assessment tools	Ensures ROI and aligns R&D with strategic goals
Crowdsourcing and crowdfunding	Crowdsourcing platforms, crowdfunding websites	Diversified funding sources engage the public
Sustainability and green R&D	Sustainable tech (e.g., renewable energy, green materials)	Attracts environmentally conscious investors
Risk mitigation	Risk assessment software and financial modeling tools	Mitigates budgetary risks and ensures project viability
Regulatory compliance	Regulatory compliance software, legal tech solutions.	Ensures legal adherence and avoids penalties

prices, and increase the flexibility of R&D finance. Additionally, by emphasizing sustainability and ESG factors in R&D financing [7], investors prioritizing environmental and social responsibility may provide funds and help R&D initiatives connect with larger sustainability objectives.

To stimulate and assist R&D projects, governments now play a crucial role in financing R&D by providing grants for research, tax breaks, and regulatory support. Organizations need to comprehend and take advantage of these activities because they are crucial for obtaining R&D funding and financial support. By embracing these technological and economic developments and adjusting R&D plans accordingly, businesses may greatly increase the efficiency of their R&D finance and ensure the availability of the resources required for successful innovation. The impacts of associated technologies, including the diversification of funding sources and risk reduction, are summarized in Table 4.4, along with trends in R&D financing.

References

1. S. Barkoczy et al., Innovation, start-ups and venture capital, in *Incentivising Angels: A Comparative Framework of Tax Incentives for Start-Up Investors*, (Springer, Singapore, 2019), pp. 11–27
2. E.C. Stoica, E.M. Iliescu, Crowdfunding-a viable alternative to finance small and medium enterprises, in *Challenges of the Knowledge Society*, (Nicolae Titulescu University Editorial House, Bucharest 2022), pp. 597–602. https://www.proquest.com/openview/0687838e14c5b ba1de58946a22b41557/1?pq-origsite=gscholar&cbl=2036059

3. M. Hassanzadeh, T.B. Bigdeli, Return of Investment (ROI) in Research and Development (R&D): Towards a framework, in *Collaboration–Impact on Productivity and Innovation: Proceedings of 14th International Conference on Webometrics, Informetrics and Scientometrics & 19th COLLNET Meeting 2018*, 5–8 Dec 2018, University of Macau, Macau, 2019
4. S. Ren, Optimization of enterprise financial management and decision-making systems based on big data. J. Math. **2022**, 1–11 (2022)
5. V. Chang et al., How Blockchain can impact financial services–The overview, challenges and recommendations from expert interviewees. Technol. Forecast. Soc. Chang. **158**, 120166 (2020)
6. O. Gassmann, M. Von Zedtwitz, Trends and determinants of managing virtual R&D teams. R&D Manag. **33**(3), 243–262 (2003)
7. L. Di Simone, B. Petracci, M. Piva, Economic sustainability, innovation, and the ESG factors: An empirical investigation. Sustain. For. **14**(4), 2270 (2022)

Chapter 5
How to Carry Out an R&D Project

This chapter provides a detailed guide on successfully conducting research and development (R&D) projects. It emphasizes aligning the project approach with organizational objectives and tailoring the project to fit seamlessly within the broader context. A key aspect of this chapter is the integration of complementary disciplines and how these diverse components can work in synergy to foster innovation and address complex challenges. It also explores various R&D project types and how to select the most appropriate one based on specific goals and requirements. This chapter covers the practical side of R&D project management, focusing on systems engineering and the vital role of project management documents. It stresses the importance of establishing clear project measurements and baselines to make informed decisions throughout the project's lifecycle. "Active Management" and "Seizing Opportunities while Reducing Risks" are core concepts, offering insights into adapting, seizing opportunities, and mitigating risks in dynamic R&D environments. This chapter highlights the need to provide unwavering support and facilitation for the R&D team, ensuring they have the resources and an enabling environment to accomplish their goals effectively. It is an indispensable resource for those involved in R&D projects, offering a comprehensive roadmap for planning, executing, and overseeing innovation initiatives.

> **Learning Objectives**
> - Recognize the key procedures for managing an R&D project
> - Recognize and choose the best project management strategy
> - Recognize important project management records in an R&D project
> - Be aware of the need to establish a foundation for the use and project measurement

5.1 Project Approach, Alignment, and Fit

In R&D, the project approach, alignment, and fit play critical roles in determining the success and efficacy of any initiative. A well-defined project approach serves as the roadmap that guides the research journey. It encompasses the methodologies, strategies, and timelines researchers will employ to achieve their objectives. The selection of the right approach depends on the nature of the project, whether it is exploratory research, product development, or process improvement. The approach should match the project's unique requirements, resources, and constraints. Project alignment is all about ensuring that the R&D effort is in harmony with the organization's overall strategic goals. The alignment process requires a thorough understanding of the company's mission, vision, and long-term objectives. For R&D to be effective, it must be tightly integrated with the business strategy. This alignment ensures that the R&D projects are not pursued in isolation but are closely connected to the organization's broader mission. It prevents resource wastage on endeavors that do not contribute to the company's core goals. Project fit, on the other hand, refers to how well the project aligns with the organization's existing capabilities, competencies, and resources. It is crucial to assess whether the organization has the required technical expertise, equipment, and infrastructure to execute the R&D project successfully. A good project fit minimizes the need for excessive investments and streamlines the execution process. Organizations often look for synergies between their existing strengths and the requirements of the R&D initiative to enhance efficiency.

The effectiveness of an R&D project is heavily contingent on the convergence of these three elements: approach, alignment, and fit. When there is a strong alignment between the project's goals and the organization's strategic objectives, securing the necessary resources and support becomes easier. Moreover, aligning R&D projects with the company's core mission helps prioritize and justify their significance within the organization. A mismatch between the project requirements and the organization's capabilities can result in numerous challenges in project fit. It can lead to delays, cost overruns, and compromised project quality. Therefore, thoroughly evaluating the fit between the project and the organization's existing resources is crucial to ensure the project proceeds smoothly. The project approach, alignment, and fit should be reviewed and refined throughout the project's lifecycle. Flexibility in adapting the approach and realigning with changing business needs is essential to keep the R&D initiative on the right track. Successful R&D projects are flexible but evolve as new insights and challenges arise.

The project approach, alignment, and fit are intertwined aspects that determine the success of R&D initiatives. A well-planned approach provides a roadmap; alignment ensures that the project aligns with the organization's strategic goals and fit guarantees that the project is a good match for the available resources. Effective R&D requires continuous evaluation and adaptation of these elements to remain relevant and achieve meaningful outcomes [1].

5.1 Project Approach, Alignment, and Fit

Table 5.1 delves into the critical aspects of "Project Approach, Alignment, and Fit in R&D," highlighting their significance in the success of research and development initiatives. The table's structured layout offers a concise breakdown of each key element. It underscores the importance of a well-defined project approach, which serves as a guiding roadmap, and how it should be tailored to the unique demands of a research project. Project alignment emphasizes the need for R&D efforts to seamlessly integrate into the organization's overarching strategic objectives, ensuring that resources are utilized effectively. In project fit, the table underscores the importance of assessing whether a project aligns with the organization's existing resources and capabilities, thereby minimizing costs and resource wastage. The "Convergence for Success" section reiterates the interdependence of these elements, emphasizing the need for ongoing adaptation and flexibility to achieve meaningful outcomes in R&D endeavors.

Table 5.1 Key elements of project success in R&D: approach, alignment, and fit

Aspect	Description	Importance	Implications
Project approach	Methodologies, strategies, and tailored execution	Guides the research journey	Influences research quality and speed
	Tailored to the project's specific needs	Ensures effective use of resources	Flexibility required for adjustments
	Determines the execution of research activities	Enhances project planning and management	Influences outcome reliability
Project alignment	Integration with the organization's strategic goals	Ensures R&D contributes to business strategy	Secures resources and top-level support
	Avoids isolation by connecting to the broader organizational mission and vision	Provides clarity on the project's mission	Enhances project's perceived value
		Fosters shared understanding among teams	Aligns R&D with long-term objectives
Project fit	Compatibility with organizational resources, capabilities	Minimizes unnecessary investments	Can impact project timelines
	The synergy between project requirements and available infrastructure	Improves utilization of in-house knowledge	Ensures optimal resource allocation
		Reduces potential disruptions in execution	Influences overall project efficiency
		Enhances adaptability and agility	Reduces the need for external support
Convergence of elements	A harmonious balance between approach, alignment, and fit	Increases the likelihood of project success	Encourages ongoing assessment and adaptation

5.2 Complementary Discipline Components

R&D is a complex and multifaceted field that drives innovation, technological advancement, and economic growth across a wide range of industries. It is a collaborative endeavor that draws upon various disciplines and components, working in synergy to achieve common objectives. One crucial aspect of effective R&D is the integration of complementary discipline components, each of which contributes to the overall success of research and development endeavors. First and foremost, interdisciplinary collaboration forms the backbone of R&D. It enables the amalgamation of diverse expertise, methodologies, and perspectives from various fields, such as science, engineering, mathematics, and social sciences. This collaborative approach is instrumental in addressing multifaceted challenges by leveraging experts' collective knowledge and skills in different areas. Scientific research is another essential component of R&D, providing the foundational knowledge upon which innovations are built. It encompasses systematic investigation, experimentation, and observation, generating new theories, empirical data, and insights that drive the development of cutting-edge technologies and solutions. Engineering and design follow closely, serving as the bridge between theoretical concepts and practical applications. Engineers are tasked with creating prototypes, optimizing processes, and ensuring that innovations are functional and aesthetically appealing. In today's data-driven world, data analysis and information technology specialists play a pivotal role in R&D. They manage and interpret vast datasets, apply machine learning algorithms, and extract valuable insights that inform decision-making and innovation. Regulatory and compliance expertise is essential in many industries to ensure product safety and adherence to legal requirements. Specialists in this field navigate complex regulations and standards, ensuring that R&D projects meet all the necessary criteria. Market and business analysis is crucial for the commercial success of R&D projects. Experts in this field evaluate market trends, identify potential customers, and assess the financial feasibility of new products or services, thereby guiding business strategies.

Ethical and social considerations are gaining prominence in R&D. This discipline component involves assessing the potential impact of innovations on society, the environment, and individuals, addressing issues such as privacy, sustainability, and inclusivity. Intellectual property and innovation protection are vital for safeguarding the fruits of R&D efforts. Specialists work on patenting, trademarking, securing innovative ideas and products, promoting innovation, and safeguarding competitive advantages.

These complementary discipline components contribute to the R&D process's overall success. Interdisciplinary collaboration, scientific research, engineering, data analysis, regulatory compliance, market analysis, ethical considerations, and intellectual property protection collectively foster innovation and ensure that the resulting products or solutions are viable and ethically responsible. This synergy advances knowledge and drives progress across various industries, benefiting society.

5.3 Facilitate the Kind of R&D Project

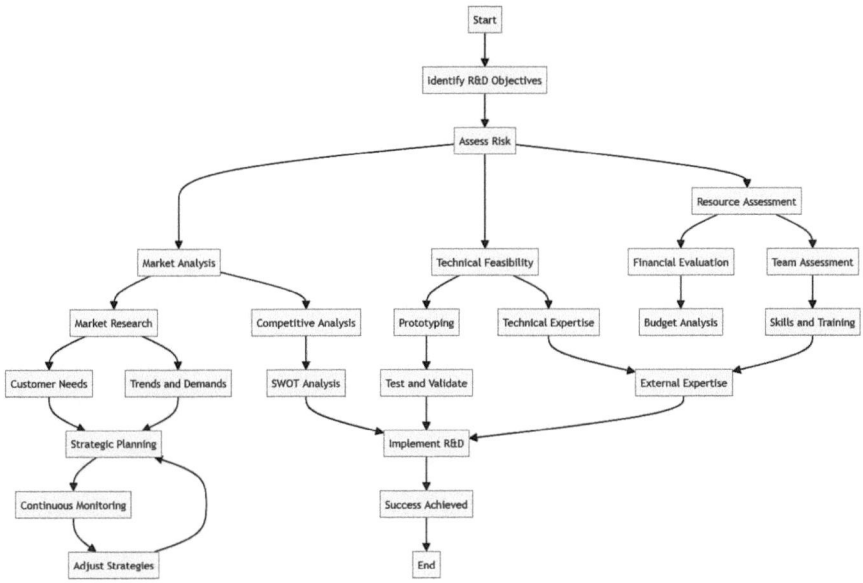

Fig. 5.1 Methods for lowering the risk of R&D and advancing it

Figure 5.1 is a flowchart that outlines a structured approach for managing R&D projects. This flowchart visualizes the steps involved in the R&D process, highlighting critical components such as identifying objectives; assessing risks in market analysis, technical feasibility, resource assessment; and conducting market research, technical prototyping, and more. It provides a comprehensive overview of the systematic decision-making and planning process essential for minimizing risks and advancing R&D efforts. The figure offers a clear and organized way to visualize the key stages and interconnected steps in R&D project management, aiding in the efficient execution of these critical initiatives.

5.3 Facilitate the Kind of R&D Project

Facilitating a R&D project involves a series of critical steps that ensure the project's successful execution (Fig. 5.2). To begin, defining the project's scope clearly is essential, outlining the objectives, goals, and expected outcomes. This initial step is the foundation upon which the entire project is built. Once the scope is established, the next step is assembling a cross-functional team. This team should comprise members from various disciplines who can bring diverse perspectives and expertise to the project. A multidisciplinary team is well equipped to address the complexity of R&D projects. Securing funding is a crucial aspect of R&D facilitation. Identifying and obtaining the necessary budget and resources are essential for carrying out the project effectively. This may involve seeking financial support from

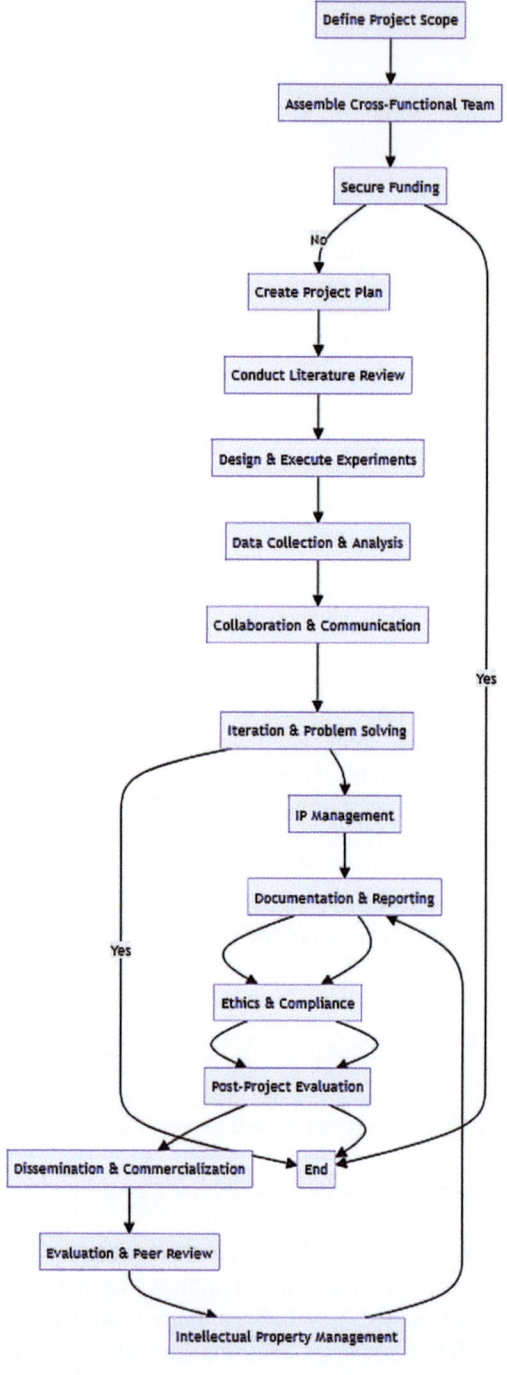

Fig. 5.2 Simplified flowchart for facilitating an R&D project

various sources, such as internal funds, external grants, or potential investors. Creating a detailed project plan is the blueprint for the project's execution. This plan should encompass tasks, timelines, milestones, responsibilities, and risk mitigation strategies. A well-structured project plan ensures everyone is on the same page and helps monitor progress. Before diving into the project, conducting a comprehensive literature review is vital. This review should encompass existing research, patents, and prior work related to the project's subject matter. It helps in understanding the current state of knowledge and identifying gaps. Designing and executing experiments or research activities come next. These activities are the heart of the R&D project, involving laboratory work, field studies, computer simulations, or other methods tailored to the project's objectives. Data collection and analysis follow the experimentation phase. This step requires appropriate tools and methodologies to extract valuable insights from the gathered data. Accurate analysis is critical for making informed decisions. Effective communication and collaboration within the team and with stakeholders are ongoing requirements. Regular updates, meetings, and information sharing help maintain transparency and ensure everyone is aligned with the project's goals and progress. R&D projects often encounter unexpected challenges. It is crucial to be prepared for iteration and problem-solving. Flexibility and adaptability are key as you navigate these challenges and make necessary adjustments to the project plan. If applicable, managing intellectual property is another significant consideration.

Patents, copyrights, or trade secrets need to be secured and managed to protect the project's innovations and ideas. Documentation and reporting should be a continuous practice. Keeping detailed records of activities, results, and findings is vital. Comprehensive reports and documentation are essential for sharing the project's outcomes and insights with the relevant audience. Ethical considerations and legal compliance must be maintained throughout the project. This involves obtaining approvals or permits and ensuring the project meets ethical standards and regulations. Finally, after the project is completed, a post-project evaluation is essential. This evaluation assesses the project's success against its initial objectives, identifies lessons learned, and highlights areas for improvement in future R&D endeavors.

5.4 Systems Engineering and Project Management Documents

Stakeholders should agree on what will be done, how it will be done, and by when as a result of the preparation of these papers. The documents are crucial since they will be the starting point for evaluating progress. Project plans, communications plans, and risk management plans are all crucial project management papers. The project plan for the R&D project should include the trajectory (or trajectories) that will be taken in the direction of the strategy in its scope. The work breakdown structure (WBS), organizational breakdown structure (OBS), related work packages, and explanations of the overall budget allocator, management reserve, and contingency will all be included in this plan, which will act as the project's guiding document. The best approach for R&D projects is to create a 1- to 3-month resource-loaded timetable, projected budget allocations, scheduled experiments, tests, and risk mitigation and reduction methods for that timeframe. A collection of subordinate schedules can be systematically constructed in 3-month increments or less depending on the pace of the experiments if the project scope and trajectory are properly documented, and a clear set of experiments, tests, and anticipated outcomes have been established. By using this technique, timelines for R&D are guaranteed to be as accurate as feasible. The previous phase's outcomes will determine the following phases' timetables, objectives, and financial constraints. By doing this, the project manager may remain flexible, which is essential for successfully managing the R&D project.

The systems engineering management plan (SEMP), the systems engineering master schedule (SEMS), the technical performance measures (TPMs), the measure of effectiveness (MOEs), the measure of performance (MOPs), and the key performance indicators (KPIs) are the papers that will be necessary (Fig. 5.2). To reduce the amount of information that is distracting but not necessary for obtaining effective outcomes, KPIs can be developed and tracked. Any interface control documents (ICDs) that are found should also be recorded to reduce the impact on associated operations. R&D projects should use technical levels of readiness (TRLs), which offer an architecture that maps well to phase reviews for assessing and regulating technological progress over the life cycle. If the project uses the TRL approach, there will be extra documentation and reviews to make sure that each step of the life cycle—from fundamental research through production—is carefully controlled and that development does not proceed until it is ready to do so (Fig. 5.3).

In R&D, systems engineering and project management are critical in ensuring that projects are successfully planned, executed, and completed. Several key documents are essential for effectively managing R&D projects within these domains. Here are some key documents used in R&D projects involving systems engineering and project management.

Project Charter. The project charter serves as a foundational document in R&D. It defines the project's scope, objectives, stakeholders, and initial requirements, providing a clear sense of purpose for the team and stakeholders.

Project Plan. This document outlines the overarching strategy for the R&D project, detailing timelines, milestones, and resource allocation. It is essential for setting a project's direction and ensuring everyone is on the same page.

Work Breakdown Structure (WBS). The WBS is a hierarchical breakdown of project tasks into manageable components. It aids in organizing work and understanding how various elements relate to one another.

Requirements Document. This document specifies the project's technical, functional, and nonfunctional requirements. It acts as a reference point for design and development decisions.

5.4 Systems Engineering and Project Management Documents

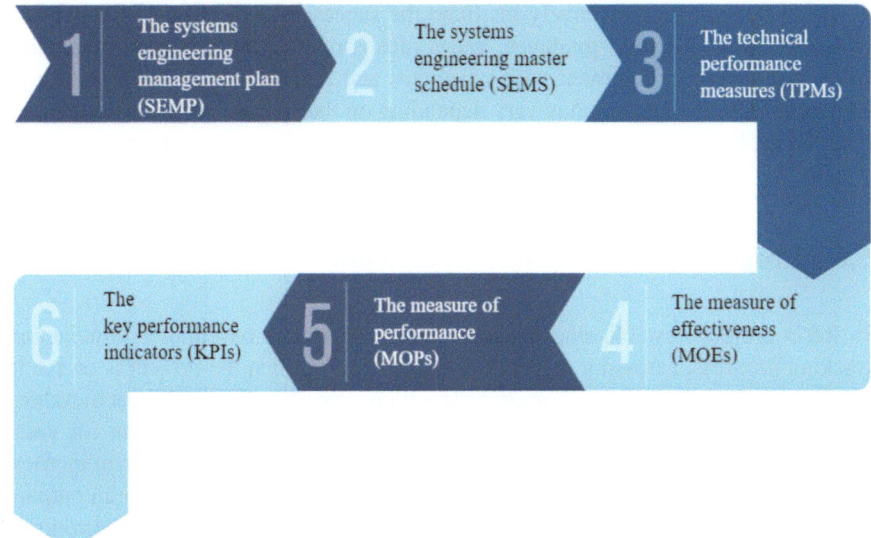

Fig. 5.3 Some aspects of project management documents

- *Risk Management Plan.* R&D projects often involve uncertainties. The risk management plan identifies potential risks and outlines strategies for managing and mitigating these risks effectively.
- *System Architecture Document.* This document provides a high-level view of the system's structure and components in systems engineering. It guides the system's design and development.
- *Change Management Plan.* Change is inevitable in R&D. This document defines the process for requesting, reviewing, approving, and implementing project scope changes.
- *Quality Management Plan.* Ensuring quality is crucial in R&D. This plan details quality standards, testing procedures, and quality control measures to meet project expectations.
- *Schedule and Gantt Chart.* A detailed project schedule presented in a Gantt chart format visually represents project tasks, dependencies, and timelines.
- *Cost Estimate and Budget.* This document outlines projected costs and budgets for the R&D project, providing the financial framework for the project.
- *Progress Reports.* Regular progress reports inform stakeholders about project status, accomplishments, and challenges, fostering transparency and accountability.
- *Communication Plan.* This plan outlines how project information is shared among team members, stakeholders, and other relevant parties, specifying communication channels and frequencies.
- *Lessons Learned Document.* At the project's conclusion, a lessons learned document captures insights, best practices, and areas for improvement, benefiting future R&D endeavors.

These documents collectively facilitate the successful execution of R&D projects, providing structure, guidance, and a means of tracking progress, managing risks, and ensuring alignment with objectives and requirements. The content and format of these documents may vary depending on the specific project's nature and complexity, but they are essential for efficient R&D project management.

5.5 Project Measurements and Baseline

In R&D projects, establishing project measurements and baselines is crucial for tracking progress, assessing performance, and ensuring that the project is on track to meet its objectives. These objectives should be specific, measurable, achievable, relevant, and time-bound (SMART). They will serve as the foundation for your project measurements. Key performance indicators (KPIs) identify relevant metrics such as project cost, timeline, resource allocation, innovation rate, research output, and product development milestones. Baseline data is essential to measure progress effectively and serves as the starting point for your project. It enables comparisons and assessments of progress. Measurement tools and methods should be carefully chosen based on project needs, including laboratory equipment, surveys, questionnaires, experimental designs, or software tools, ensuring reliability and accuracy. Data collection and recording should be consistent at predefined intervals to maintain up-to-date information. Performance targets, derived from project goals and objectives, are set to track whether the project is progressing as expected. Regular monitoring and reporting are essential to inform stakeholders of the project's status. A risk assessment should be incorporated into measurements and baselines, identifying potential risks and their impact on the project, with contingency plans in place. Records of measurements, baseline data, and progress reports must be maintained for project evaluation, audits, and future reference. The data and measurements provide feedback to the project team, fostering a culture of continuous improvement. In R&D, where uncertainties often require adjustments, the measurements and baselines must be adaptable to changing project scopes or objectives. These measurements and baselines guide project performance assessment and decision-making, facilitating successful project outcomes.

In R&D projects, project measurements and baselines are foundational to ensure that the project progresses effectively and efficiently. These measurements should be aligned with the SMART (Specific, Measurable, Achievable, Relevant, and Time-bound) objectives and goals set for the project. These objectives serve as a roadmap for the research and development efforts. Key performance indicators (KPIs) are carefully chosen to monitor and evaluate the project's progress. They encompass various aspects such as project cost, timeframes, resource allocation, innovation rate, research output, and product development milestones, enabling a comprehensive assessment of the project's health.

The establishment of baseline data is paramount for gauging progress. It involves capturing and recording the initial state of relevant parameters before the project commences. This baseline data is indispensable for comparing and determining how far the project has advanced from its starting point. Selecting appropriate measurement tools and methods is equally crucial. These tools could range from sophisticated laboratory equipment to surveys, questionnaires, experimental designs, or specialized software, depending on the nature of the R&D project. The accuracy and reliability of these tools are pivotal in ensuring the validity of the collected data. A systematic data collection and recording approach is implemented, ensuring that data is acquired consistently and at predefined intervals. This practice guarantees the project team access to up-to-date information to make informed decisions. Performance targets are established based on the project's objectives. These targets represent the desired outcomes at specific points in time and provide a benchmark against which progress can be measured. Regular monitoring and reporting mechanisms are implemented to inform stakeholders about the project's status. This includes communicating progress, issues, and potential areas of improvement. Risk assessment is integrated into measurements and baselines. Identifying potential risks and their potential impact on the project is essential. Contingency plans are developed and monitored to mitigate risks and prevent project derailment. The records of measurements, baseline data, and progress reports are meticulously maintained. These records are valuable for project evaluation, external audits, and future reference. Feedback from the collected data and measurements is used to inform the project team, fostering a culture of continuous improvement. It enables the team to make data-driven decisions and refine project strategies. In R&D, where projects often involve uncertainty and evolving scopes, the adaptability of measurements and baselines is key. They should be flexible enough to accommodate changes in project objectives and direction.

Project measurements and baselines in R&D are essential tools for tracking and ensuring the success of a project. They guide the project team, enabling them to assess performance, manage risks, and make data-informed decisions as they work toward their SMART objectives and goals.

5.6 Active Management

Active management in R&D is a dynamic approach to ensure that R&D projects progress efficiently and effectively. It commences with establishing clear project objectives and milestones, employing the SMART framework to ensure goals are Specific, Measurable, Achievable, Relevant, and Time-bound. Setting these objectives and milestones provides a roadmap for the project's progression. Resource allocation is a pivotal component of active R&D management, allocating personnel, funding, equipment, and time. Decisions related to resource allocation are made based on project priorities and evolving needs. A comprehensive project plan outlines the scope, timeline, budget, and resource requirements. Active management entails continuously adapting and updating the project plan as circumstances evolve or new insights emerge. The proactive management approach extends to risk management. Risks and uncertainties are identified, and strategies are devised to mitigate these risks. Active management involves ongoing monitoring of risk factors throughout the project's lifecycle, enabling timely intervention in case of adverse developments. Active management promotes cross-functional collaboration, fostering cooperation between different teams and departments. This collaboration is vital because R&D projects often require expertise from multiple disciplines. Monitoring and reporting are central to active management, as they allow for tracking project progress against objectives and milestones. Reporting mechanisms ensure that project status is communicated to stakeholders.

Effective decision-making is another hallmark of active management. Project managers and teams should be equipped with the necessary information and data to make informed decisions that align the project with its goals. Flexibility and adaptability are also integral to active management. Acknowledging the uncertainty inherent in R&D, active management allows flexibility in adapting to new information, discoveries, or priority changes. Continuous improvement is emphasized, encouraging project teams to learn from their experiences and apply lessons learned to future projects. Quality assurance processes are implemented to maintain the integrity and reliability of R&D activities. Ongoing checks and audits verify that quality standards are consistently met. Furthermore, active management recognizes the importance of intellectual property protection in many R&D projects, leading to strategies for safeguarding intellectual property assets. Engagement with key stakeholders, including project sponsors, senior management, and external collaborators, is part of active management, ensuring that stakeholders are informed, their concerns are addressed, and their input is considered throughout the project's lifecycle.

Active management in R&D is characterized by a proactive approach to addressing challenges, adapting to changing circumstances, and focusing on project objectives. This approach is essential for driving innovation, achieving goals, and delivering value to the organization. By actively managing R&D projects, organizations increase their chances of success and maximize the return on their research and development investments.

5.7 Seize Opportunities and Reduce Risks

In R&D, the dual objectives of seizing opportunities and reducing risks are integral to driving innovation and achieving success. A crucial aspect of seizing opportunities is the identification of emerging possibilities, whether in the form of new technologies, market trends, or evolving customer needs. Organizations must stay vigilant, monitoring the competitive landscape and staying informed about research breakthroughs that align with their capabilities and objectives.

Market research is a vital tool for understanding and seizing opportunities in R&D. It entails a comprehensive examination of market demands and the identification of gaps or niches where R&D efforts can make a significant impact. This approach ensures that R&D initiatives are market-driven and possess a higher likelihood of success, ultimately reducing the risks associated with market uncertainties.

Organizations often employ prototyping and pilot testing to mitigate the risks associated with unproven ideas. These methods involve creating small-scale models or conducting pilot trials to assess the feasibility and performance of a concept. Such strategies allow organizations to reduce technical and operational risks before committing substantial resources to full-scale R&D projects.

A systematic approach to risk assessment is essential to effective risk reduction in R&D. Potential risks, including technical challenges, market uncertainties, or resource limitations, need to be identified. Organizations should also develop robust mitigation strategies and contingency plans to address these risks as they arise.

Strategic resource allocation plays a pivotal role in seizing opportunities and managing risks. Organizations must balance investing in high-risk, high-reward projects and more conservative initiatives, maintaining a diversified R&D portfolio. This diversification strategy can help organizations achieve their goals while minimizing the impact of potential failures.

Open innovation models, such as partnerships, collaborations, and technology licensing, reduce internal R&D risks while capitalizing on external opportunities. Organizations can access new ideas and solutions by engaging with external expertise and technologies, fostering innovation, and expanding their R&D capabilities.

In summary, actively managing opportunities and risks is fundamental to successful R&D efforts. By proactively identifying opportunities, conducting thorough research, and actively managing risks, organizations can thrive in dynamic and competitive markets, driving innovation and maximizing the return on their R&D investments.

5.8 Support and Facilitate the R&D Group

The likelihood of an R&D project's success will be raised regardless of how methodically the project management methodologies are applied to the project by selecting an experienced R&D project manager; a well-diversified, committed team; and a well-organized learning environment for that team. Instead of tracking and rewarding milestone success, the ideal environment for R&D projects is centered on and fosters learning.

The project manager should assemble, maintain, and develop the team. This team may already exist, but it would be ideal if he or she could assemble one. The R&D project manager may inject fresh perspectives and ideas into the project by utilizing information and experiences from many fields, cultures, and organizational types. The project manager will have to match each team member's goals with the project's goals once they are all on board. Additionally, they should organize team-building events, properly manage workflow, and allow team members to advance in their professions.

A workforce management strategy may be created to help with team management. The R&D project manager will benefit from this plan's assistance in establishing training and career progression pathways, identifying vital skills, describing retention strategies, and providing other crucial workforce-related information to the project. It is important to incorporate techniques for rewarding and motivating team members, combining protected free time with access to tools like trip funds, machine shop access, or engineering time for concept research. Depending on the organization, incentives may also involve complete or partial ownership rights and public recognition, such as earnings, copyrights, or patents.

The ability to commit to a project's course and assume accountability for its success is one of the most crucial qualities of an R&D project manager. A relationship built on trust results from taking ownership of actions, choices, and progress while effectively conveying that information to the stakeholders. The project's stakeholders are confident that the project manager would take the actions necessary to provide the R&D project with the best chance of success. Trust and respect continue to grow when he or she is open and honest about the planned activities and progress accomplished throughout the project. Reviewing the project's progress and validating that it is proceeding as planned can be done through focus groups and reviews with outside subject matter experts.

An R&D project group has the best chance of realizing its potential if it can physically foster a climate of trust, inclusiveness, and inventiveness. This may be achieved by ensuring the team has enough free time to think about, reflect on, and discuss ideas without a target in mind. Additionally, they should be safeguarded so that internal intolerance does not stifle their ideas and allowed to engage in cross-disciplinary, forward-thinking dialogue with their colleagues. Whenever feasible, an innovative physical environment should be made accessible, offering both areas for uninterrupted concentrated thought and creativity and a considerable quantity of multidisciplinary contacts. Both of these requirements should be satisfied to maximize the potential of R&D projects. The procedure for carrying out an R&D project is outlined in Fig. 5.4.

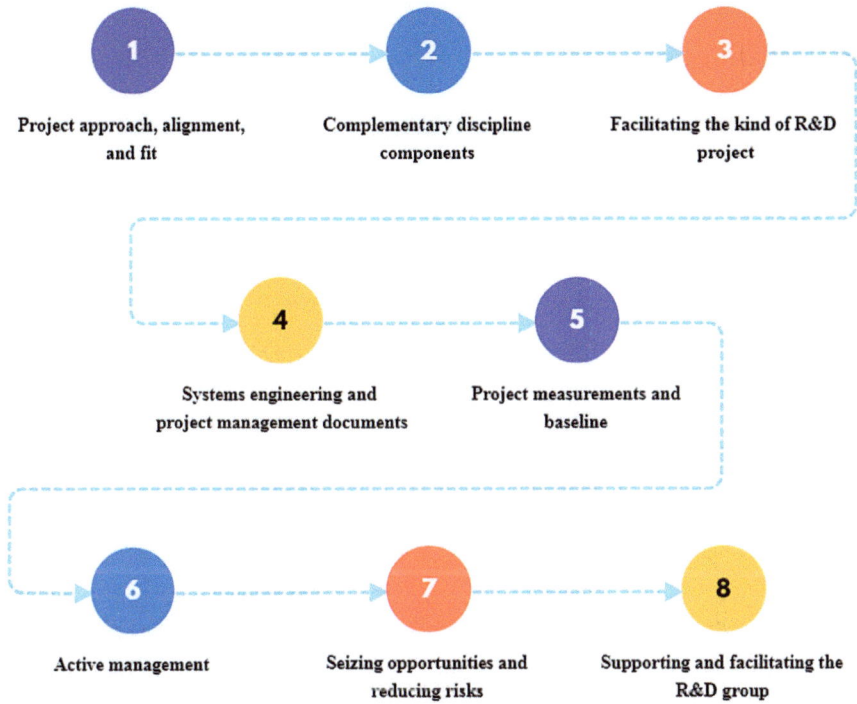

Fig. 5.4 Process for carrying out an R&D project

Reference

1. A. López-Alcarria, A. Olivares-Vicente, F. Poza-Vilches, A systematic review of the use of agile methodologies in education to foster sustainability competencies. Sustain. For. **11**(10), 2915 (2019)

Chapter 6
R&D Process

Research and development (R&D) is an essential part of innovation and development across many sectors. This chapter explores the complexities of R&D and thoroughly reviews the major steps that need to be taken to get an idea from conception to commercial manifestation. The R&D process is a dynamic journey that does not follow a linear route and is comprised of phases that need careful preparation, innovation, and adaptation. This chapter delves deeply into the critical steps of R&D, from concept genesis through commercialization. It highlights the fluid nature of R&D and discusses difficulties, pitfalls, and actual case studies. This chapter emphasizes the need for creativity and adaptation in a continuously changing environment. It is an invaluable tool for professionals and innovators aiming to transform concepts into products ready for the market.

> **Learning Objectives**
> - Understand the sequential stages of the R&D process.
> - Recognize the importance of feasibility assessments.
> - Identify common challenges and pitfalls in R&D.
> - Appreciate the value of iteration and continuous improvement.

6.1 Importance of R&D Process

Success depends not only on brilliant ideas or an abundance of resources but also on the accuracy and efficacy of the methods used. R&D is a deliberate, process-focused endeavor that requires discipline. Process management needs to be prioritized for R&D to be productive [1]. The R&D process serves as the foundation for innovation. They provide direction, help in decision-making, and guarantee that resources are used effectively. The path from idea to implementation is streamlined by a clearly defined procedure, making it easier to stay in charge, monitor the progress, and adjust to changing conditions. Process management is crucial for several reasons.

First, it guarantees efficiency by eliminating duplication and maximizing resource use. Second, it ensures consistency in job execution, which is necessary for quality control and preventing recurrent errors. Third, it makes it easier to allocate resources wisely, which is essential for the success and longevity of R&D initiatives. Additionally, risk management is crucial, assisting in identifying and mitigating possible setbacks at every level. Finally, it ideally aligns with the iterative nature of innovation since it promotes adaptation and learning by enabling feedback and insights obtained at one stage to be implemented in the next.

Regarding innovation, the R&D process is crucial in various sectors, influencing development and change. While R&D activities' specifics may differ among industries, organized processes' underlying significance is constant. The R&D process is the primary force behind development, competitiveness enhancement, and the facilitation of solutions to difficult problems [2].

The R&D process is at the heart of innovation, which is what drives the high-tech and IT industries. It supports the development of innovative hardware, software, and digital solutions. Process optimization guarantees effective software development and on-time product releases, allowing businesses to keep a competitive advantage in the ever-changing tech industry.

The value of the R&D process is felt across a wide range of sectors. Pharmaceutical businesses in the healthcare industry depend on meticulous R&D procedures to develop new medications and treatments, with accuracy and adherence to regulatory requirements being crucial [3]. R&D practices influence product engineering and design in manufacturing, improving quality and cost-effectiveness. The R&D process impacts safety, sustainability, and technical advancement in fields ranging from aerospace and military to energy and environmental preservation. Success is based on R&D processes in every industry, which has a greater effect on global trends than we often perceive [4].

6.2 Stage 1: Idea Generation

R&D begins with the creation of new ideas. In this first stage of innovation, new ideas and prospects for goods, services, and technology are generated by combining creative thinking and systematic analysis. Successful creativity is fundamental to the R&D process as a whole, determining the course and ultimate success of the project.

(a) *Brainstorming and Creative Thinking*

Brainstorming is a group activity that encourages creativity and new ideas. Its primary tenet is an open and accepting environment in which people feel safe sharing their thoughts. The brainstorming process is enriched, and new ideas are sparked when teams are made up of people from different departments and walks of life.

With the help of a professional moderator or facilitator, the meeting may be organized, productive, and enjoyable for everyone involved. While quantity is more important than quality at the outset, the conversation will stay on track and relevant to the project's aims if a clear objective or issue statement is established. Whiteboards, sticky notes, and mind maps are just a few visual tools that may help participants organize their thoughts and see the relationships between different concepts. Even the most out-of-the-box suggestions made during brainstorming sessions are written down and archived for further study. The creative process relies heavily on this interplay between divergent and convergent thought.

The R&D process does not have to be interrupted by a single brainstorming session. A project's success and originality heavily depend on the first spark of imagination that occurs during brainstorming. Brainstorming facilitates the generation of many different ideas by providing a setting for free association. This results in divergent and convergent thinking, in which many ideas are created first before a few are chosen for further development after being examined, polished, and ranked. Using visual aids and a moderator or facilitator to keep the conversation on track may help foster creative thinking. The captured concepts are the building blocks for future improvements and innovations.

(b) *Identifying Market Needs*

Identifying market needs is a crucial bridge between innovation and real-world relevance in the R&D process. Thorough market research and analysis lay the foundation for understanding the current landscape. This entails studying market size, growth trends, competitors, and customer segments through various research methods, such as surveys, focus groups, and industry reports. The goal is to acquire a comprehensive view of the market's dynamics, vital for making informed R&D decisions.

To tailor R&D efforts effectively, creating detailed customer profiles or personas is indispensable. These profiles provide insights into the target audience's specific demographics, psychographics, and behavior patterns, helping R&D teams understand what customers desire and how they make purchasing decisions. Identifying market needs goes beyond recognizing opportunities; it involves pinpointing customer pain points and challenges. By understanding the problems customers face, R&D teams can develop solutions that directly address these issues, making their products or services more appealing and relevant.

In addition to studying the present market, staying attuned to emerging trends and technologies is essential. Proactively embracing new trends and technologies can give organizations a competitive edge. Analyzing competitors is equally important; it helps to understand what is missing in the market and where competitors fall short. Regulatory and compliance considerations should also be considered, depending on the industry, as these factors can significantly shape market needs.

Building feedback loops into products or services is a smart practice for continuously identifying market needs. Listening to customer feedback, tracking product usage, and making iterative improvements based on user insights can fine-tune R&D efforts. Furthermore, it is crucial to consider immediate market needs and how these needs might evolve. R&D efforts should focus on creating products or solutions that are scalable and adaptable to changing market conditions. Cross-functional collaboration, especially between R&D, marketing, and sales teams, is crucial in identifying the market needs. These teams can share insights and data to develop a more comprehensive understanding of market demands. Table 6.1 highlights the technical aspects, methods, and benefits of market needs identification in R&D.

(c) *Leveraging Existing Knowledge and Trends*

The process starts with the creation of a concept. All new goods, services, and technology start with this imaginative and cooperative procedure. Brainstorming, a method that promotes the free flow of ideas, is crucial to this first step. It is critical to foster an atmosphere where people may freely share their unique perspectives without worrying about being labeled or dismissed. Creative solutions often emerge from multifaceted groups that combine specialists from several fields. When brainstorming, it is essential to alternate between divergent and convergent thinking to generate a diverse range of ideas, followed by evaluating and enhancing the most promising ones. Brainstorming sessions encourage original thought and the exploration of untapped avenues of inquiry by establishing concrete goals and using concrete visual aids like whiteboards and mind maps.

Moreover, innovation does not need a blank slate in every case. Using what has already been learned and keeping up with developments in the field may be very effective. Companies should thoroughly inventory their knowledge to assess their

Table 6.1 Market needs identification in R&D

Aspect	Technical aspects	Methods/tools	Benefits/outcomes
Market research & analysis	In-depth market research, data analysis, and segmentation	Surveys, reports	Informed decisions and market understanding
Customer profiles	Customer demographics and behavior patterns through data	Surveys, interviews	Targeted product development
Problem identification	Identifying pain points using data and feedback	Feedback, research	Solutions addressing customer problems
Emerging trends	Adapting to emerging tech and trends	Trend analysis, scouting	Innovation and competitiveness
Competitor analysis	Analyzing competitors for differentiation	Benchmarking, SWOT	Identifying opportunities
Regulatory considerations	Addressing legal compliance	Legal experts, audits	Ensuring compliance
Feedback loops	Implementing feedback mechanisms	Customer feedback, analytics	Continuous improvement
Long-term considerations	Planning for scalability and adaptability	Trend analysis, planning	Future-proofing products
Cross-functional collaboration	Promoting collaboration for market understanding	Meetings, data sharing	Holistic insights

intellectual property and skills. Organizations may adopt or embrace technologies to solve particular difficulties by engaging in "technology scouting," actively exploring answers from similar or unrelated sectors. The available information and experience pool may be expanded via collaborations, partnerships, and open innovation. R&D prospects can only be found by constant vigilance of market shifts, customer feedback, and competitor research. Organizations may secure a spot in the vanguard of innovation by conducting proactive analyses of new technologies, social developments, and industry changes.

Combining the spontaneous spark of brainstorming with the systematic application of current information and trends helps foster creativity. It allows businesses to strike a good risk-reward balance, speeding up R&D efforts by making use of existing knowledge while keeping an eye on the future. Finding the sweet spot between originality and emulation is critical for producing market-ready solutions that stand out. A company's strategic objectives should be considered when new technology or trends are implemented.

6.3 Stage 2: Feasibility Assessment

The R&D process continues with feasibility assessment, the second step. This step connects the brainstorming stage with the detailed planning stage of the R&D endeavor. A feasibility study is all about evaluating the technical and commercial viability of a proposed project. It is a major factor in deciding whether the project should be continued as-is or tweaked.

(a) *Market Research and Analysis*

In the R&D process, the feasibility assessment phase relies heavily on market research and analysis. In this crucial step, we analyze the market where we want to launch our new product or service. First, narrowing down the potential clientele by researching their demographics, interests, and demands is necessary. Based on this knowledge, the invention may be adapted to fit the market's needs.

In addition, it is essential to evaluate the size and development prospects of the industry. This entails making predictions about the market's future growth and assessing the level of market saturation. This helps businesses better allocate resources, develop strategies, and enter new markets. Also crucial is a detailed examination of the competition, which may shed light on the state of the market, reveal unfilled niches, and inspire novel approaches to setting apart the company's product or service.

The last element of the jigsaw is keeping up with market dynamics and developments. Foreseeing and responding to changes in consumer tastes, new technological developments, and economic conditions is crucial for good decision-making. This all-encompassing market research procedure lessens risks and improves R&D's chances of success in a cutthroat industry by ensuring that it is tailored to the demands of consumers.

(b) *Technical Feasibility*

Assessing technical feasibility is crucial to the R&D process's feasibility evaluation phase. It requires a thorough analysis of the technical feasibility of the project in question. The major objective is to guarantee that the project can be effectively implemented within the established limits.

6.3.1 Technology Assessment

Evaluation of the proposed technology or approach is crucial in the first phases of the technical feasibility study. Many important factors need to be considered during this analysis, beginning with how mature the technology is. An important question is to what extent the technology has progressed beyond the experimental stage. The time and money required to create a project are proportional to the degree of preparedness. The capacity to expand is also crucial. Stakeholders in a project, especially those involved in potentially expansive ventures, must consider whether the technology can scale to meet the anticipated production or operational needs. The suggested technology's compatibility with the organization's current hardware, software, and infrastructure set-up should also be evaluated. If addressed, compatibility difficulties may prevent projects from running late and spending more money.

Another important factor in determining technical feasibility is the availability of necessary resources. This entails ensuring the required assets are available or can be procured to advance the project. A company needs to evaluate its human capital resources to determine whether it has or can obtain the necessary engineers, developers, scientists, and technicians. Additionally, it entails figuring out what materials, tools, and equipment are needed and assessing their accessibility, pricing, and possible supply chain concerns. To ensure the project's technical needs can be met, the current IT infrastructure, including servers, software, and hardware, needs to be evaluated.

Assessing potential dangers is essential in every technological feasibility study. It entails a thorough analysis of possible technological difficulties and hazards. The first step is to catalog all potential technical risks that might arise throughout the project's development, including technology limitations, unforeseen roadblocks, and equipment breakdowns. Alternative methods, backup plans, and investments in redundancy and backup systems are all examples of mitigation techniques. Staffing the project team with knowledgeable persons who can anticipate and adapt to any difficulties that may arise is equally important. Another sensible step toward successful risk management is to think about prototyping and testing to verify the project's technical feasibility before going into full-scale development.

One of the most important aspects of technical feasibility, especially for R&D initiatives, is conformity with applicable regulations. This aspect involves several important elements, beginning with determining the regulatory requirements, certifications, and permissions necessary for the project. Such standards are essential in

6.3 Stage 2: Feasibility Assessment

fields like healthcare, aircraft, and finance. In addition to examining the technology's effects on intellectual property, data security, and ethics, the study needs to consider legal and ethical factors. It is critical to design a thorough compliance strategy to guarantee that all regulatory criteria are met throughout the project's development and post-launch stages. Figure 6.1 demonstrates the phases involved in assessing technical feasibility, from gathering information to making a final decision on whether or not to go further.

(c) *Financial Analysis*

Feasibility assessments of R&D initiatives rely heavily on financial projections and projections of future costs. At this stage, the project's economic feasibility is determined by carefully analyzing the project's financial ramifications. The primary goals are cost analysis, revenue projection, and return on investment (ROI) evaluation.

Estimating costs, including one-time development and ongoing operating costs, is the first stage in financial analysis. This includes evaluating and forecasting manufacturing, marketing, sales, distribution, and support costs, in addition to the upfront expenditures of research, design, testing, and prototyping. It is also important to set aside some money if the project's scope expands or contracts unexpectedly.

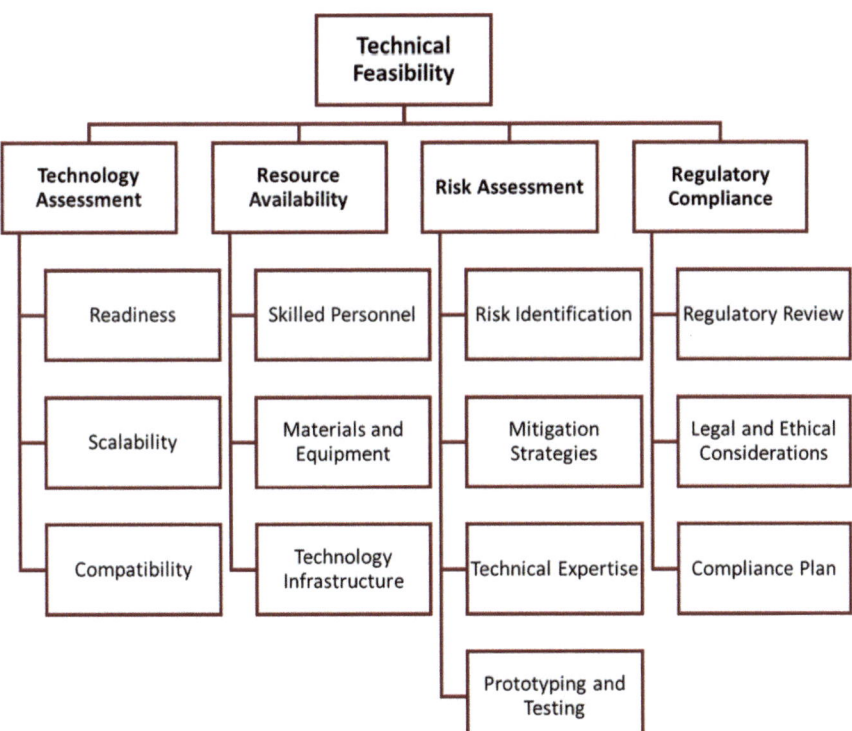

Fig. 6.1 Technical feasibility assessment process

The prospective profits of the R&D project are next evaluated by looking at revenue forecasts. This process includes formulating price plans, considering client willingness to pay, assessing market demand, pinpointing sales channels, and investigating licensing prospects. One of the most important components of revenue forecasting is setting competitive prices while yielding a profit.

The third phase of financial analysis involves estimating return on investment (ROI). Profitability may be calculated using the ROI formula and analyzing the break-even point, payback duration, and sensitivity of the projections against the actual expenses. A project's value proposition may be evaluated via a cost-benefit analysis and scenario planning once sensitivity analysis has been used to determine how changes in key financial factors affect the project's financial results. Figure 6.2 depicts the stages of economic research for R&D feasibility, including cost estimations, revenue forecasts, return on investment calculations, and sensitivity analyses.

(d) *Risk Assessment*

In the R&D process, the risk assessment phase is crucial to the feasibility assessment phase. The process includes systematically analyzing the risks and unknowns that might affect the project's outcome. Technical hazards, including technological complexity, resource availability, dependencies, and compliance concerns, need to

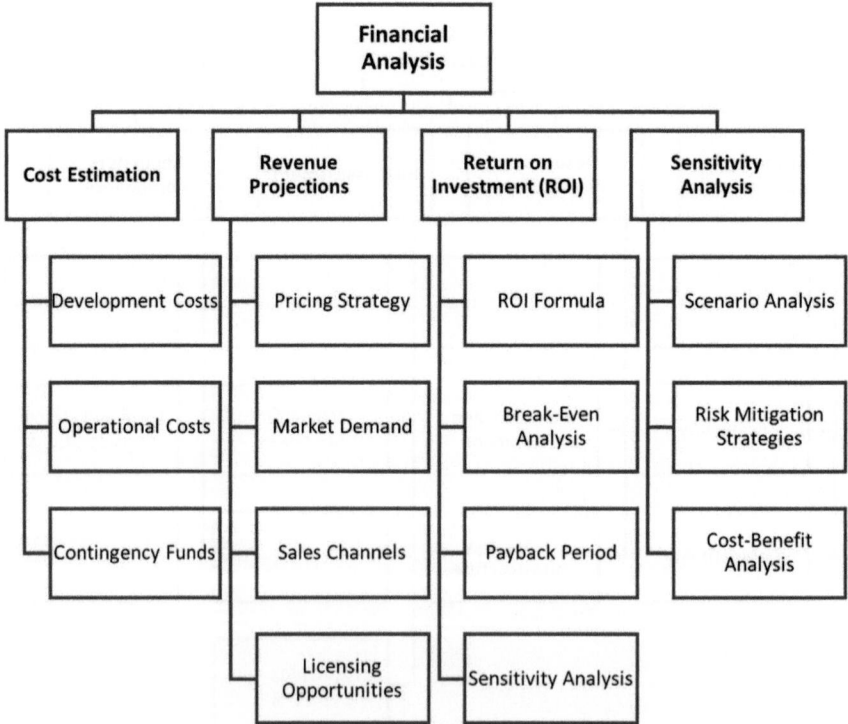

Fig. 6.2 Financial analysis flowchart for R&D feasibility assessment

be identified. In addition, it examines the accessibility of resources, which includes the necessary personnel, funds, and facilities to carry out the project.

When potential dangers are recognized, preventative measures are taken to lessen their effect or eliminate them. These methods include backup plans, means for transferring risks to other parties, constant monitoring, and extensive testing. For the R&D process, the most significant part of risk assessment is recording identified risks, prospective repercussions, and mitigation solutions. This recordkeeping facilitates open dialogue between team members and other interested parties.

6.4 Stage 3: Project Planning

Effective project planning in R&D lays the groundwork for breakthrough discoveries. Organizations may develop a structured framework for their R&D activities by identifying defined goals, allocating resources, creating timetables and milestones, and forming cohesive teams. In addition to increasing the likelihood that project goals will be met, thorough planning may help uncover obstacles and reduce risks at an earlier stage. This transitional phase between the concept and execution phases is when the innovation blueprint is brought to life.

(a) Setting Objectives and Goals

R&D project planning relies heavily on the establishment of well-defined objectives and targets. These goals serve as both a guidepost and a statement of intent for the R&D endeavor. When written carefully, they act as a beacon that keeps everyone on the same page and keeps the project on track so that progress can be tracked. Here, we will discuss why having well-defined R&D objectives and goals is crucial.

Setting goals serves as a road map for the R&D endeavor. These goals address the question, "What are we trying to achieve?" Plans are more likely to be achieved if they are SMART: SMART goals (specific, measurable, attainable, relevant, and time-bound).

Specific: Objectives should be crystal unambiguous. Team members should have a precise understanding of what needs to be accomplished. For example, rather than having a vague objective like "Develop a new product," a specific goal would be "Design a lightweight, energy-efficient smartphone with enhanced battery life."

Measurable: Measurable objectives are quantifiable, allowing for progress tracking and success measurement. For instance, "Increase market share by 10% in the next fiscal year" is measurable because it specifies a 10% target.

Achievable: Objectives should be realistic and attainable with the resources and constraints. Setting unrealistic goals can lead to frustration and failure. Assess the feasibility of each objective based on available resources and capabilities.

Relevant: Objectives need to be directly related to the project's overall purpose and the organization's strategic goals. They should contribute to the larger mission of the company. Irrelevant objectives can divert resources and focus from critical areas.

Time-bound: Establish a timeframe within which the objectives should be achieved. Timelines create a sense of urgency and accountability. For example, "Launch the product within 18 months" is time-bound.

As the larger, overarching accomplishments that give vital context and incentive for the objectives defined, goals play a key role in the world of dreams. These objectives may include creating new goods, entering new markets, becoming more competitive, or propelling technical progress in the context of R&D. Pursuing success is motivated by setting goals, which serve as both a road map and final destination for the objectives that have been established.

It is hard to overestimate the value of having clearly defined goals and objectives. Having articulated goals provides clarity and alignment, helping project teams and stakeholders work together toward a shared vision and eliminating unnecessary guesswork. They help measure progress via quantifiable results, which in turn helps project managers see problems early and make modifications as needed. In addition, having clear objectives and goals helps keep everyone on the same page and motivated since everyone knows what they are working for. The proper amount of time, effort, and money are allotted to the project, and the right personnel are assigned to the right tasks because of the clarity provided by objectives.

(b) Resource Allocation

Strategic resource allocation is a critical partner in the complex dance of R&D project planning. To guarantee the smooth running of an R&D project, it is necessary to carefully allocate resources, including money, people, time, and equipment. To accomplish R&D goals and objectives, much thought and preparation need to go into allocating available resources. Allocating resources is like putting together a complex puzzle. Optimal resource utilization is achieved by making well-informed choices regarding distributing existing assets. These tools cover the following:

Financial Resources: Budgeting and managing financial investments is critical to resource allocation. It involves determining how much capital is required for each project phase and ensuring it is available when needed. This includes allocating funds for research, development, testing, and scaling.

Human Capital: The right people with the right skills are the engine of any R&D project. Human resource allocation involves defining roles, responsibilities, and team structures. Cross-functional teams often come into play, bringing together individuals with various expertise, such as scientists, engineers, designers, and marketers.

Time: Time allocation involves establishing realistic timelines for project milestones. It encompasses task scheduling, project planning, and ensuring that timeframes align with objectives and goals.

Infrastructure and Technology: This encompasses allocating physical and technological resources necessary for the project. This could include laboratory space, equipment, software, and other tools required for R&D (Fig. 6.3).

6.4 Stage 3: Project Planning

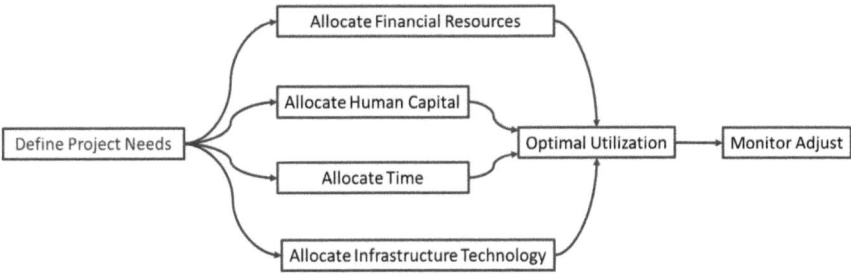

Fig. 6.3 Resource allocation flowchart

Allocating resources is a difficult task that requires careful consideration of several factors, including available funds, time, and personnel. Resource allocation is an ongoing, iterative process that requires adjustments as the project evolves. Key considerations include aligning resources with project priorities, favoring high-priority tasks, maintaining a delicate balance between sufficient allocation and preventing resource wastage, implementing contingency planning to address unforeseen challenges by reserving resources for them, and so on.

Allocating resources efficiently is essential to the success of R&D initiatives because it maximizes resource usage. This prudent allocation not only contributes to cost management by avoiding overruns and unanticipated costs, but it also helps mitigate risk by allowing for the development of a backup plan in the case of an emergency. Projects may be completed on schedule and meet their goals if sufficient resources are supplied. Teams are also more likely to produce high-quality work if they can access the necessary tools. In Table 6.2, we explore the measurements, obstacles, and project success repercussions of resource allocation on R&D success.

(c) Timelines and Milestones

R&D project planning relies heavily on timelines and milestones. Timelines are plans that detail exactly when each project step will be completed. They are a time-bound road map for the R&D project and have defined beginning and ending points. In the sometimes-unpredictable realm of R&D, it is especially important to carefully plan for the unexpected by carefully ordering tasks, allocating resources, and considering potential delays.

In contrast, milestones are beacons that direct researchers in the right direction. These landmarks serve many purposes on the project timeline. They let the project team track progress, make decisions, share information, and hold one another accountable. In addition to signaling the completion of an important job or phase, milestones serve as a source of pride and inspiration for the team. Timely milestones are especially important in R&D, where innovation can be unpredictable, because they highlight possible challenges, facilitate resource planning, and create confidence among project stakeholders.

Establishing definitive timeframes and milestones is analogous to charting a systematic path for the project in the ever-changing realm of R&D. Measurable

Table 6.2 Impact of effective resource allocation on R&D project success

Outcome	Key metrics	Potential challenges	Impact on project success
Efficient resource utilization	Resource utilization rate, cost per output	Resource conflicts, underutilization, skill gaps	Maximizes productivity and cost-effectiveness
Risk mitigation	Risk assessment, risk response plan	Changing project scope, unforeseen events	Enhances project stability and resilience
On-time delivery	Project timeline adherence, milestone completion	Scope changes, resource shortages	Ensures project stays on schedule
Quality deliverables	Quality metrics, customer satisfaction	Skill mismatches, time constraints	Delivers high-quality results
Cost control	Budget adherence, cost variance	Scope creep, resource inefficiencies	Keeps project within budget limits

outcomes and proactive risk management, resource planning, quality assurance, and project management are all made possible through this method. In addition, when project goals are met, investors, customers, and team members gain trust in the initiative. Thus, in the ever-changing landscape of R&D, deadlines and milestones are vital tools for staying on track, managing uncertainties, and eventually attaining success.

(d) Team Formation

The process of putting together a team is crucial in any R&D effort. A well-organized and cohesive team is required for R&D projects to be carried out efficiently and effectively. In this chapter, we will discuss the significance of R&D team building and the factors to consider while creating a creative group.

Putting people in a room together is not enough to establish a cohesive team capable of taking on difficult tasks; rather, it is about finding the right balance between everyone's unique strengths. There are several reasons why this strategy is essential. To begin with, it encourages a multidisciplinary approach, which is especially useful for R&D initiatives that frequently require the expertise of scientists, engineers, designers, marketers, and data analysts. Together, we can think beyond the box and solve problems creatively. Second, it ensures that everyone on the team is well rounded and can perform every step of the process, from brainstorming to marketing. Third, it fosters an environment where people feel safe voicing their opinions, receiving criticism, and working together to achieve a common objective. There is no room for misunderstanding because of the well-defined division of labor. Direction, guidance, and the ability to share and resolve information and conflicts are all made possible through establishing clear leadership roles and channels of communication.

To come up with creative ideas, it is important to build teams with a good balance of knowledge and encourage various talents, backgrounds, and experiences. Smaller groups are generally more agile and efficient, whereas larger teams can tackle more ambitious endeavors. Alignment with the company's values, mission, and vision can only be achieved by a strong cultural fit within the organization. Guidance, decision-making, motivation, and output all benefit greatly from competent leadership. For a team to function effectively, they need access to appropriate means of communication and cooperation. Finally, the team's capabilities remain aligned with project goals thanks to continued investment in training and development. Having a strong team in place is crucial to any R&D endeavor. The following results are possible with the correct mix of talent, abilities, and teamwork:

- Efficient Problem-Solving: Diverse teams can tackle complex challenges from multiple angles, leading to innovative solutions.
- Creative Ideation: A collaborative and diverse team can generate many creative ideas.
- Quality Assurance: Clear roles and responsibilities and effective leadership contribute to a high-quality standard in project deliverables.
- Adaptability: Effective teams can adjust to changing project requirements and unforeseen obstacles.

6.5 Stage 4: Concept Development

The R&D process hits a tipping point at the concept development stage when the abstract notion takes shape. This phase includes developing a prototype, testing its functionality, and honing the concept. In this phase, creators examine the concept in further depth to make sure it meets consumer demands, is feasible from a technical standpoint, and can be realized within the set budget.

(a) *Prototyping and Proof of Concept*

Prototyping and proof of concept are crucial methods that help bring an inventive idea to life during the concept development stage of R&D. Prototypes are the first iterations of a product, and they can range in complexity from rough sketches to fully functional models. They are essential since they do things like verifying technology. Risks connected with a project can be mitigated through prototypes, which help evaluate the technical feasibility of a concept by highlighting potential obstacles and areas for development. They also make it easier to get early input from potential customers, whose opinions can help better shape the product to suit their needs. In addition, prototypes are great for communicating the idea to the development team, stakeholders, and possible investors because they are a physical representation of the concept.

On the other hand, a proof of concept is a prototype designed to prove that a central idea can be implemented in practice. Addressing any lingering technical questions helps determine if the concept's core technology or idea is feasible. A proof of concept is essential for reducing risk since it allows for the early detection and resolution of any problems. If the proof of concept is successful, the project can move forward with the validated image, but if it fails, the project will need to make substantial changes or be scrapped altogether. Overall, the gap between an idea and a physical project can be bridged, user demands can be better addressed, technical feasibility can be guaranteed, and costs can be better managed by incorporating these prototype and proof-of-concept activities into concept development.

(b) *Intellectual Property Considerations*

Intellectual property (IP) concerns should be prioritized when developing a new idea. Among these factors is the need to safeguard original concepts, designs, or breakthroughs that develop as work progresses. Protecting innovations, keeping a competitive edge, and even making money from intellectual property need careful IP management. Several crucial factors should be taken into account.

To start, patents are essential for safeguarding the originality of the innovation's methods, systems, or other distinguishing characteristics. To successfully obtain a patent, it is crucial to identify potentially patentable aspects of a design as early in the process as possible, and it is often recommended to work with a patent attorney. Second, trademarks are crucial for safeguarding the reputation of the product's name, logo, and slogan in the marketplace. An intelligently selected trademark may do wonders for brand awareness and loyalty. Original works of literature, art, and computer code are all protected by copyrights. If these works are registered at the concept stage, they can be protected from being copied and distributed without permission.

In addition, trade secrets can be crucial for guarding valuable know-how like secret recipes, manufacturing procedures, and business plans. It is essential to keep confidential information protected. When working with partners, protecting sensitive information using nondisclosure agreements (NDAs) is important. When working on a project incorporating open-source code or components, it is essential to consider these factors and adhere to the terms of the relevant licenses to prevent potential legal issues. Key measures to effectively manage intellectual property include developing a thorough IP strategy linked to business goals, conducting frequent IP audits, and consulting with legal counsel. Strategic use, competitive advantage creation, investor attraction, and revenue generation through licensing or selling IP assets are all factors to consider regarding intellectual property. Therefore, taking preventative measures to secure intellectual property rights is essential from the moment a thought is conceived.

(c) *Design and Engineering*

At the intersection of design and engineering, inventiveness and technical know-how meet to make the concept a workable and useful reality. The design process considers the product's practicality, aesthetic value, and ease of use. Usability testing, user interviews, and persona creation are all integral parts of user-centered design. The visual design of a product establishes its visual identity through color palettes and typefaces. Prototyping enables early testing and iterative methods, which are crucial to maintaining a high-quality standard.

Engineering is a discipline that works with design to create a product's technical components and ensure the concept can be practically implemented. Engineers evaluate the project's technical viability, establish the project's functional specifications, develop prototypes, and deal with the project's materials and production issues. Costs are controlled, environmental impacts are minimized, and long-term sustainability is prioritized. Successful product development requires close cooperation between designers and engineers, who need to work together to translate ideas from concept to reality.

In R&D, the concept development phase is critical because it lays the groundwork for later stages like testing, scalability, and commercialization. Products that look attractive and work smoothly and efficiently result from a well-executed design and engineering process, which guarantees that the concept is original and practical to bring to market. Combining artistic and technical understanding during the ideation phase is crucial for successfully implementing novel solutions.

6.6 Stage 5: Testing and Evaluation

After establishing a product or innovation's concept and prototype, it needs extensive testing and review. This is a crucial step before mass manufacturing, and commercialization can begin to ensure that the innovation is up to par.

(a) *Product Testing*

Before a product's revolutionary potential can be fully evaluated, it needs to undergo extensive testing as part of the R&D process. To ensure the product or innovation delivers as promised, it must undergo rigorous tests. Functionality testing determines how well the product's major features and capabilities serve their intended purposes. Likewise, quality control processes examine effects under various settings meant to mimic real-world use to ensure they consistently fulfill defined criteria.

Testing for safety and compliance is also important because different industries have different safety standards and regulations that goods need to adhere to. Considerations of the product's carbon footprint and potential environmental damage are also essential. Testing the product extensively under various conditions ensures its dependability and performance. Also, testing against market standards and competition can help establish the product's value. To determine the next steps for the innovation, such as whether or not to move through with mass production, undergo additional iterations, or revert to the concept development phase in the event of serious problems, the process relies on extensive documentation and reporting of test results. The careful approach employed during testing reduces risks, quality is verified, and the invention is by expectations.

(b) *User Feedback*

Product testing is an integral part of this stage. This exhaustive analysis takes many forms, including usability testing to make sure the product does what it is supposed to, quality control to determine how well it holds up to regular use, compliance testing to make sure it abides by regulations, an evaluation of the product's effect on the environment, and reliability and performance evaluations. Comparing oneself to the market or industry norms can yield useful information, and keeping meticulous records is essential for monitoring progress.

Another important part of the testing and assessment phase, user feedback, connects the product's creators with its eventual consumers. In beta testing, a user sample can share honest opinions and suggestions for enhancements. Users' happiness, pain areas, and preferences can be systematically gathered through surveys, questionnaires, and usability testing. Users' expectations and needs can be better understood through focus groups' qualitative observations. Customers can provide continuous feedback through social media and customer service chat. Iterative product development is guided by systematic analysis of customer feedback, which ensures that the final product meets the needs of its target audience.

To bridge the gap between the development team's vision and the demands of the end-users, input from those using the product is essential. It helps direct development, improves final quality, and increases satisfaction with the result. Creating products that resonate with their target audience and provide a competitive advantage in the market demands the systematic gathering and utilization of customer feedback. Risks are reduced, and quality requirements and consumer expectations are met during this stage through extensive product testing.

(c) *Iterative Development*

Iterative development is a cyclical process that uses data from user testing and feedback to improve the product or innovation at each stage. For the finished product to meet user requirements, quality benchmarks, and market expectations, it needs to go through this iterative process. Iterative development is based on continuous improvement, with development teams constantly searching for ways to enhance the process by eliminating defects, revising goals, reducing costs, and incorporating new technology. Project management systems, such as agile methodologies, enable adaptable project administration and iterative development cycles; each iteration is validated through testing. Iterative development relies heavily on open lines of communication and cooperation between team members and stakeholders.

Iterative development is an organized and flexible strategy that accounts for the dynamic nature of R&D initiatives. Researchers and developers can keep up with the market and provide goods that meet and surpass users' expectations by listening to their suggestions, basing their judgments on data, and committing to continuous improvement. Through iterative development and user feedback, a product can be fine-tuned to meet the needs and preferences of its target audience, increasing the likelihood of commercial success. To survive in today's competitive market, being agile, efficient, and open to new ideas and technologies is crucial.

6.7 Stage 6: Scaling and Production

Once testing and evaluation of an innovation or product have been completed successfully, attention should turn to scaling and production. This phase is crucial for developing a concept into a ready product for the market. The ability of the designed solution to meet market demand while retaining quality and cost-effectiveness depends on successful scaling and production procedures. This stage calls for meticulous planning, resource allocation, and coordination among numerous organizational departments.

As innovation and ideas take shape, the move to Stage 6 marks a turning point in the R&D process. Scaling and production, which calls for careful planning and resource allocation, refines the developed concept into a product ready for the market. The first phase is capacity planning since businesses need to determine how much production capacity they will need to meet demand. As part of this phase, the supply chain is optimized, the infrastructure is expanded, and efficient, economical, and quality-assured manufacturing procedures are implemented. Supply chain management depends on supplier relationships, inventory control, and demand forecasting to maintain a steady flow of materials and components.

Cost optimization measures, such as economies of scale, are essential for the production process to be cost-effective. Contingency plans for anticipated disruptions are crucial to minimizing risks, and compliance with industry-specific laws and safety requirements is a non-negotiable need. Staffing and training are essential, and skill development and workforce planning provide a skilled workforce. Additionally, businesses need to use green production techniques, recycling programs, and waste reduction strategies to reduce their ecological impact. The key to satisfying consumer demand, creating long-term success, and achieving profitability is successful scaling and production.

Organizations must keep a close eye on cost optimization measures when they start scaling and production. These tactics seek to reduce unnecessary expenses and boost production effectiveness. It is crucial to always look for ways to cut costs, such as process optimization, buying in bulk, and raising overall effectiveness. Organizations can use economies of scale to reduce per-unit manufacturing costs as production volumes rise, increasing the product's competitiveness in the market. To make sure that manufacturing costs adhere to financial restrictions and maintain economic sustainability during this crucial stage, cost monitoring needs to be closely monitored.

The scaling and production stage is when risk management is most important. Organizations need to create thorough backup plans to deal with any disruptions. These interruptions can be caused by various things, such as problems with the supply chain, broken equipment, or even natural disasters. Additionally, it is crucial to ensure regulatory compliance. The reputation of the company and the product's safety and legal compliance depend heavily on adherence to industry-specific rules and safety standards. Technical proficiency and a dedication to sustainable and environmentally friendly methods are necessary for successful growth and production.

This commitment entails researching green manufacturing techniques to lessen the negative effects of production processes on the environment. Organizations should also consider recycling and trash reduction strategies to reduce environmental impact and improve corporate social responsibility.

Staffing and training are essential elements in the field of human resources. To achieve the scaling objectives, organizations should evaluate their workforce needs and, if necessary, add new employees. Training and skill development programs should be implemented to guarantee that the workforce is knowledgeable about the production process and quality control processes. To uphold safety regulations and maintain high product quality, skilled and prepared personnel are essential.

6.8 Stage 7: Commercialization

The crucial next phase is commercialization when the product or innovation has been effectively developed and improved. This phase entails developing the concept into a service or product ready for market and introducing it to the target market. As it connects invention with revenue generating, commercialization is a crucial stage in the R&D process. In this phase, selecting a suitable market entry strategy is vital as it will define how the new product is introduced to the market and significantly influence the potential for success.

Market segmentation, which entails splitting the target market into distinct consumer segments based on shared features, needs, or preferences, is an important factor to consider when considering commercialization. This enables individuals to craft tailored messages and product positioning, subsequently adjusting each segment's marketing and distribution strategies. Another tactic is market penetration, which aims to increase market share through aggressive pricing, marketing campaigns, and customer loyalty programs. Market development could be profitable if the product can cater to new demographics or geographic areas. A competitive market necessitates product differentiation, emphasizing distinctive features, quality control, and brand creation.

The product's qualities, the market's state, and the level of competition will all influence the decision regarding the market entry approach. A combination of strategies is frequently the best course of action. Review and modify the plan continually as market dynamics shift and new opportunities present themselves. The ability to successfully position and market the idea to the target audience will determine the success achieved in commercialization. Figure 6.4 illustrates the connections between market segmentation, market penetration, market development, and product differentiation methods, highlighting the necessity of continual market circumstances adaption.

(a) Marketing and Sales

A successful commercialization strategy is built on a solid marketing plan laying a clear path for promoting a product to the target market. This strategy includes

6.8 Stage 7: Commercialization

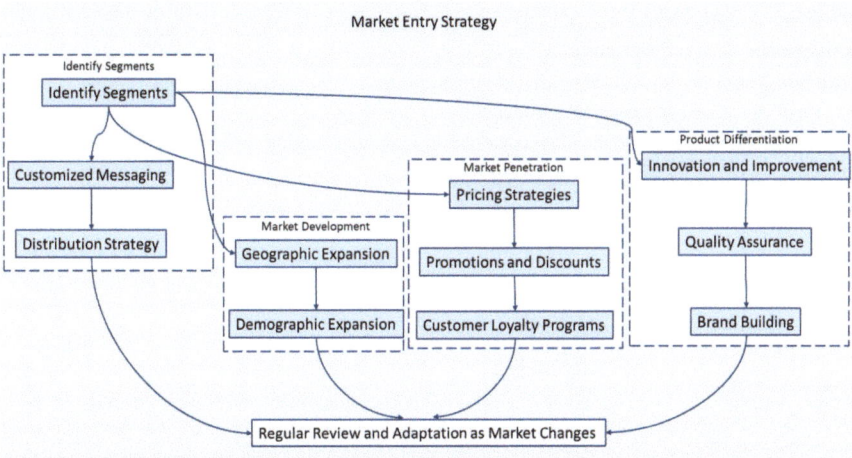

Fig. 6.4 Market entry strategies

several crucial elements, starting with identifying the target market thoroughly. Customizing marketing initiatives requires having a thorough understanding of customers' needs, preferences, and behaviors. Choosing the best marketing channels to engage the audience is also important. These channels may include traditional advertising, social media, email, content, and digital marketing. A thoughtful content strategy is essential because it makes it possible to provide interesting and educational materials that communicate the worth and advantages of the product. Budget distribution among different marketing initiatives needs to be carefully considered, prioritizing those with the best potential return on investment to make it all financially sustainable.

Sales channels, an essential component of the plan, make sure the product gets to clients effectively. Direct sales, online sales, sales through physical storefronts, and sales through a specialized sales force are all possibilities. As an alternative, using distributors or wholesalers can assist in reaching a larger market. Another option is to work together with well-known shops to get the product displayed in their stores. E-commerce platforms, such as well-known online marketplaces, offer the chance to increase reach.

It is crucial to have a well-thought-out pricing plan based on carefully examining the market environment, manufacturing costs, perceived value, and customer willingness to pay. A competitive pricing strategy can be used to place the product in the market, or value-based pricing can match the benefits the effect is thought to provide. Additionally, properly using discounts, promotions, or bundling can be a potent client acquisition and retention technique.

Brand creation is an essential part of the marketing strategy to establish awareness and trust in the market. Clear brand positioning is critical for effective branding since it creates the product's USPs and its place in the market. For brand awareness, it is also essential to have a memorable visual identity, including a logo and color

Table 6.3 Key components of a comprehensive marketing plan for successful commercialization

Component	Description
Marketing plan	A well-structured plan that guides product promotion and customer engagement strategies
Target audience	Define the ideal customer base, understanding their needs, preferences, and behaviors
Marketing channels	Identify effective channels for reaching the audience, such as digital marketing, social media, content marketing, email marketing, and traditional advertising
Content strategy	Develop engaging and informative content that communicates the product's value and benefits
Budget allocation	Allocate the budget to various marketing activities based on their potential return on investment
Sales channels	Determine how the product reaches customers and where it is made available
Direct sales	Selling directly to customers through the website, physical stores, or sales teams
Distributors	Leverage distributors or wholesalers to expand the product's reach in the market
Retail partners	Collaborate with established retailers to make the product available in their stores
E-commerce platforms	Utilize online marketplaces or e-commerce platforms to increase the product's visibility
Pricing strategy	Develop a pricing strategy based on market conditions, production costs, perceived value, and customer willingness to pay
Competitive pricing	Determine how the product's price compares to similar products in the market
Value-based pricing	Set the price based on the perceived value the product delivers to customers
Discounts and promotions	Implement discounts, promotions, or bundling to attract and retain customers
Branding	Build a strong brand identity for recognition and trust in the market
Brand positioning	Clearly define the product's unique selling propositions and positioning in the market
Visual identity	Design a memorable logo, color scheme, and other visual elements representing the brand
Brand messaging	Craft a compelling and consistent message that resonates with the target audience

palette. Establishing a connection between the product and the target market's requirements and desires by creating a compelling and consistent brand message. The components of a marketing strategy, sales channels, price plan, and branding are listed in Table 6.3 for effective product promotion and market penetration.

Continuous evaluation and modification are necessary for effective marketing and sales tactics. To make educated adjustments to the tactics, evaluate the success of the marketing initiatives regularly, collect customer feedback, and monitor sales indicators. Remember that a well-executed marketing and sales campaign can significantly influence the product's success on the market.

(b) Distribution and Partnerships

Strategic alliances and effective distribution are essential in successfully commercializing R&D-driven innovation. Products should efficiently and promptly reach clients thanks to a well-designed distribution network. To efficiently service target markets, this calls for managing logistics and the supply chain, inventory control, and distribution centers placed strategically.

Partnerships are essential for increasing resources and reach. Increased visibility and accessibility to a larger consumer base can result from working with distributors, merchants, or e-commerce platforms. Additionally, establishing strategic alliances with similar companies can help companies cross-promote their products and increase their market presence. Building ties with manufacturers and suppliers ensures a reliable and economical supply chain.

Maintaining top-notch customer service is equally crucial. Building and maintaining positive customer connections, which strengthens brand loyalty, requires effective customer service, transparent warranty policies, and means for collecting customer feedback. Establishing opportunities for development and expansion can be aided by routinely evaluating the performance of the distribution network and partnerships. Collaboration and flexibility are essential for successful commercialization in a dynamic corporate environment.

6.9 Stage 8: Post-Launch Evaluation

A thorough post-launch evaluation procedure needs to be carried out after the product or service has been effectively introduced to the market. The product's performance is evaluated, user opinions are gathered, and the project's alignment with the original goals is confirmed at this phase. The post-launch evaluation stage of the R&D process is crucial because it offers knowledge for future development and efforts at continual improvement.

Performance monitoring is a crucial component of the R&D process' post-launch evaluation phase. It entails systematically monitoring and evaluating key performance indicators (KPIs) to determine the marketability of the good or service. Sales and revenue data give firms quick access to financial information that they may use to evaluate profitability and return on investment. Analyzing market share sheds information on a product's competitive position by emphasizing growth or decline. User engagement indicators for software and online applications, such as active users, retention, and session length, provide essential insight into the user experience. Customer satisfaction surveys are crucial for understanding how well a product satisfies consumer expectations and suggests improvements. The success of a product depends on keeping strict quality metrics since it ensures dependability and consumer pleasure. Organizations may optimize the impact of their products, connect them with financial goals, and improve user experience and product quality by keeping an eye on these KPIs.

Performance monitoring is more than just tracking figures; it is an organizational strategic compass for post-launch evaluation. Sales, market share, and economic measures inform financial health and competitiveness. Metrics measuring user engagement and satisfaction aid in maintaining a user-centric strategy by encouraging loyalty and lowering churn. Quality metrics ensure customer satisfaction and product dependability.

(a) Continuous Improvement

The R&D process' post-launch review is built around continuous improvement. The goal is to improve the efficiency of the project, the product, and the procedures. Data-driven decision-making is essential since it allows businesses to use performance indicators, client feedback, and market data to pinpoint potential growth areas and improve their offerings. The method is iterative; thus, the debut of the product ushers in a journey for further improvement. Quality assurance is key, emphasizing thorough testing, frequent updates, and patches to fix flaws or vulnerabilities. Priority is given to user-centered design, where active listening to feedback and concerns informs product updates and enhances the user experience.

With the use of brief iterations known as sprints, many R&D teams use agile approaches like Scrum or Kanban to promote collaboration and adaptation. Smoothly implementing process or strategy changes depends in part on change management. It calls for dialogue, instruction, and dealing with change resistance. Documentation and knowledge exchange are essential to preserve information and best practices for upcoming projects. For measuring success and making data-driven decisions, monitoring key performance indicators (KPIs) that align with corporate objectives is necessary.

(b) Customer Feedback and Adaptation

Customer feedback can be gathered to provide priceless insights about user experiences, preferences, and market fit. Understanding elements like user experience (UX), functional recommendations, and the product's competitive positioning are made easier by this. Customer feedback can be gathered through various techniques, such as surveys, user interviews, social media monitoring, and product usage analytics. However, prioritizing this feedback based on elements like severity, user impact, and alignment with strategic objectives is crucial. Iterative development cycles are driven by this prioritizing, which incorporates input methodically to provide improvements and problem-solving. Customers' feedback is kept at the forefront of product development thanks to the effective connection with them and a never-ending feedback loop.

However, it is crucial to balance strategic goals and client feedback, as not all feedback may align with long-term aims. While considering customer wants and shifting market conditions, the iterative adaption process needs to preserve the product's primary objective. In conclusion, consumer input and adaptation create a dynamic feedback loop that directs product development, ensuring that the final product is competitive, in line with customer expectations, and adaptable to a continuously evolving market.

6.10 Challenges and Common Pitfalls

A crucial step in the innovation process, R&D has its share of difficulties and dangers. For people and organizations involved in R&D, having a thorough understanding of these issues is essential because it enables them to overcome hurdles and increase the possibility that their efforts will be successful.

Funding and Budget Management: In R&D, managing financial resources effectively is a major difficulty. The significant investments in equipment, staff, and materials needed for R&D projects frequently need help with problems like budget overruns, issues with resource allocation, and the challenging challenge of gaining early funding. These financial difficulties, especially for start-ups and smaller organizations, can burden resources and jeopardize project completion.

Intellectual Property Protection: An essential component of R&D endeavors is protecting intellectual property. By ignoring this factor, valuable innovations and competitive advantages may be lost. Intellectual property issues, insufficient data protection, and poor patent strategies are typical difficulties. Strong cybersecurity measures are essential in the digital age to preserve critical R&D data and avoid potential ownership and rights disputes.

Market Competition: R&D projects come into contact with competitive environments as they advance. It is crucial to understand competition and how to handle it. The risks of competitors copying successful ideas and the oversaturation of markets are major obstacles. Businesses need to constantly adapt, pay attention to market dynamics, and work to set themselves apart from the competition to navigate this terrain successfully.

It takes careful preparation, ongoing evaluation, and adaptability to overcome these numerous problems and avoid frequent mistakes in the R&D process. By taking a proactive approach to risk management, making wise financial decisions, and prioritizing legal and regulatory compliance, practitioners can increase their chances of successful innovation and development. With these obstacles, R&D projects can be better prepared to produce beneficial outcomes and game-changing technologies.

References

1. J. Wang, W. Lin, Y.-H. Huang, A performance-oriented risk management framework for innovative R&D projects. Technovation **30**(11–12), 601–611 (2010)
2. B. Yoon et al., Development of an R&D process model for enhancing the quality of R&D: Comparison with CMMI, ISO and EIRMA. Total Qual. Manag. Bus. Excell. **26**(7–8), 746–761 (2015)
3. E. Petrova, Innovation in the pharmaceutical industry: The process of drug discovery and development, in *Innovation and Marketing in the Pharmaceutical Industry: Emerging Practices, Research, and Policies*, (Springer, 2013), pp. 19–81
4. N.S. Davcik et al., Exploring the role of international R&D activities in the impact of technological and marketing capabilities on SMEs' performance. J. Bus. Res. **128**, 650–660 (2021)

Chapter 7
R&D as a Job

In the dynamic realm of research and development (R&D), understanding job attributes, career pathways, and organizational hierarchies is crucial. This chapter delves into these facets, shedding light on the intricate structures that underpin R&D professions. We explore the evolution of R&D jobs, ranging from traditional research roles to the advent of multifaceted positions, ultimately emphasizing the emerging dual- and triple-level hierarchies. Furthermore, this chapter investigates the multifaceted nature of R&D work. It illuminates the diverse landscapes of R&D activities, addressing the nuances of heterogeneity in R&D practices and the implications for career development. Decentralization and centralization in R&D organizations are pivotal considerations in shaping the strategic direction of R&D efforts. This chapter dissects these organizational paradigms, elucidating how they influence job roles and career trajectories. Moreover, this chapter delves into the often-overlooked aspect of career conflict in R&D professions and the strategies for its effective management. Through an in-depth exploration of career design, we provide valuable insights into creating resilient and fulfilling career paths in R&D. Lastly, this chapter offers a glimpse into the future of R&D jobs, contemplating the transformations on the horizon and their impact on career prospects. It is an essential resource for R&D professionals and those aspiring to navigate the ever-evolving landscape of R&D careers.

Learning Objectives

- Identify potential careers in R&D and consider how they have evolved through time
- Understand the benefits and drawbacks of working in R&D
- Recognize the fundamentals and expectations of working in R&D
- Describe the employment outlook for engineers and scientists

7.1　Job Attributes

Many studies support what Hackman and Oldham [1] said about the relationship between job characteristics and job well-being. An important finding from this study is that occupations with adequate feedback, occupational identity, autonomy, and diversity are more productive than those without. Feedback is information about how well a task is working. Job IDs mean that jobs can be viewed as separate units. For example, suppose you complete a specific research project or claim that a specific discovery is associated with a specific person. In that case, you can increase the relevance of a task to a specific task by allowing the researcher's name to be associated with that task.

On the other hand, the coherence of the work becomes less clear when many researchers are involved in a research project, and they are all deployed at different locations and organizational levels. Autonomy means the ability to choose what to do. Diversity means doing different types of work in the same workplace. Hackman and Oldman's theory states that work is meaningful when people feel it is important because of workplace diversity, relevance, and meaning. The higher meanings, in turn, lead to happiness.

Similarly, the autonomy that gives people control over their work and work outcomes leads to satisfaction. Finally, it is important to get feedback. People need to know how well they are doing and if what they are doing is making a difference. If someone works on something for years and no one notices (e.g., no one cites their scientific work), they will not get feedback. So, like any other kind of work, it is best to be diverse, designed, meaningful, and original and give the researcher much freedom and feedback. Much has been written about these issues [2]. Therefore, when designing R&D projects, it is important to remember that maximizing diversity, autonomy, feedback, and the importance of jobs lead to satisfaction. R&D also has challenges, but in 2005, nearly 80% of researchers surveyed said they were satisfied with where they worked. Researchers working alone was a major driver of job satisfaction [3].

Figure 7.1 illustrates the intricate interplay of essential skills and competencies in R&D roles. Technical expertise is the foundational skill, branching into analytical thinking and creativity and innovation. Analytical thinking further leads to problem-solving skills, a pivotal component of successful R&D work. Project management, a critical capability, is connected to problem-solving skills, emphasizing the importance of effectively managing R&D projects. Communication skills play a dual role, facilitating teamwork and ensuring a clear presentation of findings. Adaptability is a key attribute that connects to the core skills, highlighting its significance in the ever-evolving R&D landscape. Ethical conduct is essential and leads to regulatory knowledge in highly regulated industries. Lastly, data analysis is critical and connects to continuous learning, emphasizing the need for ongoing skill development in R&D careers.

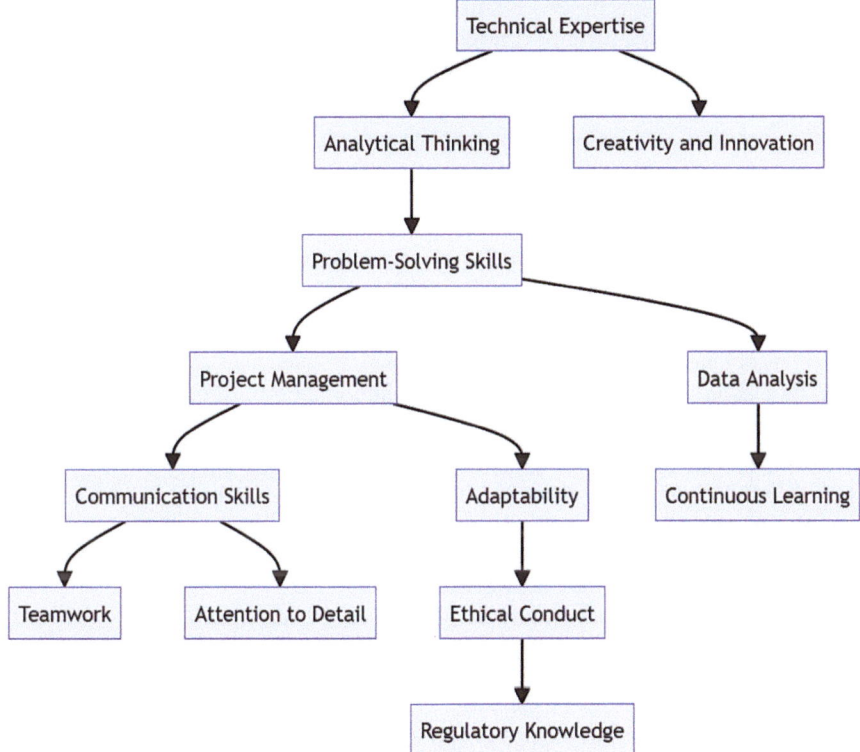

Fig. 7.1 Navigating the skill landscape in R&D roles

7.2 Career Pathways

It is important to consider a scientist's entire career when creating jobs. People who only do technical work in R&D work environments need more opportunities to rise to the organization's top regarding status and salaries. Jobs are created to make managers higher paid, but they do not match the values and needs of most tech-savvy people. When a scientist or engineer is unsatisfied with a job description, he finds happiness elsewhere, such as in his family or society. Bailyn and Lynch [4] found that this was the case for engineers who were dissatisfied with their jobs. Many organizations face this and should work to find a solution. Often, jobs should be designed to meet some of an employee's needs. When thinking about job design, it is necessary to remember that engineers' careers go through different stages. For example, Thompson and Dalton [5] interviewed over 200 managers, engineers, and scientists and found that an engineer's career spans several stages. It is said that an engineer should go through four steps to do a good job.

Engineers should work with mentors who can teach them how to plan and execute projects and how to work well with upper management and customers. In this stage, the mentor accepts the project, plans the project outline, and aligns the project with the organization's activities. Students are extensions of their mentors at this stage, so they need to be close to each other to learn how to work together.

The engineer takes charge of a specific part of a process or project, works on it on his or her own, and makes results that are easily linked to her or him. The expert starts to gain credibility and reputation as someone who knows much about a certain subject. Now, the professional is in charge of more of her or his own time and takes more responsibility for how things turn out. Relationships with colleagues and other professionals are now very important, and relationships with mentors or supervisors are less important. This means you must define the paths and place the spaces differently. For example, the expert may be far from the manager. In general, organizations consider step 2 to be important but optional. If someone stays in Tier 2 for too long, they could be fired or moved to a less important job. In other words, for an engineer to be considered "successful," he should move from stage 2 to stage 3.

This period is different in some ways. Here, engineers use their engineering skills differently instead of working on the same project. They can work with new businesses, customers, and suppliers. They also do things that benefit others and the organization as a whole. They care about how others change and grow. Many engineers stop here and think they did a great job.

Managers have a big say in how large parts of the organization move forward. She or he usually interacts with different types of people outside and inside the organization. However, it is important to remember that the technical side of a business can still be a large part of what engineers do. Therefore, rewarding those who do only technical work seems appropriate, even if there are few conductors. In other words, do not assume that managerial and technical careers are mutually exclusive.

On the other hand, professional staff in leadership positions can better develop their professional interests, giving them more freedom to make research decisions and control resources important to their research work. Technical skills and management tasks are consistent. However, there is still the challenge of providing promotion opportunities to technical staff who no longer wish to occupy senior management positions where they can no longer contribute directly to R&D.

7.3 Hierarchies with Dual and Triple Levels

One way to solve the above problem is to set up two or three levels within your organization. In a dual structure, organizations create additional levels with technical positions similar to those at the management level. These technical positions form a professional hierarchy with the same salaries, powers, and supervision as the management positions that accompany them. However, Schriesheim et al. [6] reviewed the research in 1977. They concluded that dual hierarchies could have been better at resolving conflicts between professional organizations and employees or providing more ways to be promoted and paid. It seems that the main reason these organizational structures did not work was that, by definition, moving up in the professional hierarchy meant "moving away from power." Also, those who followed the professional hierarchy failed because they felt they were on a different level than the management hierarchy and because the criteria used to judge them needed to be more fair. Because of this, Schriesheim et al. [6] suggested a triple hierarchy as a different way to handle conflict in professional organizations. The triple hierarchy gives you three different ways to move up. Those who want to move up to a managerial position can use the managerial hierarchy. The professional hierarchy is still a good choice for professionals who want to do only technical, research, and scientific work. The third level comprises professionals who have important administrative jobs in addition to their regular jobs. They have authority based on their position in the organization in areas where their professional values and organizational needs differ. The three-way hierarchy uses different types of leadership, such as champion, passive, and transformational [7]. These organizations are very similar to the structures of large research universities.

Most of the time, people with good research records run these universities. So they can connect with the teachers. They also have other qualities that help them work well with major donors, alums, trustees, government officials, etc. Some administrators are more involved with research, while others are more involved with the politics of running a university. A university that does well has a good mix of administrators.

Similarly, R&D organizations need people who can manage policy, people who can manage technical aspects, and people who can manage both. The best R&D organization has the right mix of ideas and people. It will discuss the advantages of the triple hierarchy in terms of management, professional communication, and technical hierarchies. In the case of universities, these conditions may be the president, dean, or head of the department. The president is usually a generalist who needs to study. The dean can judge the department head, and the president can judge the dean. Often, department heads know a lot about how things work and how they do research. Dean is also tech-savvy and does research from time to time. The head of the department can evaluate the professional qualifications of the teaching staff.

The argument is that a three-tiered structure can solve all three problems. Management hierarchies dominate the organization; people do not talk to each other, and there is no good way to evaluate people. Administrators need more authority in a three-part hierarchy because they need to take more of the specialization hierarchy. In the dual hierarchy, professionals and managers have different points of view and often do not speak well to each other. However, in the triple hierarchy, technical people usually talk to those in the professional liaison hierarchy rather than those in the management hierarchy. They talk to each other and get along well because they share the same values. At the dual-level, technical staff is evaluated by managers, and at the triple level, experts are evaluated by people from the professional communication level, not managers. So, the people who know them best are the ones who appreciate how well their tech workforce is doing. Some things indicate that ternary hierarchies work. Baumgartel [8] found that when people with professional or scientific backgrounds held the research administrator or director positions, researchers felt safer, and work units were more productive and had higher morale. Marcson's [9] study from 1960 and Likert's [10] from 1967 added to the evidence. Can these three approaches be used successfully by your organization? Is it even possible? In a formal organization, each person's organizational structure and responsibilities are fixed.

On the other hand, ternary hierarchies require more flexibility and cross or parallel communication. When used correctly, this three-step hierarchical approach has many unexpected additional benefits, including retaining highly skilled technical staff and preventing organizations from becoming too comfortable. (For this discussion, complacency is not being creative, feeling old, and stuck in a rut.)

7.4 Work with R&D and Its Heterogeneity

R&D consists of certain types of activities that are different from other important enterprise tasks. It also wants to show that R&D is more than just one thing. The jobs at the bottom and top of R&D vary depending on whether technicians, engineers, or researchers do them. It also depends on the business sector (pharmaceutical, aerospace, agri-food, etc.), the science and technology sector (mechanics, IT, biotechnology, chemistry, etc.), and the innovation strategy or company size. The knowledge work of R&D companies with inherent quality and internal "variability" is essential to implement and design appropriate management practices, considering the challenges faced by R&D in traditional approaches. However, Katz [11] states that management experience and knowledge were first developed in operational business activities such as manufacturing. These stereotypes make it easier for managers with an R&D background to understand what differentiates R&D and to implement management practices consistent with the logic and challenges of R&D.

7.4.1 Non-routine Work

R&D activities are not normal because they are meant to create new knowledge and objects. They are different from what other business functions do, which are more predictable, stable, or cyclical and, therefore, easier to plan for and handle. This is something that all R&D businesses have in common, even though the "R" and "D" works are different. The upstream side of R&D (the "R" side, which focuses on thinking and exploration) or the downstream side (the "D" side, which focuses on development and implementation) works on innovation for new ideas and results. In fact, like D, R's work happens when an idea is born and shows how it works at the skill level. The company's other operations then take over the day-to-day activity of developing new processes or products.

7.4.2 Uncertainty in Activities

Another aspect of R&D is its inherent uncertainty. This uncertainty is related to the search for new things and is exacerbated when reviewing previous explorations. The research has many surprises. It needs to be made clear when, what will be found, how much and how much will be sold, etc. The idea of coincidence suggests that we can get different results than is thought. Uncertainties such as development value, level of economic technology, and possibilities are low. Therefore, Le Masson et al. [12] define development as "a controlled process that uses existing technology and knowledge to produce a system (organization, process, or product) that meets well-defined criteria (schedule, cost, quality). The value has already been established, conceived, or assessed." However, unforeseen issues can change the work's details, cost, and schedule.

7.4.3 The Work in R&D: Engagements and Interactions with the Environment

R&D professionals are an important external oversight critical to R&D and the company's ability to develop new ideas. Their work also generates knowledge that is "redistributed," used, questioned, evaluated, and redistributed by actors in external technological and scientific environments. You must recognize the skills and knowledge working in your field that others worldwide are creating. Also, some of what R&D creates is for sharing with the outside world (standardization, reputation building, impact on publications, patents, etc.). In addition to helping companies innovate, this is one of the main goals of R&D [13].

This door to the outside world has been open for a long time. It is essential for business R&D. The company's R&D staff spent time conversing with other scientists, reviewing patents, and reading papers presented at universities and industry technical and scientific conferences. These connections with the external technological and scientific environment vary internationally, regionally, and nationally. The history of science shows how important these exchanges were and how globalized the competition in this field was at the time. Communication with the scientific community is established not only through interaction but also by genuine collaboration. Also, R&D people do not just read other scientists' papers or share information at meetings. They work together, too.

7.4.4 The Job with a Certain Amount of Occupational and Autonomy Norms

R&D activity is also characterized by a high degree of independence. Unusual work aimed at creating new objects and knowledge and knowledge-intensive work that requires subject matter experts and is performed by people who are part of the knowledge community. It goes beyond the organization. As mentioned above, they are known for their qualifications. However, they are also known for their independence, dedication to their profession and work, sense of belonging to a professional group, work ethic, and cooperation in learning and establishing a profession.

Researchers have more freedom than other professionals working in creative industries (for example, advertising or design) or managers. This is because autonomy is an important part of professional identification. However, R and D are not the same. In research, autonomy can affect the purpose of a task and how it is carried out (although no one can predict what will happen or what they will find). How would you describe the process leading to an undefined goal? In addition, professional rules allow industry researchers at various career points (promotions, skills, job reviews, selection, etc.). Different organizations, management, and professional programs put different pressures on development.

So, when a new product is being made, the timeline and goals are often set within a clear set of guidelines for the deadline, cost, and quality. The manager monitors the job closely if it cannot be completely controlled.

7.5 Decentralization and Centralization

Whether to use a decentralized or centralized framework is another big question regarding design work. Allen [14] states that in a centralized project or project structure, most people working on a specific activity report to a single project manager who creates the task. They are judged and close to this person for their performance as managers. Allen [14] and Marquis and Straight [15] all say that centralized and decentralized structures can work only in certain situations. In decentralized structures, most people can access all the information they need to do their jobs. In centralized structures, you have to get information from one or two people who know most of what you need to know. The decentralized structure works well when people need to share information quickly, and projects take a long time to finish. On the other hand, a centralized structure works best when the project is of short duration, and the flow of information could be faster. A decentralized structure is better when there is much new information coming into a project area that requires a flexible way of organization and a lot of communication and collaboration between the people involved in the project.

Peters and Waterman [16] stated that good firms have decentralized and centralized structures. Even at the lowest level, these companies gave many parts of the organization much freedom. However, when it comes to the company's core values, it is committed to the core. According to their study, a hybrid organizational structure fulfills three basic requirements: the need for efficiency, the need for innovation, and the need not to be too rigid. This book covers only a few other things about decentralization and centralization in large and important organizations. The hybrid (decentralized and centralized) approach works well at the lower levels of organizations that are getting bigger. There is still a need to divide the large workforce more. Divisions could be based on a project, location, or product type.

7.6 Career Conflict and Design

Organizational conflict can sometimes be caused by the way jobs are set up. This is especially true because scientists' goals sometimes differ from the organization's. The goals of scientists are based more on scientific values, while organizational goals are often based on how they make money [17]. The freedom you are given at work is an important part of your design. Employees should be able to make strategic decisions as they gain experience. It should be assumed that workers will be free to create jobs. The types of freedoms offered may vary depending on the level of people working.

Much has been written about the conflict between organizational goals and professional values. Professional values reflect the provision of scientific progress. Scientists can be modern if they act in a way that reflects their professional values. On the other hand, Thompson and Dalton [5] argue that this focus on professional growth often conflicts with the primary goals of an organization. Scientists are generally more defined by their work than by where they work. As they rise, research managers become more familiar with the company than their profession. Researchers and scientists may not understand each other because these trends are different. Because the relationship between an organization's scientists and research managers is an important determinant of the success of an organization's research activities, research managers should find ways to reduce these apparent conflicts. Ross [18] suggested four ways to improve managerial roles (Fig. 7.2).

Peltz [19] identified four types of conflict in technical organizations. Type I typically addresses technical discrepancies between peers in understanding data and achieving specific technical goals, milestones, and objectives, whereas type II addresses conflicts with others (e.g., dislikes and likes, trust, and concerns about peers' goals). Type III is disagreements between supervisors and subordinates about technical or managerial issues, such as schedules or technical approaches. Type IV is disagreements between a supervisor and a subordinate about something interpersonal, like procedures, rules, authority, or power.

When different parts of an organization have different goals and try to maximize different things, this is a common source of conflict. An example is a struggle between the groups responsible for R&D and marketing. The R&D team may want to create a product that meets certain technical standards, and the marketing team

Fig. 7.2 Four ways to improve managerial roles

may want to create a product that will sell well. Although the two objectives are not mutually exclusive, coordinating the work of these two groups can be difficult and sometimes requires special management attention. Arvey and Dewhurst [20] state that it is important for management to establish comprehensive goal clarification and planning activities to avoid conflicts between goals. Below talks about the "task-dominant," "process-dominant," and "phase-dominant" approaches.

7.6 Career Conflict and Design 145

In a process-based approach, there is no clear transfer point, and each party escalates to the next task without formality. There is no case where one group is formed, and another is disbanded so that multiple people can perform a particular task simultaneously. Instead, people engage in activities and groups and leave as needed. The marketing and technology sides almost always talk to each other, and no one says, "That is your job" or "I do not care." Instead, everyone is constantly working to improve their products. Of course, the people responsible for this process are experts. However, some people, such as engineers, know little about marketing, and those who work in marketing know little about the technical side of the project. There are no visible documents when work moves from the study group to the marketing group or from the marketing group to the study group. The product needs to move back and forth between the two groups, and it does.

The task-dominant approach is more flexible than the other two approaches. In this case, existing employees focus on the task and the end product and discuss "our" products rather than "their" features. "I am interested in this product" is better than "I am a technical person" or "I am a marketer." In this case, no changes are made to the contact person or team while the product is being built. People are experts in many things but do not act like experts. They are part of the team making the product, and it does not matter if they are scientists, engineers, economists, or experts in public opinion. All they are doing here is working as a team on a project. So, in the task-dominant approach, people work together as a team and always talk to each other. This differs from when the research team met the marketing team and had a formal structure with titles like "marketer" or "researcher."

In a stage-dominant approach, the groups responsible for R&D and marketing form a formal structure with specific functions. These groups have clearly defined roles and can only perform specific tasks. In organizations, people have jobs directly related to what they are good at. For example, while an engineer's duties and responsibilities are limited to technical aspects, a marketer who cares about how the public feels about a product has certain responsibilities related to how the public responds to it. How tasks shift from one group to another reflects how these formal structures work. There are formal and institutional branches where R&D people delegate tasks to marketers, or marketers delegate tasks to R&D people.

Souder [21] says these three ways have some benefits in some situations, so it would be wrong to say one of these three ways is always better. People with a professional organizational structure are distinguished using special technology and functional activities. Experts often have good things, but you may need to learn more about what is happening in other groups. One benefit of having many specialists is that they can do some tasks very quickly and very well. On the other hand, there needs to be more coordination with other activities, which is a drawback.

Souder has listed some important things for making one of these three organizational structures work as well as possible. Those interested in how to organize a team should review the original article [17]. This article suggests that the best type of organizational structure depends on the criteria you want to maximize. These criteria include environmental factors (e.g., environmental uncertainty and change), work factors (e.g., type of technology or innovation), and organizational factors (organizational characteristics, organizational complexity, types of communication patterns, shared responsibility). Emotional event theory states that what happens at work are personal events that affect good and bad moods and emotions. Partly because of these moods, people are happy with their jobs. When a person's values match those of the organization, they are more likely to be happy at work, committed to the organization, stay there for a long time, and want to stay [22].

7.7 Future of R&D Jobs

Some R&D innovations of the past 25 years have been well "absorbed" and can give R&D professionals a sense of achievement, at least for some of them (given the diversity of professional profiles and aspirations in this group). However, other things create challenges and make R&D difficult [23]. Table 7.1 presents an overview of the key factors that will significantly influence the future landscape of R&D jobs. As the world continues to evolve in response to technological advances, shifting global dynamics, and emerging industry needs, professionals in R&D must adapt and prepare for the challenges and opportunities these factors present. The comprehensive table highlights how various elements, from technological advancements and interdisciplinary skills to regulatory changes and the emphasis on sustainability, will shape the R&D sector. These factors underscore the importance of staying current, embracing innovation, and cultivating a diverse skill set to thrive in the future of R&D careers.

Table 7.1 Key factors shaping the future of R&D jobs

Factor	Future impact on R&D jobs
Technological advancements	High demand for professionals in AI, ML, and data analytics
Interdisciplinary skills	Cross-disciplinary collaboration is essential
Data-driven R&D	Proficiency in data analysis becomes vital
Global collaboration	Distributed teams require cross-cultural skills
Sustainability and green technology	R&D focuses on eco-friendly solutions
Regulatory changes	Understanding evolving regulations is necessary
Remote work and virtual c	Ongoing influence on work culture and collaboration
Intellectual property and patents	Navigating the complex IP landscape is important
Industry-specific changes	Varies by industry, e.g., pharmaceuticals vs. tech
Lifelong learning	Continuous upskilling is essential for staying current
Global economic factors	Economic conditions impact R&D funding and jobs
Healthcare and biotechnology	Significant growth in R&D due to medical innovations

7.7.1 How Long-Lasting Is This Concentrated Focus and Heightened Pressure, Both Privately and Communally?

R&D has come under much pressure quickly due to many things mentioned above and has lost much independence. Prove your return on investment (ROI) upfront. Increased dependence on business, strong time bias and competition from market timing, periodic evaluation, and the organization of the entire project. Instead of resisting change, the shift from R&D to marketing and manufacturing led to significant results, such as more projects and shorter design times. However, one wonders how long the new R&D models and activities (in terms of individuals and groups) will last (corporate, sectoral, R&D, etc.).

Evaluation is work performed by many people, including business unit managers, external partners, project managers, project team members, and colleagues from other departments. Project management makes it easier for people to communicate and collaborate because they are in the same room. However, it also puts much pressure on everyone because clients, project managers, and colleagues always watch.

It can also be frustrating for some R&D professionals to cling to projects that do not have a strong goal of generating new knowledge. It does not align with their professional goals. It also raises the question of the importance of focusing on the steps leading to product development and using a company's strategy (and, therefore, monetization). Many companies talk about how important innovation is and how much it can give an edge over the competition. However, using research resources for short-term development projects means using doubts about R&D resources. After these problems were found, some organizations came to solve the problem. Tools and practices that promote the exploitation of knowledge in knowledge management are good examples of this. Human resource management has also changed. For example, when candidates are recruited, or project managers are selected, key new skills to work within the project structure and willingness to manage uncertainty are sought.

7.7.2 How Will Relocation, Globalization, Outsourcing, and Open Innovation Affect R&D Work Future?

R&D work is now undergoing several changes, some of which are either underway or are clearly on the horizon. R&D and investment are going to nations in the south, for instance. Large corporations in the north have long maintained R&D facilities in other northern nations, but the push toward southern nations is more recent, and it is still being determined how far it will go. It is unexpected to see businesses close

R&D facilities in their home nations before reopening them, particularly in China. The first series of questions is about where and who will conduct R&D in 30 years. How much R&D employment will also remain in northern nations like France? The second series of questions focuses on the different sorts of work the laboratories conduct depending on their location. R&D activity was divided up across nations for a considerable amount of time.

For the business to remain near to its clients, they also made it simpler to employ bright individuals and obtain a better grasp of how regional marketplaces function. The laboratories established in the southern nations concentrated on customizing the center's goods for foreign markets and providing technical assistance to struggling industrial clients. What will the function of R&D laboratories in northern nations be compared to those in southern countries in 30 years? How would these several laboratories from the same firm collaborate? Expanding international R&D will increase the amount of distant cooperation, with all the drawbacks and issues that include national and technical cultures, languages, time zones, etc. Additionally, it results in geographic mobility, which, depending on how long the project is, may be either long term (international careers) or temporary (for the duration of a project).

R&D outsourcing is another important trend. Large industrial companies "refer" some of their R&D work to the outside world. This is due to various factors, including avoiding activities, taking fewer risks, and having lower fixed expenses when it is difficult to determine how lucrative they are or how much money they will recover. One set of procedures is paying service providers (such as technical platforms, government research facilities, small firms, or Altran Technologies) to complete tasks formerly completed internally. This enables others to work on R&D initiatives (or carry out tasks like making show samples, etc.) that are not carried out internally. Additionally, it enables the organization to access human resources without having to make an employee hire. The second set of practices involves keeping an eye on the innovation and research ecosystem to see what is interesting before thinking about how to capture it (startups or competitors, purchasing of licenses or patents, recruitment of engineers and researchers).

Another change has to do with the fact that partners and entities are spread out all over the world. Digital devices make it possible to work together remotely or collect ideas for improving a product from people worldwide via the Internet using crowded sourcing platforms. In recent years, how people work together has changed from bilateral to multi-lateral, involving many groups. Participating in the innovation and research ecosystems for large enterprises is the goal. They expect to gain respectability, access to talent pools for hiring, information, ideas, and other things like access to technology and equipment. They also want to split the expenses and risks of funding large R&D projects and equipment. In this type of ecosystem, R&D activities are carried out in stakeholder chains or networks that span numerous organizational boundaries, and coordination, interaction, and interface work are all becoming crucial components of R&D activity.

7.7.3 What Effects Will the Digital Revolution Have on R&D Work?

First, it is important to note that, unlike other occupations or economic sectors, R&D has always been linked closely to digital technology. Many digital technologies are not just created via R&D; in some cases, such technologies were created first for R&D (including, but not limited to, in the military area) before being made available to the general public. The Internet, supercomputers, computers, CAD software, and many others support this. People are concerned that the coming digital revolution, particularly artificial intelligence, will result in employment losses. R&D has a more "natural" and grounded relationship with these technologies, which are also perceived as empowering, even though it is also affected by these risks.

R&D has long been experimenting with the incredible effects of various digital technologies. This is true for instruments like social scientific networks, several online scientific publications, electronic databases, the Internet, e-mail, etc., that make it simple to exchange knowledge swiftly and affordably. With just three clicks, it can now access much new knowledge worldwide, making tracking and reviewing the literature easier. How research and design research is done has changed greatly because digital tools can do complicated calculations. They have cut the time it takes to calculate and process data from experiments by an unbelievable amount.

Additionally, they enable the simulation of complicated systems and occasionally replace all actual testing. They occasionally alter how research is organized. For example, in biology and chemistry, where molecules have been studied based on their structure for a long time, digital methods now allow huge libraries of molecules to be tested without knowing anything about them. This changes how scientists do these things and what skills they need to do them. These new, scientific, multidisciplinary fields (like bioinformatics) show great promise for technological and scientific advances.

Digital tools also improve the efficiency of R&D. Additionally, the repeated studies and the fact that computer-aided design techniques are more accurate than hand-drawn designs all contribute to the results being more trustworthy. However, these tools also alter how they function and support the development of new knowledge, concepts, and technologies.

Online forums, specialized social networks, wikis, and other virtual online development communities offer new venues for knowledge exchange. They are incredible instruments for enhancing the dynamics that are already rather frequent in these universes of trading and sharing knowledge, even though this may raise concerns about how corporations secure their intellectual property.

Digital technologies are changing the way projects and R&D teams work together. Collaboration tools make it easy to work remotely or near you. These tools are useful for emerging international R&D networks and multilateral R&D partnerships. Digital technologies, which offer new opportunities for prototyping, are changing how they work (low cost and fast). Software applications are becoming increasingly important for many products. However, prototyping is easier with tools and software such as 3D printers, making it easier to turn ideas into reality. It can be seen how easy it is to use an object, how well it works with other things, etc. Other tools, such as augmented and virtual reality, also enable early testing of how users interact with future products. Using digital tools in this way allows you to view production steps more often and faster for a fraction of the cost. It is easier to work because you can communicate with other companies and customers in an agile way. Digital technology offers several ways to attract virtual or physical prototypes, such as virtual or physical prototypes, virtual or internal prototypes, distant or physical prototypes, and remote and abuse platforms.

References

1. J.R. Hackman, R.J. Hackman, G.R. Oldham, *Work Redesign*, vol 2779 (Addison-Wesley, Reading, 1980)
2. B.T. Loher et al., A meta-analysis of the relation of job characteristics to job satisfaction. J. Appl. Psychol. **70**(2), 280 (1985)
3. T. Studt, World's best R&D companies. R&D Mag. **49**(10), 12–15 (2007)
4. L. Bailyn, J.T. Lynch, Engineering as a life-long career: Its meaning, satisfactions, and difficulties. J. Occup. Behav. **4**(4), 263–283 (1983)
5. P.H. Thompson, G.W. Dalton, Are R&D organizations obsolete? Harv. Bus. Rev. **54**(6), 105–116 (1976)
6. J. Schriesheim, M.A. Von Glinow, S. Kerr, Professionals in bureaucracies: A structural alternative, in *Prescriptive Models of Organizations*, ed. by P. Nystrom, W. Starbuck, (North-Holland, New York, 1977), pp. 55–69
7. D.A. Waldman, L.E. Atwater, The nature of effective leadership and championing processes at different levels in a R&D hierarchy. J. High Technol. Managem. Res. **5**(2), 233–245 (1994)
8. H. Baumgartel, Leadership style as a variable in research administration. Adm. Sci. Q. **2**, 344–360 (1957)
9. S. Marcson, *Scientist in American Industry: Some Organizational Determinants in Manpower Utilization* (Industrial Relations Section, Princeton University, Princeton, 1960)
10. R. Likert, *The Human Organization: Its Management and Values* (McGraw-Hill, New York, 1967)
11. R. Katz, *The Human Side of Managing Technological Innovation: A Collection of Readings* (Oxford University Press, 1997)
12. P. Le Masson, B. Weil, A. Hatchuel, *Les processus d'innovation: Conception innovante et croissance des entreprises* (Lavoisier Paris, 2006)
13. S. Frickel, F. Arancibia, Environmental science and technology studies, in *Handbook of Environmental Sociology*, (Springer, 2021), pp. 457–476
14. T.J. Allen, *Managing the Flow of Technology: Technology Transfer and the Dissemination of Technological Information with the R&D Organization* (MIT Press, Cambridge, 1977)
15. D.G. Marquis, D.M. Straight, *Organizational Factors in Project Performance*, Working Paper No. 133-65 (M.I.T., School of Management, Cambridge, 1965)
16. T.J. Peters, R.H. Waterman, In search of excellence. Nurs. Adm. Q. **8**(3), 85–86 (1984)
17. W. Souder, A. Chakrabarti, Managing the coordination of marketing and R&D in the innovation process. TIMS Stud. Manage. Sci. **15**, 133–150 (1980)
18. M.H. Ross, Opportunities for maximizing the effectiveness of the administrator/researcher relationship. J. Soc. Res. Admin. **22**(1), 17–23 (1990)

19. D.C. Pelz, Some social factors related to performance in a research organization. Adm. Sci. Q. **6**, 310–325 (1956)
20. R.D. Arvey, H.D. Dewhirst, J.C. Boling, Relationships between goal clarity, participation in goal setting, and personality characteristics on job satisfaction in a scientific organization. J. Appl. Psychol. **61**(1), 103 (1976)
21. W. Souder (ed.), *Stage-Dominant (SD), Process-Dominant (PD) and Task-Dominant (TD) Models of the New Product Development (NOD) Process: Some Straw-Men Models and Their Contingencies* (Technology Management Studies Group Paper, 1975)
22. A. Joshi, J.J. Martocchio, Compensation and reward systems in a multicultural context, in *The Influence of Culture on Human Resource Management Processes and Practices*, (Psychology Press, 2007), pp. 199–224
23. P. Gibert et al., *Innovation, Research and Development Management* (Wiley, 2018)

Chapter 8
Team Building in R&D Projects

Team building in research and development (R&D) projects delves into the intricate interplay between R&D and human resource management (HRM), comprehensively examining the multifaceted relationships between these two critical aspects of organizational success. This chapter begins by scrutinizing the dynamic interrelations between R&D activities and HRM strategies, illustrating how talent management influences innovation and vice versa. The discussion then proceeds to explore the roles and relationships of researchers within the R&D environment, emphasizing their connections with peers and management. It illuminates the importance of fostering collaborative and harmonious interactions to facilitate optimal project outcomes. This chapter further elaborates on the critical aspects of creating and leading effective R&D teams, addressing the challenges and strategies for workforce management in this specialized domain. It also offers insights into enhancing the productivity of R&D initiatives, emphasizing the significance of innovation culture and process optimization. Finally, it guides managing projects tailored to specific R&D specialties, offering a well-rounded approach to navigating the intricacies of research and development within the contemporary organizational landscape.

Learning Objectives
- Understand the importance of team building in an R&D project
- Recognize major factors to consider in the success of a team
- Realize how to choose the right project manager and team member
- Identify how to build the most productive R&D project
- Describe how to facilitate a learning environment for the R&D team

© The Author(s), under exclusive license to Springer Nature Switzerland AG 2024
H. Taherdoost, *Innovation Through Research and Development*, Signals and Communication Technology, https://doi.org/10.1007/978-3-031-52565-0_8

8.1 R&D and HRM: Complex Relationships

Although human resources departments (HRD) have previously been unable to influence the R&D sector, this is no longer true. HRM practices have grown in importance over the past 20 years in this more sophisticated and scientific society. Numerous tensions are created as a result. These conflicts result from fundamental contrasts in their philosophical outlooks and a reciprocal failure to understand these two realms. As a result, one of the points of structural tension is the conflict between the need to standardize and differentiate HRM techniques used in R&D from those used by other staff groups of the organization. One of the main factors that call into question the applicability of conventional HRM techniques and necessitates the development of novel ways to regulate work in R&D is the advent and rising popularity of project organizations in R&D.

8.1.1 Project Management and HRM Adaptation in R&D

Project organization, which is increasingly typical in R&D activities, implies that the requirement for certain HRM approaches has been acknowledged. The latter is a result of acknowledging both project mode specificities—which are now applicable to a wide variety of business operations and are not only reserved for R&D—and R&D specificities. Since good human resources are needed for successful project performance, particularly in R&D, and since projects are a vector for skill development, HRM should pay attention to the leadership of people participating in projects. The conventional (hierarchical, vertical, and functional) structure that HRM was developed inside and by was disrupted by the features of projects, causing its traditional methods to become unstable. Indeed:

1. Projects involve dual project/professional authority to varying degrees.
2. They cast doubt on the legitimacy of the professions, as represented by traditional functional frameworks.

The challenges highlighted by this new organizational structure and its possible effects on HRM were the topic of studies and the pertinent remedies that HRM may offer to these particular instances of temporary assignment of workers to cross-functional projects. Loufrani-Fedida [1], Zannad [2], and Baron [3] are a few of these. The project structure poses several difficulties for HRM, including career management, pay, assessment, training, and hiring project managers, as demonstrated by Garel et al. [4]. The many HRM techniques now used in R&D are evaluated in the next section, where these challenges are highlighted.

8.1.2 Integration and Recruitment

Several variables determine the requirement for recruitment. As noted by Akhilesh [5], they result from the diversification of operations, integration or development of new technologies, creation of new product lines, partnerships, or other forms of collaboration. Staff turnover in R&D also affects recruitment demands since R&D experts might leave their jobs for various reasons, including promotions within R&D, departures, firings, and transfers to other departments within the business. Cross-functional movements are often observed in major industrial businesses nowadays; they happen early in the professional career and mostly include egressive movements from R&D toward other functions. Reverse mobility is relatively uncommon. When it comes to R&D, businesses typically opt to hire outside. Therefore, the latter is not advantageous for people and is rarely sought. The R&D recruiting process has additional features beyond the fact that it primarily focuses on the external labor market. These qualities are brought about by the unique nature of R&D operations and the uniqueness of applicant profiles.

Finding the correct profiles far enough in advance is a strategic component of recruiting management. This entails identifying and luring potential individuals before rivals do in the context of the "battle for talent." So, building connections with engineering schools, universities, and research groups is important. As a result, HR departments have developed the "campus managers" job to find, recruit, and pre-screen young individuals with fascinating profiles while still in their academic careers. Managers of HR and R&D are responsible for locating regional, national, and worldwide higher education and research institutes that serve as potential candidates for the organization by offering specialized training or carrying out pertinent research. The institutions should then be informed of the company's operations, employment prospects, hiring practices, etc. Beyond the communication initiatives, several actions could strengthen the relationship between the business world and academic institutions, such as providing student internship scholarships, co-supervising Ph.D. theses and research papers, various grants, post-doctoral internships, and, more recently, participating in developing academic curricula. Such measures assist in adjusting academic learning content to meet future business demands, which will benefit both students and businesses. The hiring demands and applicant profiles may change depending on the business area, the scientific and technical field, and the company's innovation strategy.

Even though youthful engineers and researchers who have just finished their training make up the majority of R&D hiring, the organization may occasionally desire to add more seasoned profiles. These researchers frequently work for clients, partners, or rival businesses. They often have extensive industry knowledge, important connections, and even hierarchical management or project management expertise, which could be advantageous for a corporation looking to hire senior R&D personnel. The growing standards for the competencies required of new hires is another feature of the recent era. The so-called "behavioral" talents are gaining more weight in recruiters' evaluation grids beyond the scientific and technical skills that are still crucial. Companies are getting increasingly demanding, and they may even provide an (excessively) ideal image of an R&D person who should be skilled in many areas and able to perform a variety of responsibilities. The concern is that the posts will not be filled because stringent hiring standards drastically shrink the pool of possible candidates.

These issues are made worse by the conflicting expectations of developing R&D management techniques. This occurs when an organization demands that R&D personnel abandon routines, exercise creativity, see the future, and abide by more regimented work procedures. It would be more meaningful to reason in terms of both individual and group competencies, given the aggregation of skills and functions expected of R&D professionals. This would enable more rational consideration of how the group, rather than each member individually, can assume all the necessary abilities and tasks. Beyond the appropriateness of duties and talents, it is also important to examine the new hire's compatibility with the team in which he or she will be included. However, since hiring R&D specialists always has a longer time horizon, thinking beyond the job will be filled shortly. The candidate's career goals and professional growth skills should be questioned, especially regarding knowledge, project coordination, hierarchical management, and other duties.

It can be seen a much higher presence (and power) of HR actors in recruiting processes, when before they might be entirely excluded, along with the growth of these new expectations (behavioral capabilities, ambidexterity, capacity for development, etc.). A candidate's soft skills, motivation, career goals, and capacity to integrate and advance within the organization, even outside of R&D, are better understood by HR actors when compared to the company's ideal career routes. On the other hand, peers continue to be involved in sourcing (detection) and selecting applicants with HR managers and the direct hierarchy due to the highly specialized nature of technical and scientific talents. A more thorough evaluation of the various dimensions of candidates' talents is made possible by various assessors. Additionally, it helps students see their potential short- and medium-term working environment. To foster a realistic vision of the nature of their future jobs, projects and missions they would be focusing on, career chances, the company's operating norms, etc., it is vital to convey a properly considered message to applicants. This helps us avoid disappointments for the new hires once they join the organization.

While taking on project assignments may be temporary mobility, R & D personnel should consider pursuing their career goals inside and outside of R & D. The trends in this area mirror those seen in the other activities. The HR department encourages more mobility on all axes, including organizational, geographic, and functional. The defined processes and mobility incentives, in addition to the communication efforts, encourage this mobility. For instance, it was demonstrated at Rhodia that, starting in the early 2000s, there would not be any more internal promotion and that a candidate for management roles in R&D would need to have completed several mobilities. Sometimes, it was also necessary to have overseas experience.

Similarly, moving through operational positions in subsidiaries is a crucial requirement for professional advancement in the oil business [6]. The HR department creates a variety of tools to ease these transfers. Some tools encourage individual initiative, including internal employment databases, job mapping, and potential career routes. Others are more managerial tools, such as "people evaluations" conducted by "career and mobility committees," made up of HR and line managers, who assess each R&D professional's situation and suggest potential advancements for those who are seen to have remained long enough in their present role.

There is significant pressure on R&D engineers in certain organizations to quit R&D and contribute their innovations to other business divisions, motivated by the notion that individuals should only work in R&D for a while. This is the pool approach [7], where R&D acts as a hiring agency and training facility for technical managers across the board. The leaders of other functions can also recognize bright R&D engineers within the context of projects than they were in the past, and they try to entice them with appealing ideas. Many R&D experts are aware of these evolutionary possibilities. In reality, professional advancement is frequently quicker, pay is better, and non-cash compensation aspects (responsibility, proximity to clients and authority, attractive workspaces, etc.) are more exciting than in R&D.

Even though these are powerful tendencies, they bring up certain issues. When considering R&D, where its operations occur over longer periods, is the temporal pattern of movement every three years—which is nearly established as a norm—relevant? Is it problematic to insist on the traditional norms of mobility for R&D experts, given that the goal of R&D is to create a foundation of unique scientific and technical skills that takes around ten years to develop? Would it be preferable to have a pool of highly qualified engineers and scientists in R&D who can create completely original products due to scientific discoveries? Here, the limitations of the use of a typical HRM model that does not take into consideration the unique characteristics of R&D can be observed. On the other hand, some businesses work to promote R&D as a desirable career field where experts may advance without needing to leave the field of science and technology.

8.2 Relationship of the Researcher to Peers and Management

The directors or managers of research organizations are seen negatively by many researchers. Sharing authority is crucial in an R&D organization. Researchers desire to share managerial and executive power with the administrative system. Particularly when it comes to administrative and executive power, researchers want to be able to discuss the parts that have an impact on their research operations. According to Naveh [8], the interaction between formalization—which refers to structure and stability—and discretion—which refers to spontaneity—affects several elements of research, including project team size, project duration, and technical competence. As a result, all parties concerned should implement some form of power sharing. On the other side, researchers need to be aware that such power-sharing also comes with certain administrative responsibilities and responsibilities. It is impossible to avoid this. Researchers who share these abilities should do some administrative duties, adhere to deadlines, consider the opinions of others, reach agreements when opinions disagree, and refrain from engaging in guerilla warfare after choices have been made.

Researchers have certain unfavorable opinions of management based on their interactions with businesses. When managing R&D projects as of 1998, approximately 90% of Fortune 500 corporations prioritized formal endeavors focused on stability, structure, and adhering to regulations [8]. In certain instances, the administrative structure of universities and research institutions appears to expand proportionately more quickly than the number of faculty or research groups. Many scientists working for R&D businesses and university faculty members believe that the administrative staff is growing geometrically in size, which makes research and teaching operations look like ancillary concerns. Managers should regularly review the administrative systems in their workplaces and consider if every position is necessary.

8.3 Creating Teams

When assembling a team, numerous factors should be taken into account. Effective teams, according to the research of Pelz and Andrews [9], are characterized by individuals who support one another's work, have high regard for other team members, and have complementary abilities, tactics, and approaches. The similarity in talent level, as well as in fundamental attitudes and ideals, is a prerequisite for respect. Team members, however, need to have a variety of abilities and distinct, complementary mindsets. They could all be top-tier researchers with a shared appreciation for autonomy, but they might have different specialties and viewpoints on certain data-collecting techniques.

Research is becoming increasingly multidisciplinary, and many research teams have people from many academic fields and cultural backgrounds. This not only causes coordination issues but may also boost creativity. Including multiracial team members might boost inventiveness. According to a study by Leung et al. [10], the multicultural experience can boost creativity because it gives people the ability to (a) widen the range of ideas and concepts that are understandable, (b) realize that the same form can serve multiple purposes, (c) disrupt preexisting associations, (d) become more open to ideas from various sources, and (e) develop greater cognitive complexity. People should not view their coworkers as too cognitively dissimilar; therefore, matching ability is crucial. Cooperation will deteriorate if coworkers are perceived as "dummies" or "geniuses," which makes people uneasy. Complementarity is desired for numerous personality traits at the same time. For instance, a person who wants to control a team may get along with others far better than those who like to be in command. People who enjoy listening get along well with chatty people.

8.3 Creating Teams

What is the ideal team composition? According to studies, "five" is considered the ideal number for a discussion group. However, McComb, Green, and Compton [11] argue that there is no fixed amount. Instead, the degree of flexibility within a team influences the staffing quality. The leader may become more dictatorial and dominate the available time in larger groups, members may feel they need more time to give their thoughts, and rival subteams may form. Smaller organizations frequently need a clear leader, may not have clear objectives, or may need more diverse viewpoints to prevent groupthink. The caliber of the team members and the coordination of their efforts determine a team's effectiveness. The assignment will, however, determine how these factors affect the team's efficiency. There are three different kinds of jobs to take into account.

Unitary vs divisible tasks. Different people can perform actions that can be broken down into smaller parts, such as approving references on a reference list (checks can be assigned to as many officials as there are reference pages to be checked). A single activity, such as understanding a text, cannot be broken down into smaller tasks.

Optimizing vs maximizing tasks. Finding as many references as possible is an example of maximizing task criteria with no upper bound. Determine how much space the project needs using a sample optimization job with ideal-level criteria (too much space is wasted, too little, and the project could be more efficient).

Conjunctive vs disjunctive tasks. In solving problems, if one participant has an exact solution, the others should agree, like the root of a quadratic equation. In integrated assignments, each member should agree, for example, on an arbitration panel or a panel where everyone has veto rights. In discrete problems, if the probability that a member gets the correct solution is P and the probability that someone fails to solve the problem is Q, then if $Q = 1 - P$, the theoretical probability that the group solves a problem is equal to $1 - Q^n$ where n is the number of members.

The more members you have, the better the group's chances of success. In the case of complex tasks, the opposite is true. If the group is smaller, the solution is higher than when it is larger. By the way, the groups are empirically made less theory ($Pg = 1 - Q^n$), especially because of the damage to those who discuss different solutions. He also expects people to create human results when splitting tasks, but that does not happen. The group often needs to produce more. This lack of individual responsibility, called "social loafing," results when there is no clear identification of each person's output. If the production of each person as a whole is defined, then there is little or no social bread. Finally, unit tasks are better performed by individuals than by groups.

Teams engaged in R&D are different from other teams in that they execute non-routine work. Teams in R&D firms have many benefits. Typically, R&D projects cope with complicated tasks that call for technical expertise in several different fields; as a result, cooperation is necessary and advantageous since jobs are difficult and call for a variety of talents [12]. Teamwork has significant drawbacks as well in the R&D industry. Because technical specialists are chosen for their scientific and technical abilities rather than their communication and social skills, problems might arise in a team environment. Research has revealed that technicians may need more qualities to perform effectively in a team atmosphere, which is why many technical professionals choose this employment, so they will not interact with others on a social level [13]. On a different level, as many R&D professionals desire to be recognized for their accomplishments at work, the reward system can also discourage effective collaboration. Teamwork is not a solo activity, which may demotivate people. Conversely, difficult research projects necessitate cooperation and teamwork, which may offer the required drive.

Based on this knowledge, Levi and Slem [12] conducted a study to evaluate the effectiveness of teams in R&D firms. They created a 30-item poll to gauge the employees' values and worldviews. Three R&D centers for the electronics sector in California received the survey. Key professionals at each site were interviewed in addition to taking the survey to determine the authenticity of the survey results. According to the analysis of this study, it can be beneficial to encourage cooperation in R&D firms. From an individual standpoint, the absence of incentives in a team environment was considered a serious issue. This is because people frequently need help to recognize the connection between performance in a team setting and their success inside the company. It was also discovered that conflict and disagreement-causing team members might undermine the efficacy and cohesion of the group.

Additionally, there is no ideal method of leadership or decision-making due to the diversity in the traits and composition of teams. Organizations may encourage a team's leadership style even when it does not suit the purpose of the work. Teams could only be successful as a result of this.

Creating self-managing teams in an R&D environment has been quite challenging, as described in this section's beginning. Self-managing teams are emphasized by all three of the organizations in the poll, even though there is no evidence to support their superiority over other forms of cooperation. Finally, Levi and Slem [12] discovered that corporate culture impacts how well a team works. This makes intuitive sense. When placed in a collaboration situation, employees will be far more eager and capable of success if a firm has a corporate culture that promotes employee participation and involvement while highlighting teams and their advantages.

8.4 The R&D Team Leadership

Managing an R&D team entails exhibiting effective leadership abilities, effective workforce management, and providing an educational setting. This chapter explains the steps and techniques to oversee R&D and creative project management. The maximum likelihood of successful outcomes may be ensured by using these methods and procedures. The team members and the project manager will significantly affect the success or failure of any R&D project, regardless of how methodical the project management techniques are used. Figure 8.1 shows how organizational

8.4 The R&D Team Leadership

Fig. 8.1 Factors for enhancing product R&D and innovation

design, which in turn promotes workforce management and workforce empowerment for product R&D, influences leadership and, ultimately, drives innovation. The components of leadership, organizational design, and workforce management are examined in this section. These three factors are potent enhancers or force multipliers for product R&D and innovation.

8.4.1 Project Management for R&D

R&D is a generally ill-defined, complicated, and ambiguous field. Only some people have the temperament, education, or experience necessary to manage well in that setting. A successful project manager is naturally inquisitive, self-assured, adaptable, and typically has a higher tolerance for uncertainty and taking risks than an ordinary person. The R&D project manager should be a team player who is at ease leading highly creative group members, especially principal investigators who are frequently completely devoted to the projects in which they have a financial stake. Most of the time, the project manager serves as the project facilitator, freeing the principal investigator and other subject-matter experts to concentrate on the R&D rather than on the management activities of scope, schedule, budget control, workforce management, and other essential project tasks.

Project managers will need to use a combination of project management, organizational management, and leadership strategies to lead R&D initiatives successfully. When selecting a project manager for an R&D project, it is critical to understand where the project is in its life cycle and its strategy and trajectory to ensure that the person's abilities align with the project's objectives. Where extreme R&D is required, for instance, a strong leader is nearly always required to dismantle obstacles and challenge the existing quo. When incremental R&D is undertaken, a capable team and a skilled project manager may be sufficient.

Projects are frequently clearly phased in companies, and project managers are switched from one stage to the next. A creative project manager who excels at initial planning may be assigned to the project's early R&D stages. When the project is prepared to begin progress, a project manager with execution skills may be added. An organization that is aware of the skill set of its personnel will use that understanding to place them best. This makes it possible for the project, the project manager, and the company to succeed.

8.4.2 The Role of Leadership

There is a sizable body of written knowledge on leadership and several examples of outstanding leaders. They are frequently credited as inspiring a successful company or a famous individual in the arts, technology, or science fields. These tales inspire businesses to search for the ideal leadership recipe that will boost the likelihood of project success. Finding and keeping excellent leadership becomes essential, given how much its leadership affects R&D efforts. A great leader might have many different qualities. However, effective leaders often have a few characteristics.

Leaders are frequently described as being devoted and motivated; they have a firm understanding of the organization's plan and frequently have a clear vision of how to accomplish it. They are prepared and equipped to seize new strategic possibilities when they present themselves. They are powerful decision-makers who can compel behavior without using direct authority. They are renowned for influencing and motivating rather than managing. Leaders use unconventional thinking and original solutions to tackle challenging issues. They often possess skilled communication abilities, which is crucial for executives in charge of R&D initiatives. Most leaders feel at ease communicating with people at all levels of the business, including colleagues, customers, and other supporting organizations.

Additionally, they frequently have good soft skills. They frequently rely on intuition, inventiveness, and positive energy. They often lead the team by charm and persuasion rather than via direct reporting connections.

A project manager with strong leadership abilities should be chosen if there is a high requirement to remove obstacles, alter the status quo, or make dramatic changes [14]. A project manager with good implementation abilities would be a more sensible choice if there is a high requirement to accomplish established objectives, achieve incremental modifications, or effectively execute current procedures.

Whether strong leaders are created or born has been debated for a while. Successful R&D team leaders respect and trust their team members, communicate effectively, and recognize the value that each team member offers. These characteristics are frequently listed as those of leaders and are frequently considered intrinsic or not necessarily "teachable." A sizable body of information also suggests that leaders could be taught and provide programs to do so, which is why the argument exists. Those with a higher than average likelihood of becoming effective leaders via the use of targeted training are those who can adapt and can apply to learn.

All prospective R&D project managers should go through assessments that will show whether or not they exhibit these qualities, and if they do not, further thought should be given to whether or not additional experience and training will result in a significant enough improvement to warrant implementation or whether the candidate should be given a different opportunity. Both self-assessment and honest appraisal by stakeholders, like a person's supervisor, immediate superiors, and peers, are often heavily emphasized in leadership training. The application of the information acquired and shown modified personal behavior is intended to be the outcome of the training. Strong self-reflection develops the individual's self-awareness. These evaluations and training should, at the very least, emphasize (1) discovering an individual's motivations, (2) identifying their strengths and areas for development, (3) developing their capacity for accurate perception and intuition, (4) making decisions and judgments based on limited information, and (5) how effective they are at making decisions, communicating with others, and directing those who are in charge of lower-level outcomes. The finest training opportunities include accountability and implementation assessments to see if the teachings have been used effectively. Leadership development takes time. As a result, those willing to adapt to change and are conscious of how their actions affect others are the ideal candidates and should be given training chances. These are also the people who should be considered for upcoming leadership roles in R&D project management.

An R&D project manager should be able to drive change, innovate, inspire, and accomplish more than is expected. They should also have a solid understanding of the strategic vision and the project objectives. To accomplish the intended outcomes, she or he will often have the creative freedom to modify processes and procedures drastically. The skilled project manager will be able to apply project management standards, methods, and practices to her or his R&D project significantly and will feel at ease with the accountability that results from doing so. To reach maximum performance, the project manager should be assured and comfortable enabling the role of subject matter experts, technical leaders, or principal investigators. Following the discipline, she or he will make sure there is contact with the important parties who have endorsed and supported the R&D throughout the whole life cycle.

An R&D project manager who is in tune with the organization's senior leadership, knows and supports the strategy, and supports the organization's objectives will be better able to get the funding for the study. This alignment will eventually advance the research trajectory by creating a relationship based on trust with important stakeholders. The company will then benefit from knowing that the R&D project is progressing an acceptable life cycle. The project manager's responsibility is to carry out the project's objectives and advance the strategic objectives with which it is aligned. Comparatively speaking, the technical or scientific lead is in charge of directing the project's goals and advancing along the R&D pathways [14].

8.4.3 The Management Role

Any person in a position to manage personnel below them will also be expected to carry out management tasks. Organizing the work, growing the workforce, and ensuring that the job is being accomplished on time, as scheduled, and within the proper scope are some of these actions. Corporate goals should create team goals. General supervisory activities include organizing work, recruiting and employing employees, handling employee relations, conducting merit reviews, motivating employees, delegating duties, communicating, and providing staff development and training. Additional supervision duties may be necessary for a company, such as authorization for labor billing, benefits, communications, and information systems operations to provide staff access to the systems. When necessary, management training often concentrates on developing teams, managing projects, leading change, comprehending economics and accounting, bargaining, ethics, workforce planning, performance metrics, coaching and conflict management, and communications. Additionally, the particular organization could need management training that tackles the risks and culture of the organization.

8.5 Workforce Management

Controlling an R&D staff entails:

- Assuring the presence of a talented and well-rounded team
- Organizing team-building and motivational events
- Coordinating an individual's goals with the project's goals
- Getting agreements for using common resources
- Adapting the group

Although selecting the proper project leadership style for the team is crucial, there are other factors to consider. A person seldom accomplishes great things by themselves and without the help of others. A project manager's capacity to create and have an impact on an effective team becomes increasingly crucial as the world gets more integrated, complicated, and interrelated. The project manager frequently takes on the role of an orchestra conductor, giving instructions, creating possibilities for success, and inspiring his or her team to excel. Most books and articles highlighting incredible achievements in engineering, science, or other creative fields ultimately mention the team that made them possible. Teams can accomplish a great deal more than the abilities of their members combined. When all the pieces fall into place, the group can qualify as a "high-performance" team, capable of achieving great things and exceeding expectations. Most project managers build their teams with the idea of having a high-performance team contribute to an R&D project in mind.

Choosing the "correct" team is essential for R&D initiatives. They often consist of small groups of knowledgeable, skilled, and passionate people about their work. Adding subject matter experts with complementary skill sets to the team can impact the supplementation or boosting of weaker skill sets. This is particularly true if the team consists of a highly motivated workforce enthusiastic about the project and the possibility of joining the team. This kind of team typically has confidence in each member's skills, is well-aligned with the project's goals, and is at ease with its mission.

8.5.1 Staff Diversity and Selection

The ideal situation is for the project manager to select and employ the team of their choosing. The most difficult situation is when a team is allocated to a project, and there is no way to add more people. This is so that the project manager can select team members who are first and foremost motivated to join the team and genuinely dedicated to the purpose. The project manager who is given the authority to create the team can do so by carefully considering and selecting a team with a wide range of backgrounds, analyzing the project's requirements, and adding team members with complementary abilities essential to the project's success. The R&D project manager may combine abilities, fill in skill gaps, and assemble a team with the best opportunity to contribute fresh ideas to the project by combining knowledge and experiences from various disciplines, cultures, and organizational types.

Leading an existing team requires comprehensive skill evaluation and knowledge of the allocated staff's attitudes toward the project. The team should be assessed quickly to determine its strengths, shortcomings, aspirations, and dedication to the team and the project. Despite how challenging it may be, decisions should be taken regarding current employees who are unwilling or unable to perform at the level required by the team. To fill the team role with someone who will contribute to the project's success, every effort should be taken to enable any employees who are not fit for the task to find a suitable position outside of the team. Those who do not want to join the team but still want to contribute to the project because they support the purpose ought to evaluate their total skill set and match them with tasks that best utilize their abilities. This can entail using previously underutilized abilities or tackling brand-new tasks for which they have yet to be given the chance. The objective is to ensure that each team member is devoted to the project, contributes distinct viewpoints, and has the fewest abilities in common with the other group members. While certain overlapping skill sets provide cross-training and backup, the ideal approach for small R&D projects is often to ensure that the widest array of complementary skill types is present.

8.5.2 Activities for Building Teams

An R&D project is frequently interesting and captivating. It is simple to feel a part of a particular group since it comprises a tiny core of highly gifted people who are cut off from the main operations or manufacturing business. A compact team may be brought together quickly for an R&D project. Action is essential to unite and focus freshly created teams of distinctive and different individuals. The goal is to create an atmosphere where each team member feels a feeling of pride in being a part of the group and a particular connection to the other team members. Each member of the team should take ownership of the project's outcomes and be prepared to intervene, stand up for, and safeguard other team members. Personal ties are permissible and result in better performance than when team members keep their distance.

The next most crucial task for the project manager is to bring the team together through team building after initial hiring or team building. The team comes together through this activity, which is a firm foundation for all subsequent work. Every member of the team should be made aware of the project's purpose and the expectations for them during the team-building exercise. Review the project's precise goals and have all your questions addressed. A common team-building activity involves putting the group through a series of really difficult activities where they should support one another. It works because teamwork is boosted, and trust and relationships are formed more quickly when uncomfortable actions are undertaken. Regularly using these activities in team meetings will improve team dynamics and frequently result in lifelong friendships. These exercises may take the shape of a quick assignment, such as creating competing teams to finish a difficult task in a limited amount of time. Alternatively, they may be something casual outside the office, like a group luncheon at an unfamiliar ethnic or local eatery. The team's ability to work together to support one another while dealing with stress is the most significant influence of the scenario.

Although there are several methods for creating teams, the main requirements for a successful team are that all members share the same knowledge of the project's goals and are dedicated to accomplishing them. It is important to be aware of each individual's special abilities and skills as well as their planned contributions to the project. This should clearly state what special skills each team member brings and what is expected of them in terms of how those skills will ultimately affect the project's goals. Every team member should consider how each person fits into the overall project team.

8.5.3 Alignment of Objectives

Making sure people are engaged in their jobs and have the necessary abilities and expertise is one of the simplest methods to guarantee project success. When people work on a project that piques their true interest and offers educational chances, they will perform above expectations. They should be allowed to improve their knowledge and pick up new skills. The project manager needs to be aware of the team's strengths and limitations, and the group members need to be aware of their own goals for their careers and level of self-awareness to achieve the optimum alignment.

The most efficient way to guarantee this alignment is to gather the team, review the project's goals, and give everyone a chance to choose their roles and responsibilities. The flexible ways of project development, in which team members choose development tasks from a bundled set of requirements being worked on over a specified period, are compatible with this manner of ensuring group members choose the work they are most interested in executing. The idea may be used successfully in

R&D initiatives as well. Team members are free to select jobs not only in which they can excel but also in which they are interested in getting more experience.

Redundancy should be avoided, and people who have the aptitude and desire to fill the function should be permitted to do so. Self-selection with the possibility for cross-training, mentorship, and succession planning frequently produces a committed team member. If team members are taking on duties outside of their typical skill set, they should be partnered with a subject matter expert to enable proper mentorship until they are all on the same page with the project's goals.

8.5.4 Workforce Sharing

The ideal staffing scenario for R&D projects occurs when labor resources and essential skills are tightly integrated into the project and are not shared with other projects. However, this is only sometimes the case. R&D projects frequently exchange crucial talents, and operations do so occasionally. The project managers should carefully bargain over the use of shared resources. The worst-case time-sharing situation is a split in which time is allotted to each project during the day. The greatest scenario for sharing time is when it is done in large time chunks, like a week or a month at a time.

In these setups, there is a propensity for the workload to exceed the time allotted for the task. Therefore, regular monitoring of job performance is necessary to guarantee that the commitments are being completed once the agreed-upon amount of shared time has been set and the team member is at ease with the shared arrangement.

8.5.5 Evolution of Team

The management of the personnel is a crucial duty for a project manager. Projects should be rebalanced, requiring the decrease of some abilities and the augmentation of others as they progress through the project life cycle (from basic to applied to R&D to production). Each team member will seek to advance personally. Some people might desire to take on more duties, acquire new skills, or work on a different aspect of the project.

It is the job of the R&D project manager to manage the workflow and give team members efficient chances to advance in their careers. The project manager should thoroughly understand the team member's competencies, specific capabilities, and career goals to be effective. Replacing team members with critical talents for the project is difficult since they are frequently in great demand. Creating a workforce strategy will assist in determining the necessary skills and establishing a procedure for managing transitions. A workforce management plan may include the capabilities or requirements of the workforce, strategies for identifying and retaining vital skills, career advancement, performance management, career development, and other topics deemed crucial to the project. It should address the problem of personnel with crucial talents being dispersed among too many projects or departing the project at a crucial stage.

Ensuring that the project crew is suitably linked with the duties they can outperform is an efficient strategy for evolving the team. Team members will be actively involved in the project until the conclusion by giving mentorship and training on those activities, then offering possibilities for new growth experiences, and ultimately revisiting and updating the alignment of the goals at regular intervals. For the project's life cycle, the needs for personnel management evolve from having to guarantee that vital talents are accessible to discussing how to move them onto new projects. While each R&D project is unique, a good project manager can manage the team and transition people on and off in time for changes from one life cycle to the next. This requires expertise, including the ability to negotiate with other internal groups and a solid grasp of the organization's strategic direction and trajectory.

The project manager should assess the team as the team develops to decide which team members have the expertise and skill set necessary to oversee upcoming R&D projects. Although many different assessment methods are available, the goal is to utilize the person's past performance as a deterministic indicator of how they will perform going forward. Opportunities for extra training and experience should be made available to people with a penchant for leading R&D. Wherever learning may boost performance, it ought to be made available, and project managers should be asked to show their use of newly acquired skills as part of their performance review. The major processes and interactions in personnel management within the framework of research and development projects are depicted in Fig. 8.2, with the project manager's function being highlighted.

8.6 Improving the Productivity of R&D

Enhancing the R&D project's productivity entails ensuring that the organization's R&D project is appropriately aligned; the project and the organization create a learning environment, and team members get accountability and responsibility.

The likelihood of an R&D project producing successful results is high if the methods and procedures for managing R&D are in place, the project manager is efficient, and the team is devoted, competent, and adequately diverse. However, at the organizational and project levels, a few crucial foundational components should still be implemented. These are so all-encompassing and forceful that the impact can significantly lower the likelihood that the project will succeed if ignored. This section will examine the effects of a well-aligned organizational structure, a structured learning environment, and a setting that gives people the right amount of responsibility before holding them accountable.

8.6 Improving the Productivity of R&D

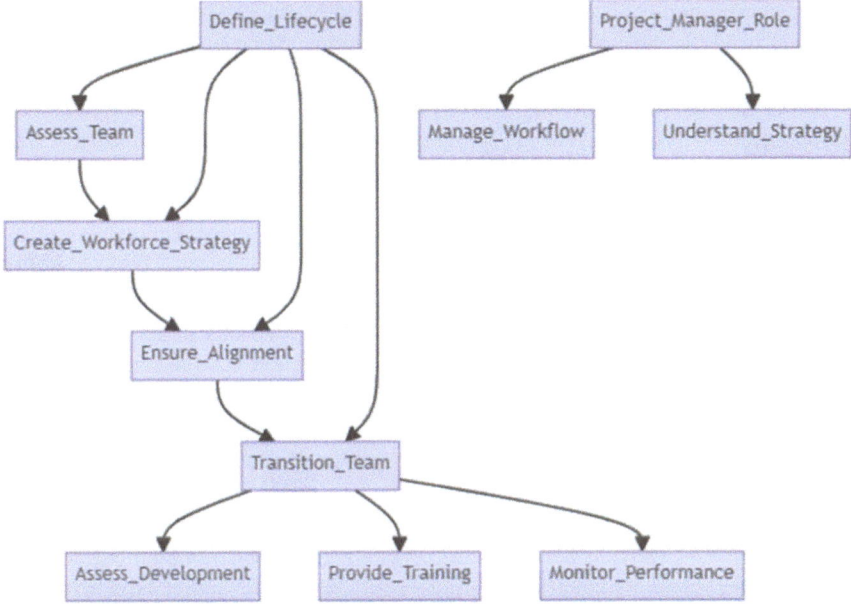

Fig. 8.2 Managing personnel in R&D projects

8.6.1 Alignment of Organizations

R&D projects need to be properly linked organizationally to guarantee the maximum possibility of success. Operating an R&D project inside a bigger organizational structure can be done in various ways. The organizational alignment may vary depending on the project's precise objectives and where it is in its life cycle. For instance, R&D initiatives that prioritize delivering gradual change could be integrated into an organization's operations and share resources with other initiatives. Through this alignment, the R&D project team and the group in charge of day-to-day operations are guaranteed to work closely together and clearly know how design choices will affect the operating model. Because they are founded on difficulties faced, ideas developed via this close cooperation can create incremental change.

Radical innovation-focused R&D initiatives frequently perform better when they are isolated from the operations activities. The development of novel concepts that have the potential to alter paradigms and potentially result in the creation of goods or services that could replace present operations can be facilitated by assembling groups of people who are not bound by the current operational environment. In this situation, the team should be cut off from the operations so they can freely consider their choices. This kind of R&D effort should be more closely coordinated with outside partners that can contribute fresh ideas. Being near a significant transdisciplinary innovation hub might encourage innovative thinking and present chances for collaboration with outside partners more frequently than if the team is remote from the wider community. Therefore, even if this kind of team would gain from being liberated from the pressures of day-to-day operations, it can be anticipated that the influence of external partnerships will lead to improved performance.

Incorporating crowdsourcing is a current trend inside corporations for both radical and gradual progress. This is an open approach of cooperation when a firm shares information with a willing group of volunteer contributors, generally material kept as company proprietary. The "crowd" is made up of eager individuals who volunteer their time and talents for free. They have a personal stake in accomplishing this. Through crowdsourcing, they have the chance to contribute their thoughts, expertise, or material to improve and advance the good or service and reach a broad objective. Adding to the efforts of others and creating a capacity larger than the sum of its parts are two ways that every contributor may benefit from the endeavor. This can be an effective organizational structure for businesses with a public-domain good or service to accomplish both gradual and radical development.

Irrespective of where the R&D project is physically located inside the firm, it should be considered one of the crucial tasks to answer to the top management. R & D should be an integral part of the business by the company as a whole. Executive leadership should provide the R&D organization its utmost, unflinching support. When organizational budgetary restrictions arise, R&D is sometimes the first activity to be removed since it is frequently viewed as discretionary. However, an organization's future typically lies in its R&D initiatives. The organization will ultimately stagnate and be outperformed by emerging skills in the external environment if R&D expenditures are not made. The R&D efforts that it approves should be supported by executive leadership, which should comprehend, respect, and accept this. The executive leadership should safeguard R&D operations against cuts that might jeopardize the organization's long-term existence if new financial difficulties arise, which they always do.

In practice, the project manager frequently needs more control over how the organizational leadership supports the project or where it is aligned. Although the project manager may not be in control of the actual alignment, he or she should be able to identify the ideal configuration and, if it cannot be executed physically or organizationally, should establish the right stakeholder connections to accomplish the same outcomes. Project managers have the authority to recommend that specific adjustments be made by the department or division that owns the project if they believe doing so will increase the likelihood of the project's success. Additionally, the project manager has direct authority over the project and may make adjustments if they stay within organizational boundaries.

8.6.2 Environment of Learning

Creating an environment of learning in Research and Development (R&D) is essential for fostering innovation, staying competitive, and ensuring that your R&D team continually improves and evolves. By fostering an environment of learning within your R&D department, you can harness the collective intelligence and creativity of your team, leading to continuous improvement, breakthrough innovations, and a competitive edge in your industry.

Free Time

Free time is among the most prized and significant components of R&D. It is common to think of having time to think, reflect, and talk about ideas without having a specific objective as a luxury. A learning environment that encourages free time allows people to converse together without a predetermined agenda. A vital part of R&D, the creative process, and innovation is free discussion time to explore concepts and brainstorm with people inside and outside of the company. There should be plenty of chances for cross-disciplinary collaboration since the most innovative breakthroughs frequently cut across fields.

Some project managers say that unrestricted free time is necessary for all R&D initiatives and that R&D should be the spontaneous pursuit of ideas. Only a tiny percentage of R&D projects, often in the fundamental and applied research fields, can efficiently employ spare time to explore ad hoc and opportunistic ideas. When a purpose is not understood, free time presents an opportunity to explore the uncharted, which might push the pursuit in a new direction. For these kinds of initiatives, spare time should be planned out in advance. Where the goal is clear but the means to achieve it are not, spare time allocations should be allowed at key times in the plan. The location of the spare time inside the project is crucial to consider. When a project enters a more formal R&D process, spare time in the timetable is replaced with established development activities.

It would be challenging, if not impossible, to stay on course to accomplish a goal without some discipline and concentrated attention. The likelihood of effective outcomes is likely smaller. The R&D project requires a goal to remain generally focused on the future and the main route to achieving that future objective. Trajectories, as opposed to a single, long-term project route, are used to measure progress since the road to achieving the objective may be hazy and might change depending on the outcomes of trials and testing. Moving further along the trajectory requires some flexibility, emphasizing the necessity of one- to three-month scheduling chunks. It is essential to be open to learning new things and to seize chances when they come along. Human nature dictates that we lose interest in tasks as they get more challenging, choose the route of least resistance, or stop doing them when they no longer hold our attention or excite us. This may result in wandering in the R&D area when short-term interests take the lead. Roaming can be a waste of important resources that could be used to tackle the toughest issues along the trajectory. Having this objective or course of action reduces the roaming risk, and allowing for spare time within the appropriate period of the timetable can meet exploratory needs with the least amount of risk to the R&D trajectory.

Free Speech

A company should build on its core skills and keep progressing if it wants to remain competitive. An innovative R&D environment based on fresh thought—sometimes radically fresh and different—is necessary for such progress. Free speech is the idea that people have plenty of opportunities and the capacity to engage in interdisciplinary, forward-thinking dialogue with their peers, that their opinions are not constrained by intolerance inside the organization, and that information about the R&D activity is regularly communicated to the stakeholders.

Collaboration across disciplines is a familiar idea. For the expected advantages of close cooperation with other academics, the enhanced opportunity for ad hoc talks that can spark new ideas, and the availability of an R&D-oriented staff, many top R&D businesses expressly choose to locate in innovation parks. The fundamental idea of guaranteeing regular spontaneous and informal communication with one's peers is not in question, even though there is considerable controversy about the real benefits garnered by this sort of setup from an organizational standpoint. A correlation between this kind of research and communication results may be shown in significant successful partnerships over time. Each R&D project manager should periodically attend interdisciplinary seminars, colloquia, lectures, debates, or other forums for knowledge exchange.

It can frequently be incongruous with the cultural environment, where people are expected to keep R&D work private to be able to discuss research across divisions and outside of the normal communication chains. When there is a high level of competition and a compelling product or service, stringent secrecy may be necessary in some situations to limit the danger of losing a patent. Each circumstance should be assessed to establish the amount of cooperation that can be permitted, and any restrictions or boundaries should then be communicated to the R&D employees. People need to feel at ease, forming connections, working together, and sharing ideas within the limitations of any restrictions. Diverse relationships foster creativity, thus opportunities for those connections should be encouraged.

An atmosphere conducive to learning will be open, transparent, and devoid of any antagonism or perceived restrictions. Of course, adhering to the rules and regulations is necessary to guarantee a workplace free from discrimination and harassment. Furthermore, the right to express radical or contrarian opinions should be safeguarded. There is an aversion toward fresh ideas in many well-established organizations. Deploying new ideas occasionally makes employees who were instrumental in creating the successful company feel intimidated because they worry that it would undermine their achievements or destabilize their success. Free information interchange will stop if intolerance is permitted to control conduct. To ensure that the flow of new ideas and free speech are unrestricted and unobstructed, it is crucial that any acts that stifle new ideas or seek to divert them into efforts that support the status quo be detected. Holding evaluations for incoming R&D projects at the organization's most strategic level as opposed to lesser ones is one strategy to guarantee that this occurs.

It is crucial to convey the goals and developments of R&D activity. Setting expectations, establishing relationships, and building respect all depend on honest and open communication with all parties involved in the R&D activities. For R&D to be successful, it is crucial to discuss the risks associated with following a trajectory, comprehend experimental results and where they may demand full pauses, accept the loss of sunk costs (expenditures that have already been made and cannot be recovered), and reevaluate future course. Personnel working in R&D will not cooperate if they face consequences for disclosing knowledge that would be considered unsuccessful in a typical project. For any R&D project manager to provide information concerning outcomes that are less than positive, without consequence, efforts to develop impact-free communications channels should be promoted.

Focus on Outcomes

A successful conclusion for a normal project satisfies the cost, time, and scope requirements. Progressing toward the strategic objectives would be a good result for an R&D project. Focusing on results that will result in strategic goals can help avoid the atmosphere that comes from creative thought without a clear path to closure. The project will likely start along a road, pause, change direction, pick up another forward route, etc., as it is being carried out. In contrast to standard projects, where this would not be desirable, R&D projects are anticipated to include this. As a result, emphasizing R&D results rather than accomplishing a list of predetermined milestones is a highly recommended approach that shows a dedication to learning and delivering results.

The execution of an outcome-based project significantly relies on post-experiment or post-test informal interactions, examining the results of those tests and experiments, and talking with the stakeholders about the next steps. When an experiment or test yields unanticipated results that do not support moving further along the same course, this should not be viewed as a failure but rather as a success. Trying and verifying the outcomes matters since it enables the project manager to determine what will and will not work and to go on even when things do not go as planned.

Many R&D project managers believe these early results will only be seen as essential to the process and prefer to report on progress after they have accomplished a substantial accomplishment. It can occasionally be challenging to communicate how R&D initiatives are progressing. This can be a result of the difficulties in explaining the technical aspects of the task to nontechnical workers. Sometimes, it is difficult to communicate progress on a route not driven by milestones. The project manager may wish to review the result before sharing it with interested parties if they are unsatisfied. However, it is crucial to inform the stakeholders of the R&D path's development regardless of the justifications for not doing so to maintain their support for the project. Lack of communication about progress and results causes stakeholders to believe that nothing is moving forward or everything is going according to plan, which may be accurate.

Due to a simple lack of knowledge, it frequently happens that individuals would believe that more needs to be done. In other words, people can think that nothing has been done if they are unaware of it and cannot perceive that it has been done. That may be the furthest thing from the truth, but how would they understand if there were no regular and active communications? To effectively manage R&D projects, the project manager should incorporate informal conversations with the stakeholders after each experiment or test and include them in decision-making at each gate. These two easy actions will help the impacted stakeholders become more aware of and involved in the R&D project.

Physical Environment

The actual architecture of workplaces should be considered while establishing a learning environment. Two fundamentally opposing requirements should be met for creative pursuits. The first one is to provide a setting that encourages a lot of cross-disciplinary contacts. The next is to provide areas where people may reflect and create without interruption. Both of these requirements should be satisfied to maximize the potential of R&D projects. By choosing to be in closely adjacent buildings with common public areas, interdisciplinary encounters may be made physically possible. Employees taking advantage of common public places creates the potential for collaboration. Physically open areas in buildings may be planned for collaboration. People will flock to them and be inspired to engage in interactions that foster creativity and learning if they are appealing enough. Additionally, open spaces that offer learning resources, access to internet databases, sizable areas and instruments, and machine shops for process redesign offer intriguing choices for investigating concepts on an as-needed basis.

It is also a needed place for uninterrupted, creative reflection. People frequently gather to share ideas, but when it is time to write them down, there has to be a quiet, undisturbed location to do it. This may be handled by using office space that is either permanently allotted or could be borrowed for short periods as necessary. These rooms' aesthetic attractiveness should accommodate the demands of those who work there. Most people given full-time private studios or offices style them to fit their tastes, reflect their personalities, and show who they are as people. The significant drawback of using a borrowed office space is that it does not offer that creative outlet and may even make it more difficult for someone to be creative there. It is simple in theory but challenging in practice to understand the needs and preferences of the R&D project employees for their physical locations and then design an environment that would maximize their chosen learning modalities. Businesses frequently find themselves bound by leasing agreements, committed to office arrangements determined by organizational culture and past preferences, and hesitant to adopt more creative workspace arrangements. To execute architectural modifications to workplaces that would encourage the learning environment required for the success of the R&D project, it frequently needs a strong leader to overcome cultural resistance.

Performance Incentives

Working on novel thoughts and ideas, setting up tests and trials, and developing new procedures, goods, or services are all exciting. Working on initiatives directly relevant to one's area of interest or being a part of a successful R&D project team can frequently be motivation enough. However, several rewards may excite an R&D team and provide even greater outcomes. The provision of blocks of spare time either at the start of a project or following major events during the project's course is one of several concepts. However, resources like machine shop accessibility, engineering time, or travel funds for concept investigation in applied research can be major incentives besides offering free time. For complete or partial ownership rights and public acknowledgment, such as about patents, copyrights, revenues, etc., the majority of organizations have also made agreements. It would be feasible to offer some incentives by sharing ownership of the patents, in an instance, or by giving the creator a share of any profits generated by employing the patented item or method. Depending on the project manager's and team's priorities, incentives like this will vary in strength. Since R&D is frequently put at risk during economically challenging times, offering strong senior management support and secured funding is frequently considered a powerful incentive.

8.6.3 Responsibility and Accountability

Being responsible and accountable entails the following.

Commitment. The ability to commit to a project's course and assume accountability for its success is one of the most crucial qualities of an R&D project manager. The project manager is also responsible for making sure that the R&D progresses fairly and is prepared to enter the following stage. The R&D progress is the project team's responsibility, including the principal investigator and subject matter specialists. The likelihood that the project will be completed successfully increases markedly if the team is committed to its success.

Responsibility. Building a relationship of trust with the stakeholders requires the project manager to take charge, accept accountability for actions, progress, and choices, and then effectively communicate that information to the stakeholders. The project manager's capacity to carry out the actions that will provide the R&D project the best chance of success gives the stakeholders peace of mind. Trust and respect continue to grow when a project manager is open and honest about the planned activities and progress achieved throughout the project. The project manager may implement such openness by putting together a list of goals that will be achieved for the project. These goals should be precise, quantifiable, time-limited, and achievable. The inclusion should be agreed upon by the project manager and their immediate supervisor.

Fortitude. It is not enough to have a responsible, knowledgeable, and capable R&D project manager who is accountable for her or his activities. Being prepared to submit to inspection and assessment of one's work and results is a necessary component of accountability. A reliable technique to ensure that the project is operating as anticipated is to conduct reviews with outside subject matter experts in the field of R&D. Another technique to confirm that the procedures and results are as claimed is to conduct focus groups. When a project is very complex or difficult for stakeholders to verify, these kinds of assessments and conversations are especially crucial. It is helpful to validate how other people see the team's progress, even if the project manager is getting the expected project results. Although the customer's viewpoint on the team's performance is typically clearly known, it is only sometimes obvious how successfully the project manager communicates with all other stakeholders participating in the project. Multi-rater assessments are used to get a range of opinions regarding the project manager's capacity to oversee, lead, and manage project team members.

Additionally, they can gather data on whether the project manager is effectively influencing supporting organizations and satisfying client demands. These evaluations reveal to the team members how well they can engage with others and shape the environment. These evaluations offer a glimpse into how the project manager and team members engage with the larger environment. The assessments give team members and others connected to them a way to express their opinions and report on their impressions. Beyond the real progress being shown on the project, a variety of measures can be employed to assess how successful the R&D team is. One of these tools is multi-rater reviews. However, if there is no plan to address and remedy any identified performance shortcomings, these assessments may be pointless. Resolution should involve a series of corrective steps taken over time, like training, that either fix the problem or result in expulsion from the team. Each organization's processes and procedures for evaluations and follow-up activities will be unique.

8.7 Managing Projects for Certain Specialties

The project manager should think carefully about applying specialized strategies for various disciplines and how to effectively support principal researchers and other subject matter experts if they are to increase the likelihood that an R&D project will produce good results.

Project managers support R&D initiatives across all disciplines by implementing project management methods and procedures. A project manager can also place more emphasis on the project management procedures related to their understanding of the processes and procedures for distinct R&D disciplines. For instance, a scientist and an artist could have varying levels of comprehension of what they desire the final output to look like, or their knowledge of the outcome may be either an unknown or imagined result. A scientist could have a general idea of what they want to achieve, but an artist might know what they want to produce but need to know how to get there. Iterative design is more common in some fields, whereas sequential design is more common in others.

The project manager can create a schedule if the R&D lead researcher has only a general idea of the project's end goal. This schedule will ensure sufficient time frames for structure and exploration and brief timeframes for developing and carrying out tests and experiments and timely decision gate reviews. The project manager can then aid in outlining the next steps. This sequential method offers more information that may be used to clarify the vision and purpose.

Conversely, an iterative method could be most effective if the R&D principal investigator needs clarification about how to start the development but clearly understands the final aim. The project manager can create a timetable to concentrate on brief design iterations with active participation and time blocks for free-form thought and result contemplation. For instance, an artist could conceptualize the end outcome, but they would likely attempt several different approaches before settling on their preferred approach to development. The schedule may be created most effectively to guarantee that the principal investigator has the time and resources to establish the procedures required to accomplish the desired outcome, and the project manager can track progress to keep things moving in the right direction. The project manager will be able to constantly monitor the schedule elements and the critical route once the overall time frame is defined to ensure the project is completed on time.

8.7.1 Scientist or Researcher R&D

Additional attention should be given to

- Determining what stage of the research's life cycle it will be in
- Creating project strategies that prioritize learning and exploration
- Using an adaptable project management technique
- Assisting with evaluations following each experiment
- Informing the stakeholders of the outcomes

Scientists and researchers are goal-driven and eager to push the boundaries of science and technology. If they had taken the traditional educational path, they would have received a solid appreciation for the steps involved in formulating hypotheses, designing experiments and testing, recording discoveries, and cooperating with colleagues. They should adhere to procedures to submit journal papers, give speeches at symposiums, publish books, etc. However, they may oppose or object to procedures that do not appear to be directly connected to the success of their study. They often work independently and concentrate on their scientific or research endeavors. The project manager should ensure that short sprint timeframes and tests or experiments as endpoints are used while developing project plans, emphasizing discovery and learning. After each test, reviews should be arranged, and communication should be given top attention.

References

1. S. Loufrani-Fedida, La gestion des ressources humaines au service de l'articulation entre management des compétences et organisation par projets. Revue de gestion des ressources humaines **1**, 24–38 (2011)
2. H. Zannad, La gestion des ressources humaines dans les projets industriels. Revue de gestion des ressources humaines **68** (2008)
3. X. Baron, Gestion des ressources humaines et gestion par projet, in *La Fonction ressources humaines*, ed. by D. Weiss, (Les Editions d'Organisation, 1999), pp. 611–653
4. G. Garel, V. Giard, C. Midler, *Management de projet et gestion des ressources humaines* (Université Paris I, Panthéon-Sorbonne, Institut d'administration des entreprises, 2001)
5. K. Akhilesh, *R & D Management* (Springer, 2014)
6. O. Lelebina, *La gestion des experts en entreprise: dynamique des collectifs de professionnels et offre de parcours* (ENMP, Paris, 2014)
7. L. Gastaldi, *Stratégies d'innovation intensive et management de la recherche en entreprise: vers un nouveau modele de recherche concourante* (Université de Marne-la-Vallée, 2007)
8. E. Naveh, Formality and discretion in successful R&D projects. J. Operat. Manag. **25**(1), 110–125 (2007)
9. D. Pelz, F. Andrews, *Scientists in organizations: Productive climates for research and development* (Wiley, New York, 1966)
10. A.K.-Y. Leung et al., Multicultural experience enhances creativity: The when and how. Am. Psychol. **63**(3), 169 (2008)
11. S.A. McComb, S.G. Green, W.D. Compton, Team flexibility's relationship to staffing and performance in complex projects: An empirical analysis. J. Eng. Technol. Manag. **24**(4), 293–313 (2007)
12. D. Levi, C. Slem, Teamwork in research and development organizations: The characteristics of successful teams. Int. J. Ind. Ergonom. **16**(1), 29–42 (1995)
13. D. Lea, Managing the high-tech professional. Personnel **65**(6), 12 (1988)
14. L.M. Wingate, *Project management for research and development: Guiding innovation for positive R&D outcomes* (CRC Press, 2014)

Chapter 9
R&D Project Management

Research and development (R&D) projects drive organizations toward success and competitiveness. This chapter delves into the intricate world of R&D project management, shedding light on the critical principles, strategies, and tools that govern these projects. From traditional methodologies to adaptive approaches, this chapter explores the dynamic nature of R&D projects and their complexities. It also addresses the vital aspects of risk, change, and the project life cycle, which are crucial for achieving R&D objectives. The importance of project management in R&D cannot be overstated. Organized and efficient project management becomes paramount as organizations invest substantial resources into research and development. In this regard, the first section of this chapter outlines the fundamental importance of project management in the context of R&D, emphasizing its role in aligning objectives, optimizing resource allocation, and ensuring that innovative ideas translate into tangible outcomes. To navigate the intricacies of R&D project management, it is essential to establish a common understanding of key terms and control mechanisms. Definitions and controls for project management are discussed in the second section, providing readers with a foundational understanding of the terminology and practices that will be explored in subsequent sections. This foundational knowledge is essential for making informed decisions and ensuring that R&D projects are structured and organized, ultimately leading to successful innovation and growth.

> **Learning Objectives**
> - Understand basic definitions of project management.
> - Outline major research and development project management processes.
> - Differentiate project, program, and portfolio.
> - Recognize change management in research and development projects.
> - Identify project life cycle.

9.1 Project Management's Importance

Since the 1980s, project form, which was first only used in engineering (nuclear, defense, building, space industries, etc.), has extended to all spheres of endeavor. The project-based organization is frequently employed in R&D to encourage the innovation process. An innovation project should be managed once the first choice has been made. Then, this makes up the project management object, which has several different components. Methodologies mostly pay attention to time and money. It seems important to consider human resources, which are crucial to a project's success [1].

In the corporate sector, project management is used in many areas, including operations and process management. Because all companies create goods, services, or some mix of the two, all companies have operations. R&D results in a new product or service, and a marketing plan is needed to introduce that product to the market. Ansoff [2] has offered a straightforward version of such a tactic. This section demonstrates that most corporate activities require project management to effect change and make things happen, independent of the marketing approach. Ansoff's product/market matrix is reliable for locating market expansion prospects.

Ansoff's grid delineates four universal growth options, each strategically designed for different scenarios. First, we have "Market penetration," which involves increasing market share for the current product within the existing market. Next is "Market development," which explores new markets for existing products. "Product development" comes into play when enhancing or introducing new products within the current market. Finally, "Product diversification" is the strategy of choice when both the market and product are entirely new to the organization, paving the way for innovation and expansion into uncharted territory. These four growth approaches offer a comprehensive framework for organizations to chart their course in pursuing growth and diversification.

The sole quadrant of Ansoff's [2] product/market matrix related to operations management is labeled "current product/new market," the other three are related to project management and entrepreneurial management. Change management must be avoided, even in operations management. Project management disciplines should thus be used in any corporate plan. A literature survey on current thinking in innovation and NPD activities firmly supports the "stage-gate" approach. This encapsulates the stages of a project lifecycle and suggests that NPD projects embrace project management techniques. An increasing amount of work has been completed in business during the past few decades using projects and project management. However, projects need to improve regarding outcome (quality) and scope (i.e., specification) despite the widespread use of time and cost measurements. This is despite the rapid increase in project management literature and bodies of knowledge. The emphasis on enhancing project planning and stakeholder and people management has increased significantly due to rising competitive pressures, uncertainties, globalization, and the resulting scope (specifications) creep.

People issues, including their expertise, are among the most important factors for successfully delivering a project's end products. They should be given special consideration during implementation and planning activities. This view is supported by project management bodies of knowledge, such as the PRINCE2 system and the PMBOK Guide [3]. This chapter begins with project management processes and definitions, followed by project management characteristics, such as complexity, regulatory requirements, governance, risks, stakeholders, and the project lifecycle.

9.2 Definitions and Controls for Project Management

The complexity, length of time, scope, and scale of projects can range greatly from a quick, one-person project to a multidisciplinary, multiyear project involving hundreds of people in multidisciplinary engineering construction or new product development. The Project Management Body of Knowledge (BOK), which was created by the APM [4] in the UK, gives the following definition of project management:

> A project's control, monitoring, organization, planning, and everyone involved's drive to complete it safely and within performance and predetermined budgetary parameters.
> The sole point of accountability for doing this is the project manager.

The Institute of Project Management [5] offers the following project management definitions:

> A project is a brief endeavor to produce a special outcome, service, or good.
> Applying procedures, tools, skills, and knowledge to project activities to achieve project requirements is known as project management. Through the application procedures, this is achieved.

Here is another way of describing project uniqueness in PRINCE2 [6]:

> A project is a transient organization formed to supply one or more commercial goods by a decided business case.
> Project management is the control, supervision, delegation, and organization of every part of a project and the inspiration of individuals engaged in accomplishing the project's goals within the anticipated performance parameters.

The "iron triangle" of quality, time, and cost is the focus of engineered performance objectives. Scope schedule and budget are shown in the three corners of the triangle in another representation of the "iron triangle" [7]. The project triangle has largely been removed from the latest editions of the PRINCE2 [6] and Project Management Institute [5]. This is because a project has many more restrictions to consider besides its budget, schedule, and scope, sometimes known as quality, times, and costs. In PRINCE2, six project control variables are recommended (Fig. 9.1).

Regarding PRINCE2 [6], quality appears to have a part in the definition and functioning of the deliverable, and this role overlaps with scope and benefits.

Fig. 9.1 Six project control variables

9.3 Traditionally Managed Projects

Due to its similarity to the flow of a waterfall, conventional project management, also known as the waterfall approach, mandates that specific procedures are followed and tasks are finished in the correct order. One is institutionalizing a strategy outlining the actions that will be carried out, by whom, when, and within certain technical and financial constraints. Documentation and communication are necessary both following and during the actions. The discipline offers methods for comparing performance and quality to the established documentation. Establishing a baseline definition of what will be done, when, and for how much, and then monitoring against that baseline is the basis of how the discipline operates. The project's budget, resource-loaded timeline, outcomes, requirements, scope, and any other project-related paperwork are included in the baseline, the completed version of the project. The edition has received agreement and endorsement from all project participants.

Additionally, it is the edition against which effectiveness, quality, and improvement are evaluated. When all of these actions are combined, they give a framework or structure that aids in giving an idea form and provides insight into the steps necessary to bring the concept into practice. Communications, progress measurement, risk management, budget, schedule, organization, approach, requirements, outcomes/deliverables, and scope are the elements of conventional project management.

9.4 Adaptive Project Management

Adaptive project management in R&D is a dynamic approach that acknowledges the inherent uncertainties and complexities of R&D projects. Unlike traditional project management methods that rely on fixed plans and predefined goals, adaptive project management adapts to changing circumstances, allowing teams to navigate the unpredictable nature of R&D work effectively. In an R&D context, projects often involve exploratory phases where the outcomes are unclear. Adaptive project management embraces this ambiguity by encouraging teams to iterate, experiment, and pivot when necessary. This approach promotes flexibility, critical in R&D, as unforeseen discoveries can lead to exciting opportunities or unexpected roadblocks. One of the key principles of adaptive project management in R&D is continuous stakeholder engagement. It encourages open communication between the project team and relevant stakeholders, including researchers, engineers, and executives. This ongoing dialogue ensures that the project remains aligned with the organization's strategic objectives and adapts as new information emerges. Adaptive project management tools and techniques are also vital in R&D projects. These tools often include agile methodologies like Scrum or Kanban, which enable teams to prioritize tasks, collaborate efficiently, and adjust their strategies as needed.

Additionally, adaptive project management incorporates feedback loops and regular retrospectives, facilitating continuous improvement. Another advantage of adaptive project management in R&D is its ability to manage risk effectively. R&D projects frequently involve high levels of uncertainty and risk, and adaptive management provides a structured approach to identify, assess, and respond to these risks. It promotes risk mitigation strategies that evolve as the project progresses. One significant aspect of adaptive project management in R&D is its focus on data-driven decision-making. Teams collect and analyze data throughout the project to inform their choices. This data-driven approach allows for the rapid adjustment of project direction based on evidence, enhancing the likelihood of success and reducing the chances of costly failures. Leadership is essential in implementing adaptive project management in R&D. Leaders must foster a culture that encourages innovation, collaboration, and adaptability. They should also support teams to experiment and explore, recognizing that not all endeavors will yield immediate results.

Adaptive project management in R&D is an invaluable approach for organizations that aim to excel in innovative and dynamic environments. It enables R&D teams to thrive in uncertainty, facilitates ongoing stakeholder engagement, employs agile methodologies, manages risk effectively, and prioritizes data-driven decision-making. With strong leadership and a culture of adaptability, organizations can harness the power of adaptive project management to drive successful outcomes in their research and development endeavors.

Figure 9.2 illustrates the key components of adaptive project management in R&D. It includes elements like continuous stakeholder engagement, the use of agile methods, data-driven decision-making, risk management, and the role of leadership in implementing this approach. This flowchart provides an overview of how adaptive project management adapts to the dynamic nature of R&D projects.

9.4.1 Portfolio, Project, and Program

"In practice, the phrases are frequently conflated or used somewhat indiscriminately, leading to a great deal of confusion and miscommunication." Making the difference between a portfolio, a program, and a project can sometimes be helpful. They differ in how they manage the scope and measure results, although they have the same rigorous, organized approach and fundamental procedures in common. Again, a project is described in the PMBOK as "a temporary activity intended to generate a unique outcome, service, or product." According to the PMBOK, a portfolio, and a program are defined as follows:

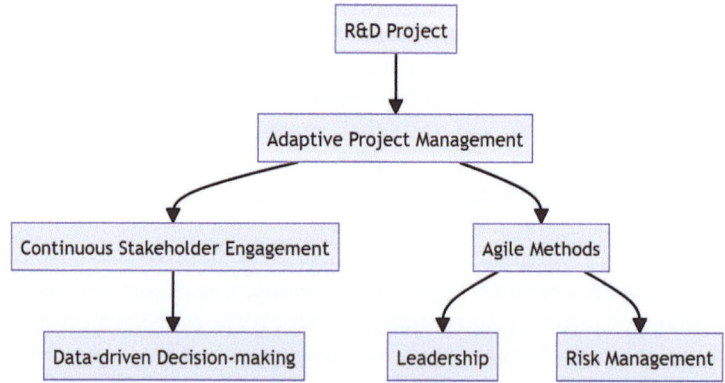

Fig. 9.2 Flowchart representing adaptive project management in R&D

9.4 Adaptive Project Management

The portfolio includes a grouping of projects, programs, and other types of work that enable efficient work administration to achieve long-term company goals.

The program is a collection of connected projects handled and coordinated to get advantages and control that handling them separately would not provide.

The term a program in the PMBOK is generally consistent with definitions found in other literature. A program is "a series of initiatives handled together for extra advantages," according to Turner [8]. According to MSP [9], a "particular group of initiatives established by an organization that combined will fulfill some stated aims, or a collection of objectives, for the organization" constitutes a program. Murray-Webster and Thiry's [10] emphasis on the "strategic and tactical advantages" of program management conflicts with the PMBOK description of a portfolio's reference to "strategic business objectives." More people are realizing that programs can be the vehicle for attaining an organization's corporate and strategic goals. What are the extra advantages of program management over project management? Removing risks associated with project interfaces, the prioritization of resources, and a decrease in management effort are further advantages. Is there a noticeable change in how we approach things, too? The crux of the issue is that program management may concentrate on problems with organizational changes and program results. To achieve the business's strategic goal, a program may involve components of related activity that fall outside the purview of the individual initiatives in the program.

Managing a project portfolio is seen differently, probably more prevalent, as analogous to maintaining a financial portfolio. This kind of project portfolio management (PPM) starts with the company creating an exhaustive inventory of its projects with sufficient descriptive details to enable analysis and comparison. The PPM process involves department heads or other unit managers reviewing each project and prioritizing it by predetermined criteria after the project stock is prepared.

Last but not least, the portfolio management team regularly reviews the project portfolio (quarterly, monthly, etc.) to decide which projects are succeeding, which may require further assistance, and which may require a reduction in scope or elimination. A project manager's life may be significantly eased, and their career can be more fulfilling with effective PPM. More significantly, it may assist a company in coordinating its project burden with its strategic objectives to maximize the utilization of its scarce resources. There are certain variances in focus with using three dimensions, even if the basics of project quality are relevant to managing a portfolio, a program, or a project. While program management focuses more on organizational quality, project management emphasizes product and process quality. Program management's main goal is to hasten the change brought about by stakeholders and the evolving requirements of systems, scope, and organizational culture. The primary foundation of projects is the development of products and their timely and cost-effective delivery. Like managing financial portfolios, project portfolios will likely put product quality in the driver's seat.

9.4.2 Managing Stakeholders

The project manager uses a variety of project management procedures and tools to resolve possible disputes that may arise throughout a project. Many project management tools, such as critical path analysis, Gantt charts, and earned value management, are available to help project managers and team members in addition to the recommendations in PMBOK and PRINCE2. The use of project management methods and tools impacts the continuous control of the project stakeholders. "Stakeholders" are people or organizations who actively participate in a project or whose interest might be impacted—both favorably and unfavorably—by how well or poorly it is carried out. For the team and a project manager to effectively communicate and manage expectations, it is essential to identify and understand the stakeholders.

Team managers, project managers, function managers, and top management are examples of internal stakeholders. Clients, rivals, suppliers, regulators, and environmental organizations are examples of external stakeholders. Major infrastructure projects also display various distinctive characteristics, such as itinerant labor, the need for different talents, the planning of the project, and the work's limited timelines. Additionally, many supplies, tools, and services must be planned for and implemented flexibly. Maintaining a connection with stakeholders to produce a good project outcome is known as stakeholder management. A stakeholder may have an impact on the project or product scope. Therefore, a stakeholder analysis should be conducted right at the start of the project and be directly included in the first scope and vision document. When determining the scope of the assignment, a stakeholder assessment can reveal a variety of helpful information, such as who is responsible for several tasks, whose aims conflict with those tasks, or which business processes are redundant and which are frequent.

This analysis can be prepared in a variety of methods. Following a few easy actions will help.

- List all the stakeholders and group them due to their impact and level of interest in the project.
- Develop each person's duties, objectives, descriptions, influence, and knowledge.
- After conducting preliminary interviews, classify every stakeholder group based on their support levels for the corresponding project goals. Passive opposition, strong opposition, neutral, passive support, and strong support are a few examples of these types.
- Specify the needed position as well as the existing position.
- Summarize the arguments against it.
- Create an action plan to get each set of stakeholders to the necessary position, such as a concession, a shift in roles, mentorship, and training (Fig. 9.3).

Stakeholder analysis should be done as early in the project as possible, but this is a dynamic process. Each set of stakeholders will adopt a different viewpoint and levels of commitment as their familiarity and involvement with the project

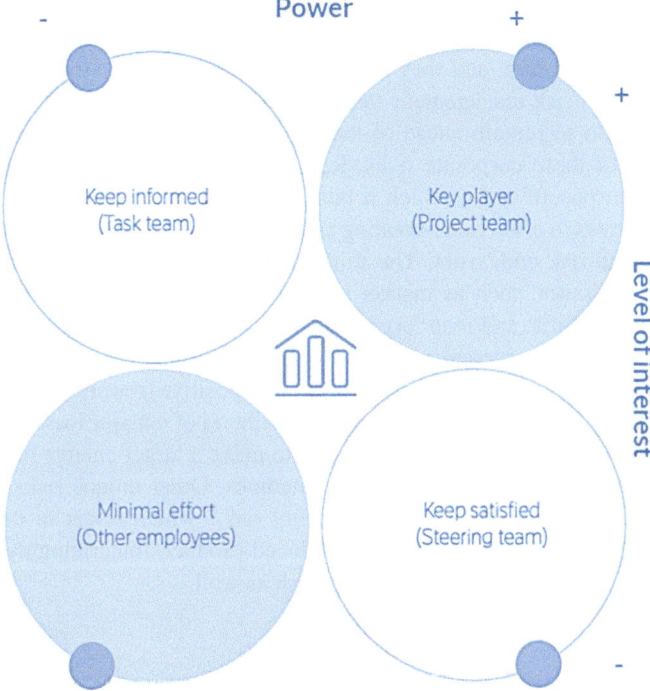

Fig. 9.3 Matrix of stakeholder [11]

develops. In addition, more stakeholders might join the project at a later stage. Accepting that you might not "win them all" is sane, and assembling an alliance of disparate viewpoints is frequently creative. The main goal should be to obtain the support of the "key actors" who have the most impact on project results and control significant resources.

9.5 Risks of Project

Projects are dangerous by definition and by nature. Cost, timeliness, and quality all demonstrate how project risks have an impact. Risks include going over budget, finishing late, or getting results that need to meet expectations. In the worst-case scenario, there may also be a significant violation of safety or environmental regulations. Project management basics are designed to deal with these hazards. To help the risk management of projects, there are extra methods and instruments. The reasons behind or sources of risks in initiatives should be comprehended to minimize their impacts. Hillson [12] proposes three different and independent causes of project hazards: shared traits, intentional design, and the external environment. All initiatives, including R&D, share traits that naturally make them dangerous. These traits include uniqueness, complexity, unknown assumptions, newly created organizations, stakeholders, and change in scope. All projects have these dangerous qualities by their very nature, and they can only be adjusted by altering the project or taking the proper risk management steps. A business strategy is frequently created with added risks to remain ahead of the competition, and the project reflects the performance of these corporate risks. R&D initiatives are closely related to this category of purposeful risks, which it falls within. Projects' function in a project-based company is to offer value-creating skills. Projects are, therefore, purposefully created as high-risk endeavors. The third cause of project risk involves external environmental issues, such as market volatility, the state of the global economy, rivalry, political shifts, and inner organization changes. In the contemporary world, these elements are also increasingly susceptible to change.

A risk can result in either a negative (risk) or a positive (reward) consequence. A bigger risk has a higher potential reward or likelihood of a major loss. According to Hillson [12], when an organization attempts to make a larger change more rapidly, it risks experiencing both good and bad outcomes. Other unique risks associated with R&D initiatives include "appropriability risk," which refers to the easiness with which rivals could copy a freshly produced product. Integrating the research, development, and delivery cultures carries risk as well.

9.6 Complexity of Project

The complexity of a project is a metric for the inherent complexity of completing projects, and it can vary greatly depending on the factors that influence the amount of complexity and the meaning of the term. Many metrics and deliverables, including schedule, financial, organization, process, and design complexity, provide evidence of a project manager's impression of complexity. The term's misuse exacerbates the difficulty in controlling project complexity. A project manager's sense of complication has been classified scientifically into dynamic and detailed complexity. There are several definitions of complexity, and Sussman [13] provided a list of 20 of them. Some highlight the intricacy of a system's behavior, while others focus on its internal organization.

Senge [14] may be credited for coining the terms "detail complexity" and "dynamic complexity," which are said to have popularized the idea of complexity in corporate processes. The many-to-one (many resources applied to a project) detail complexity requires a thorough plan specifying who does what and when and a cascade milestone chart demonstrating interdependencies and key routes. Dynamic complexity is managed through the ongoing, dynamic triage process and is characterized by subtle cause-and-effect relationships and less visible long-term repercussions of treatments. According to Senge [14], true power over most managerial problems comes from a knowledge of dynamic complexity rather than detail complexity. The majority of businesses need to be made aware of dynamic complexity. They devote their whole lives to data processing, thinking that improvement in

9.6 Complexity of Project

corporate performance can be achieved via painstakingly laborious data collection and analysis. Projects often place a similar emphasis on specifics. While paying attention to the little things is crucial, what they mean for the big picture is more crucial. Dynamic complexity is simpler to comprehend, quantify, and manage when adopting a dynamic management process rather than a content-focused strategy. Organizations often mature their project governance and control systems over time to a point where they can handle the intricate intricacy of the projects they frequently embark on. However, these restrictions are frequently ineffective because of excessive bureaucracy and the need for comprehensive knowledge of dynamic complexity. Baccarini [15] proposes two forms of complexity in a project, namely, technological and organizational complexity, and defines project complexity as consisting of numerous diverse and connected pieces produced from differentiation and interdependency. A complicated project organization has distinct components: the more differentiation, the more complicated the organization. The level of interdependency among the project organization units is another characteristic of organizational complexity. The level of diversity or variation in work is called technological complexity via differentiation. The interdependence between tasks may be included in technological complexity through interdependency. According to Baccarini [15], "integration," which refers to coordination and management, manages differentiation's and interdependencies' complexity.

Project complexity provides challenges but should be turned into opportunities by understanding its components, causes, consequences, or their distinction and interdependence. The insight complexity provides into organizations fundamentally distinct from other prevalent management discourses may be the most important component of complexity. The model that follows, which was condensed and modified from Stacey's [16] work, should assist in illustrating how complexity can be used to one's benefit. It can be seen in Fig. 9.4 that the model considers two dimensions, agreement and certainty, and based on these, four project-related zones are given: "anarchy," "complex," "difficult," and "simple." The "anarchy zone" denotes circumstances without clarity or consensus on plans, which leads to a breakdown. Avoidance is the only viable tactic in this situation. There can be agreement on the results in the "complex" project zone, but there may be disagreement over which outcomes are preferable. Politics could be crucial in this situation, necessitating coalition building and compromising. There is great potential for innovation in "complex" projects, breaking with the past, where novel modes of operations are expected to diverge from conventional management practices. The classic management strategy is used to make reasonable judgments in "simple" initiatives when there is consensus on goals and clarity. Through efficient coordination, communication, and stakeholder management, this is the area of potential and innovation. Efficient management of the complicated project area and the "integration" of difference and dependency are related concepts.

In the end, dynamic complexity refers to ongoing changes in people's behavior, whereas detail complexity refers to the difficulties resulting from product design and process details. It is crucial to use a mixture of project control tempered by

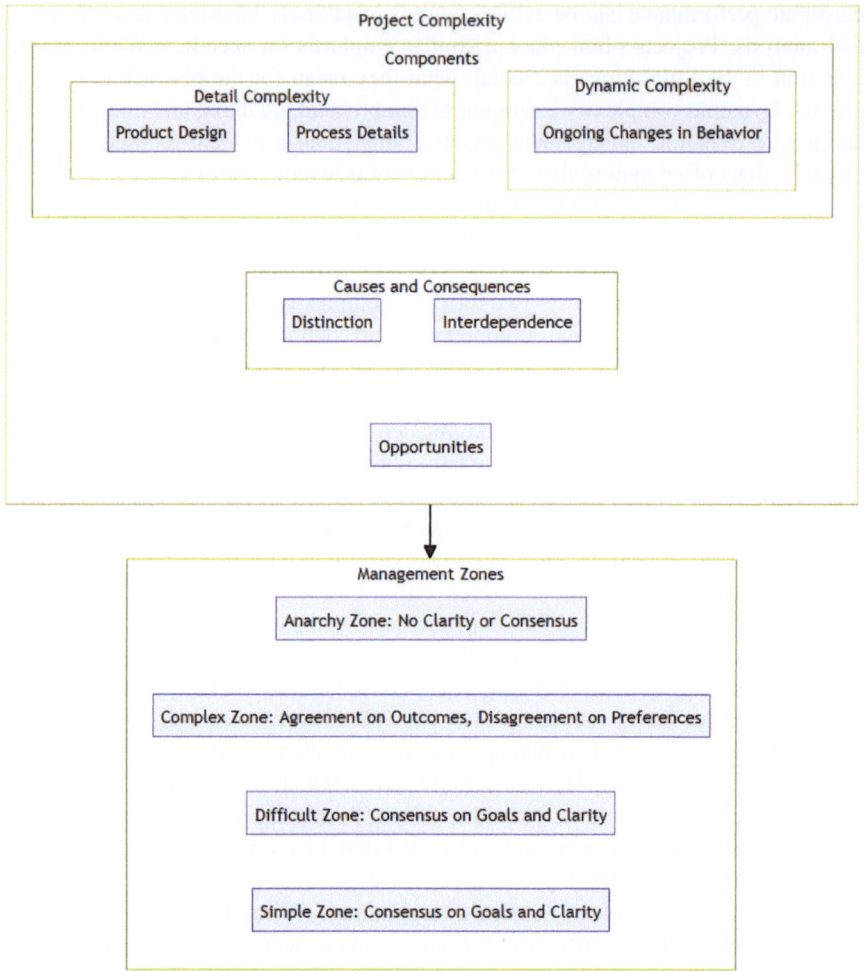

Fig. 9.4 Visualizing project complexity and management zones

regular interactions and reviews supported by ongoing communication and learning to handle dynamic and detailed complexity.

Figure 9.4 visually represents the intricate concepts discussed in the text regarding project complexity and its management. It delineates the primary components, including detail complexity, associated with product design and process details, and dynamic complexity, tied to ongoing behavioral changes. The chart also emphasizes the causes and consequences of complexity, focusing on the distinction and interdependence of project elements. In addition, it highlights the numerous opportunities that project complexity presents when well understood. The flowchart introduces the management zones, from the "anarchy zone," where clarity and consensus are absent, to the "simple zone," characterized by consensus on goals and clarity. This visual helps elucidate the multifaceted nature of project complexity and the strategies required to navigate it successfully. It offers a concise overview of the key concepts outlined in the original text.

9.7 Change and Project Management

Changes in company strategy, national and global economies, and surroundings all influence project management. In change projects, project managers should manage the change in an organized manner and develop a project organization to accomplish the change. Numerous change management projects (product development, system development, and overall quality efforts) begin with unclear goals and strategies for achieving the change. Management by projects should be used for change projects whenever a client needs and means of attaining them change.

The delivery, execution, and planning of well-structured projects are also marked by quick changes that result in new demands for stakeholders and a need for appropriate information for practitioners to adapt plans to respond to the changes. Project scope is known to expand in a changing environment. Furthermore, as project management evolves, it moves beyond a basic definition of practice to elicit distinctions across different project kinds and situations. Even within a similar project, maturity might range among professions and between stages and procedures. Information technology and tools are frequently necessary throughout the decision-making process, particularly in bigger businesses, for processing information that may differ between firms. All of this has an impact on the project quality dimensions.

Cost, time, and scope are project management criteria that create a system, and changes in one affect the others. For example, any large change in scope that was not planned for may impact other criteria, particularly important cost and schedule variables. However, eliminating scope creep is impossible. Similarly, uncertainty or risk is viewed as a normal state to be considered while managing change. The organizational quality dimension pertains to people-related traits critical in managing and maintaining change with important stakeholders. Some academics [17] have emphasized the significance of "team relationship management." Developing these elements in change management and difficulties connected to project management basics are useful in exploring the importance of project quality ineffective results. With the core fundamentals of project management in mind, the next part narrows down to the particular procedures and phases of the project life cycle.

9.8 Project Life Cycle for R&D

Projects have life cycles, much as living things do. They start small, then expand, reach their peak, then begin to fall, and eventually close. A project's life cycle generally includes the commencement, design, execution, and closure (Fig. 9.5). This

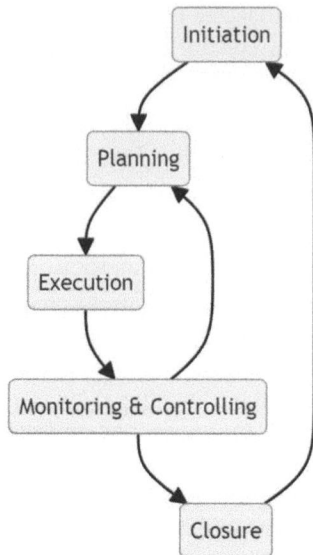

Fig. 9.5 The four stages of a project's life cycle

is equivalent to a product's life cycle's launch, growth, maturity, and decline stages. It is vital to notice that these four project life cycle components align with the eight PRINCE2 procedures. Turner's names for each project life cycle step are frequently different in various project applications (e.g., handover, implementation, design, and definition), or the phase may contain additional stages. However, alternative project life cycle phases are simple to match with Turner's given names.

The project life cycle is a group of typically consecutive project phases in the PMBOK, whose names and numbers are established by the control requirements of the organization or organizations engaged in the project. In a project life cycle or project, phases are divisions requiring additional control. As a result, there are considerable differences in how PMBOK or PRINCE2 express the numbers and names of stages in a project life cycle. The project type and its intended use affect the language and phases of the project life cycle. Some application areas like IT have project life cycle structures with common phase names well documented and applied throughout the respective sector.

A project's life cycle should be divided into stages for a variety of reasons, including the following:

- It offers reasonable portions.
- It makes it possible to manage risks related to the project at each step.
- It assigns each stage's relevant management jobs and abilities.
- It fosters communication by providing a unifying topic at each stage.
- It promotes growth over time.

It is significant to highlight that the above benefits of the project lifecycle are effectively realized by ensuring that the individuals working in two successive phases cooperate closely to guarantee that the product delivered fulfills the demands of the subsequent stage. The pertinent papers should also be signed. Additionally, proper attention should be placed on integrating product design-focused procedures with project management processes throughout a project life cycle. Six Sigma projects are now implemented and closed with the rigor of the project life cycle thanks to the Control, Improvement, Analyze, Measure, and Define methodology.

9.9 Navigating R&D Challenges: The Way Forward

R&D is pivotal in driving innovation in scientific and industrial domains. It is the engine behind progress, the source of groundbreaking discoveries, and a catalyst for economic growth. However, the R&D landscape is fraught with challenges that necessitate a strategic and forward-thinking approach to success. Funding constraints are one of the most significant hurdles in R&D (Table 9.1). Securing adequate financial support for projects can take time and effort. Organizations can overcome this challenge by exploring alternative funding sources such as public-private partnerships, venture capital, and crowdfunding. Another obstacle is the need for more talented researchers and scientists. A worldwide shortage of skilled individuals poses a critical problem.

Mitigating this challenge involves forging strong academic-industry collaborations, investing in STEM education, and fostering international talent exchange programs. The rapid pace of technological advancement presents a challenge of technology obsolescence. Innovations can become obsolete in the blink of an eye.

Table 9.1 Strategies for overcoming business challenges

Challenge	Suggested way forward
Funding constraints	Explore alternative funding sources like public-private partnerships and venture capital
Talent shortages	Foster academic-industry collaborations, invest in STEM education, and promote talent exchange
Technology obsolescence	Cultivate adaptability and continuous learning within the organization
Regulatory hurdles	Engage with regulatory bodies early and advocate for regulatory reform
Intellectual property	Utilize strong patent strategies and partner with legal experts
Market uncertainty	Employ data analytics and market research to reduce uncertainty
Global competition	Foster international collaborations, seek global partnerships, and tap into emerging markets
Environmental sustainability	Implement sustainable practices and embrace green technologies
Data security	Ensure robust cybersecurity measures and educate the workforce on data protection

Organizations must cultivate a culture of adaptability and continuous learning to address this, allowing them to pivot as technology evolves. Stringent regulations often slow down the pace of innovation. Engaging with regulatory bodies early in the R&D process and advocating for regulatory reform can help streamline processes, enabling innovation to flourish within the bounds of the law. Protecting intellectual property is critical in R&D. Implementing robust patent strategies and collaborating with legal experts are essential for safeguarding innovative ideas.

Market uncertainty poses yet another challenge. Predicting market trends and consumer preferences is a complex task. Reducing uncertainty involves leveraging data analytics and conducting thorough market research. Global competition has intensified in R&D. To remain competitive, organizations must foster international collaborations, seek global partnerships, and expand into emerging markets to access a broader talent pool and diverse insights. With environmental sustainability gaining prominence, R&D must become more eco-friendly. Implementing sustainable practices and embracing green technologies align with global initiatives and cater to the growing demand for environmentally responsible solutions. Finally, data security is an issue, especially as R&D projects often involve sensitive information. Robust cybersecurity measures, including encryption and secure data storage, and educating the workforce on data protection are essential steps forward.

In summary, navigating the complex terrain of R&D challenges requires a multifaceted approach encompassing financial creativity, educational investments, adaptability, regulatory engagement, legal protection, data-driven insights, global engagement, sustainability, and cybersecurity. These strategies can pave the way for more successful and innovative R&D endeavors.

References

1. P. Gibert et al., *Innovation, Research and Development Management* (John Wiley & Sons, London, 2018)
2. I. Ansoff, *Corporate Strategy: Revised Ed./Ansoff I* (Penguin Books, London, 1987)
3. Institute, P.M., *A Guide to the Project Management Body of Knowledge (PMBOK Guide)*, 4th edn. (Project Management Institute, Chicago, 2008)
4. APM, *The Body of Knowledge* (Association for Project Management, Princes Risborough, 2006)
5. A. Guide, *Project Management Body of Knowledge (Pmbok® Guide)* (Project Management Institute, Newtown Square, 2008)
6. PRINCE2, (2009)
7. D. Seaver, *Respect the Iron Triangle* (IT Today, 2009)
8. J.R. Turner, J.R. Turner, T. Turner, *The Handbook of Project-Based Management: Improving the Processes for Achieving Strategic Objectives* (McGraw-Hill, London, 1999)
9. R. Sowden, *Managing Successful Programmes: Office of Government Commerce*. ISBN-13, p. 978–0113310401 (2007)
10. R. Murray-Webster, M. Thiry, Managing programmes of projects, in *Gower Handbook of Project Management*, vol. 3, (Gower, Aldershot, 2000), pp. 47–64
11. L.M. Wingate, *Project Management for Research and Development: Guiding Innovation for Positive R&D Outcomes* (CRC Press, Boca Raton, 2014)

12. D. Hillson, *Management Risk in Projects: Fundamentals of Project Management Farnham* (Gower Publishing Company, Surrey, 2009)
13. J.M. Sussman, *Ideas on Complexity in Systems-Twenty Views* (Massachusetts Institutre of Technology, Cambridge, Masachusetts, 2000), p. 2000
14. P. Senge, *The Fifth Discipline: The Art and Practice of the Learning Organization* (Random House, London, 1990)
15. D. Baccarini, The concept of project complexity—A review. Int. J. Proj. Manag. **14**(4), 201–204 (1996)
16. R.D. Stacey, *Strategic Management and Organisational Dynamics: The Challenge of Complexity to Ways of Thinking about Organisations* (Pearson Education, Harlow, 2007)
17. A. Kadefors, Trust in project relationships—Inside the black box. Int. J. Proj. Manag. **22**(3), 175–182 (2004)

Chapter 10
Managing Performance in R&D Projects

Within the realm of research and development (R&D) projects, this chapter delves into the nuanced facets of performance management. It explores the intricacies of performance in R&D, addressing the unique challenges and opportunities it presents. Section 10.2 discusses the delicate art of quantifying R&D performance. It emphasizes the need for accurate, relevant metrics in a domain where innovation often defies conventional measurement. Section 10.3 reveals how traditional project management practices can be tailored to suit the fluid landscape of R&D, providing strategies to strike the right balance between structure and adaptability. Section 10.4 underscores the pivotal role of collaborative teamwork in R&D projects. It offers insights into fostering an environment conducive to creativity, knowledge sharing, and synergistic efforts. The chapter also delves into Sect. 10.5, showcasing how technological tools and advancements can streamline R&D processes, reduce time to market, and enhance overall performance efficiency. By the end of the chapter, readers will possess a comprehensive understanding of the strategies, challenges, and opportunities associated with managing performance in the dynamic world of R&D, providing a roadmap for organizations to optimize their innovation endeavors.

Learning Objectives

- Be aware of how crucial performance management is for R&D initiatives
- Recognize the principal obstacles to carrying out R&D initiatives
- Identify key factors to take into account while managing R&D initiatives
- Describe cutting-edge methods and tools for managing R&D initiatives

10.1 Performance in R&D

Performance in research and development (R&D) is a critical aspect of a company's or organization's success, particularly in industries that rely on innovation, technology, and product development. Evaluating and improving R&D performance is essential for staying competitive, driving growth, and achieving strategic objectives.

10.1.1 A Hard-to-Define Concept

Though performance is fundamental to management in general, and possibly because of this, this notion has yet to be ascribed to a singular definition. Bourguignon [1] discusses the polysemy of the word "performance" and its controversies. The portrayal of success is relative to performance. What does R&D success entail? An invention does not necessarily result in innovation (the positive sanction of the invention user), such as a more lucrative process or a successful commercial transaction, in designing activities. And, if the idea is effective, who receives credit for it?

Efficiency (achieving an aim), effectiveness (using resources to accomplish the objective), and relevance can all be seen as performance components. In the area of R&D, management by goals could be more successful. It is possible to concentrate on how resources are used (budget compliance), but it is obvious that there are other goals than this activity. The process of putting abilities into practice is included in the performance. R&D would benefit more from this viewpoint. Additionally, it aligns with modern management control orientations and some writers' focus on process- and competency-based management. Understanding action processes and possessing abilities are related ideas, according to a study by Lorino [2]; informed by this strategy, it may be assumed that if researchers work together efficiently to collaborate with the downstream business and with marketing, then there is a chance that the outcomes will be financially beneficial.

Several parallels to R&D go beyond the general difficulties in defining performance and can be attributed to various variables. First, there are the many responsibilities that people play, frequently reduced for contributions to innovation initiatives. R&D also has multiple other objectives, including establishing the organization's credibility in the scientific community, luring talent, evaluating impractical scientific and technological paths, influencing standards, directing university research, and assisting the corporation in gaining legitimacy with the government.

Because the idea of total performance is impractical, it is challenging to uphold it by balancing efficiency, effectiveness, and relevance. As a result, some authors make distinctions between various performance kinds. Hooge and Stasia [3] define four forms of innovation and R&D (Fig. 10.1).

When considering the economic assessment of projects, it shall be talked about the first two categories of performances. The other two performance kinds are more challenging to comprehend since they are qualitative. Hooge and Stasia [3] outlined

10.1 Performance in R&D

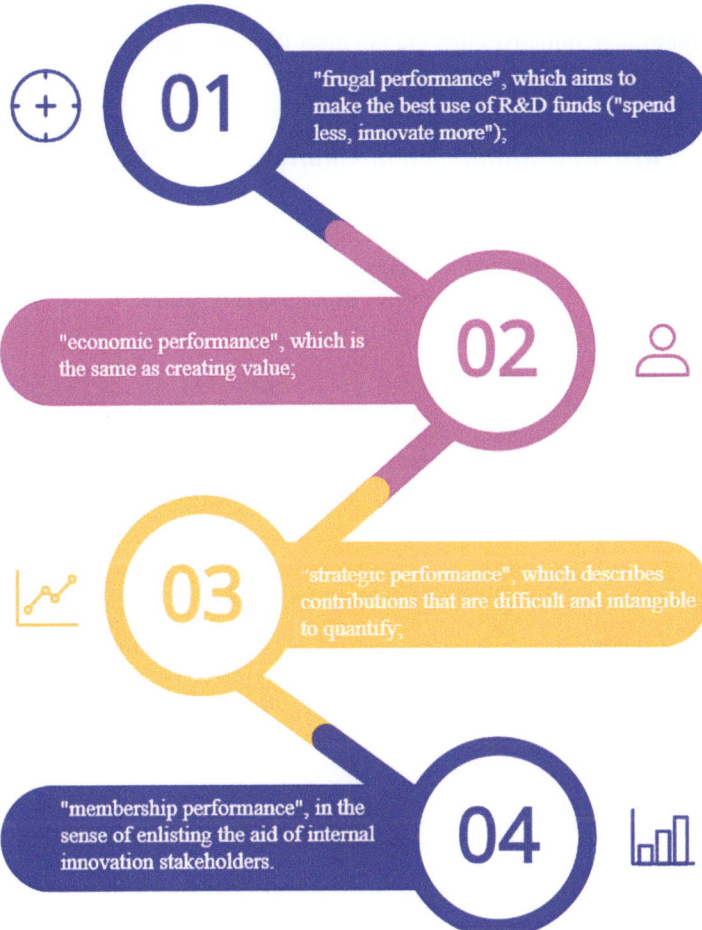

Fig. 10.1 Four forms of innovation and R&D

criteria for planned performance. The degree to which R&D is consistent with the company's plan and the contribution of R&D to business image and communication are two of the most distinguishing factors it will continue to use. How R&D reacts to problems and expectations of innovation stakeholders inside the organization indicates membership success.

In addition to categorizing the various R&D performances, another challenge is the sometimes "invisible" character of R&D achievements. For instance, in the technical area, it is challenging to comprehend the advantages or impact of a significant actor's reputation on the environment. How can we assess the benefits of research that shows that such a technological route is a dead end and prevents bad investments? Unfortunately, because academic literature prioritizes success above failure, it does not promote learning from loss or from the failures of service or product innovation initiatives that did not meet their anticipated victory.

10.1.2 Particular Management Challenges in R&D

R&D performance management is a difficult task due to a variety of elements. Some are typically related to functional studies, while others are explicitly associated with the design industry. This brings up many problems (Fig. 10.2).

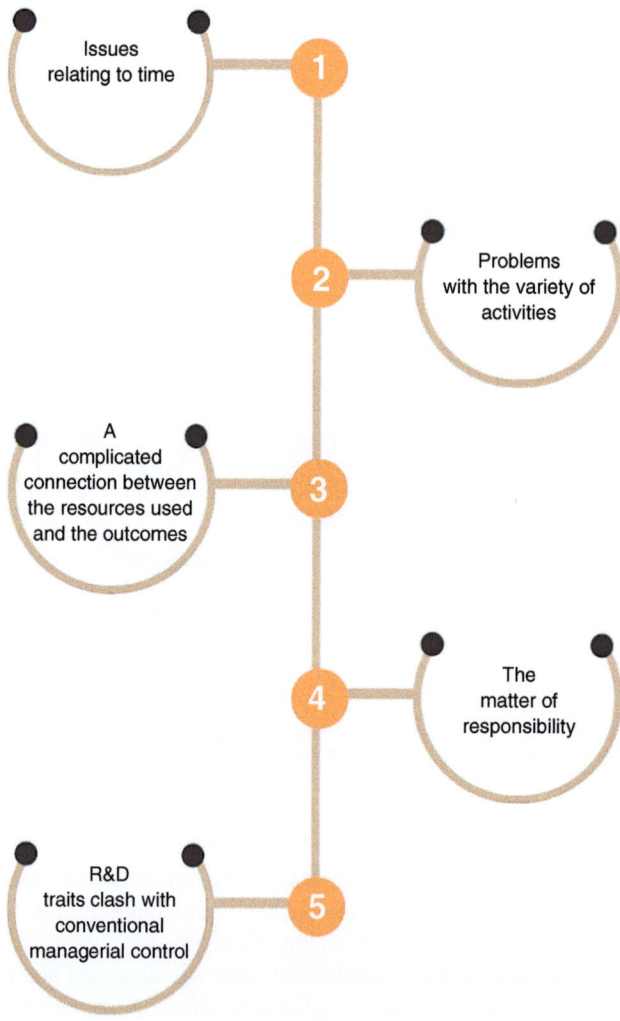

Fig. 10.2 R&D performance management problems

Issues relating to time. When it comes to managing R&D efforts and understanding the outcomes of such work, the problem of temporality is challenging. Indeed, it cannot determine a management technique's efficacy by only looking at its immediate results. There may be a significant lag between expenses and the accumulation of advantages from the company's outcomes. Therefore, breakthrough innovation initiatives that would create a more erratic turnover over a longer horizon are heavily penalized by the dominance of short-term profits over other value factors. Short-term managerial pressure can have negative outcomes.

Problems with the variety of activities. The wide range of activities under the umbrella of "R&D" presents another barrier. At the very least, tertiary or technical support will not be included as they already have a standardized solution register. It is important to distinguish between "controlled design" (or engineering-related) activities and "unregulated design" (or creative) activities, which relate to research. The period/costs/quality triptych continues to be the reference in the first scenario, which entails activating existing knowledge in line with set criteria. The canonical performance model is under scrutiny in the second instance, which deals with creating new information.

Complicated connection between the resources used and documented outcomes. The ambiguity of its findings distinguishes research. A credible link between the economic and the resources mobilized results attained, in particular, cannot be shown via simulation.

The matter of responsibility. Managers now concur that research's successful breakthroughs elevate them above the technologies they produce. The issue is that R&D is rarely fully responsible for the failure or success of an innovation. Nixon [4] states: "One of the primary obstacles in assessing R&D operations resides in the reality that a successful invention demands structural and performance features that no one function can fully control." R&D's relationships with other corporate departments are crucial, yet as Nixon points out, interactions are only sometimes active during the invention process. It changes with time.

R&D traits generally clash with conventional managerial control. The latter is made for operations where activity is frequently planned over an annual time frame and is essentially standardized, with structured procedures and predetermined goals managed by a clear hierarchical structure. Additionally, "control systems have, at best, a modest role" [5] despite the strategic relevance of development and design activities. However, performance issues have not been spared by R&D, which has sparked a newfound interest in the field.

10.1.3 The Challenges of Performance

While management control professionals have sought to offer (additional accounting and accounting) solutions to the problem, the financial crisis has shifted emphasis to performance management in R&D, which has traditionally been seen as a discretionary activity. The strategic backdrop has also become tense due to the shortened product life cycles, more rivalry, and the quickening speed of technology breakthroughs (rapid product obsolescence due to corporate rivals and plans, quicker imitation). Therefore, there is a powerful scissor impact between the rise in product design costs on the one hand and the sharp decline in the time frame within which it may be hoped to get a return on investment on the other.

10.2 The Sensitive Matter of Measurement

Everything deemed vital in management is meant to be governed by a set of metrics. The same holds for performance. Key performance indicators, sometimes abbreviated as KPIs, convey this concern. These metrics (rates, quotients, percentages, and averages) are compiled in a scorecard to be utilized in the field's performance management process and include pertinent justifications and remarks for conducting follow-up activities.

10.2.1 Traditional Indications

Three primary indicators are employed to quantify the concrete outcomes of R&D: patents, publications, and innovation, based on Gallié et al. [6] study, which surveyed the worldwide literature and considered international practices.

The most popular method of knowledge production is patenting. This measure has several drawbacks. Not every innovation can be patented. Since not all information can be codified, it is called tacit knowledge. Such knowledge cannot be saved in databases, yet they are frequently the most important. Finally, patents are expensive (both in filing and in maintaining them) and may need to be more successful at defending an innovation. Some businesses may also want anonymity.

The volume and stature of scholarly publications are other indicators of research effort. This serves as the fundamental metric for evaluating the efficacy of public research structures. Researchers in the private sector can also be assessed using this criterion, namely for upstream research and high-technology activities. However, this measurement uses peer-controlled recognition systems and disseminates information to the general public. However, the corporation must be made aware of it and consequently reluctant to employ it.

Innovations are the third category of indicators. They convey the R&D outcome of commercially available products that drive consumers in a certain market group to purchase. They especially fall within the category of economic traits, and as such, organizations naturally place value on them. However, several definitions exist of what constitutes an invention, and this idea does not reflect a uniform reality. Additionally, when innovation is accepted as such, internal process advances that indirectly enhance economic performance are ignored.

Curiously, customer happiness indicators are not among the top five most frequently employed KPIs, as highlighted by Fugier-Garrel et al. [7]. More generally, some academics have proposed that performance evaluation should be made available to a larger variety of external and internal players by further enforcing control inside the company's strategy. The balanced scorecard (BSC), which Kaplan and Norton [8] devised, serves this function. The BSC strives to respond appropriately to the requirements of each of our innovative stakeholders in R&D.

10.2.2 Considering the Different Stakeholders

Bremser and Barsky [9] demonstrate how businesses can connect their resource commitments in R&D to the company's operations and goals by using the BSC as an integrating framework. This strategy aims to advance the logic of strategic objective alignment between actor group goals and actor group goals. Researchers provide advice for businesses to follow while implementing this integrated performance design for the R&D measuring system.

In the example given (based on examining a research center's strategy), the authors suggest four "strategic objectives" that represent the viewpoints of various stakeholders, including staff development and growth, internal professionals, customers, and finances. Every department in the firm has KPIs that correlate to these goals. Metrics are connected to one or more KPIs in turn. For instance, Bremser and Barsky [9] propose the computation of the volume of training hours about the "retention" and "development" indicators or the skill covering ratio by a strategic skill category.

As evidenced by a study by Agostino et al. [10] on an Italian technological research center, complexity can result from considering many stakeholders. They provide a collection of 23 indicators broken down into five categories as part of an action research project to create an R&D performance assessment system meant to manage various expectations.

Since the presented indicators are related to a particular organizational setting and research procedure, it is highly challenging to argue their significance. Like the fundamental usefulness of KPIs, the issue of stakeholder engagement has piqued our attention. Following Simons' [11] contrast between diagnostic control (the traditional idea of management control) and interactive control (based on vertical contacts between subordinates and managers and horizontal interactions between persons of the project group), Gautier [5] promotes the relevance of interactive control, theorizing that this approach is more suited to development projects since it allows for the resolution of the unresolved issues. Insofar as it may be created and utilized in a participative and transversal method to encourage connections between many groups of players participating in the various phases of projects, using the BSC might be seen as extending this logic in certain ways.

10.2.3 Limitations of Technological Sophistication

Research on innovation performance is broad. However, the numerous research findings are yet to, as of yet, lead to a performance metric that is universally acknowledged. The variety of indicators based on sectors, nations, and measurement types is substantial. The number of indicators has increased, and the complexity of control systems has increased since the end of the 1990s, but the outcomes could have been more compelling.

The research by Hagedoorn and Cloodt [12] on the technical performance of 1200 businesses in various industries (pharmaceuticals, aerospace, etc.) demonstrates that, on the one hand, such indicators tend to strengthen one another while, on the other hand, each of them considered individually gives a pretty reasonable estimate of innovation performance.

10.2.4 R&D Departments' Financial Management

Not all businesses that engage in R&D activities should have a department devoted to doing so. However, this is typically the case in major corporations, except those who place R&D as their primary activity, as is the case, for instance, in advanced electronic activities, which are arranged in a matrix-like fashion across two divisions: by geographic region and by activity sector. Budget management, like other mechanisms for allocating resources to support an activity, is a crucial component of management control. All of the projections considered criteria to be adhered to are intended to be clarified in the budget, which shows the expense of a program as part of a plan. Let's examine each of these significant words in turn: a schedule, a plan, and a budget:

- The plan outlines the activities the R&D department should conduct over several years. Establishing important parameters for the future is stated in qualitative and rather generic words.
- In contrast to the plan, the program is a short-term prediction (typically for a year), detailing in-depth and considering various scenarios, activity levels, agreed-upon quantities (especially the workforce), etc.
- The budget serves as a translation of the strategy by estimating the cost of carrying out the program, as specified by various cost elements, in monetary terms.

Equipment required, supplies consumed, and salaries for this activity make up most of the R&D department's expenses. Their relative shares vary significantly depending on the industry in which they are employed, for example, between the digital sector and more capital-intensive industries like the steel industry that use expensive machinery.

In many cases, wages and related expenses are the most important factor. The fundamental prerequisite for this calculation is to ascertain the precise number of workers required (in terms of qualification and quantity). Equipment for experimenting, testing, and trial may equally represent a significant item, depending on the activity. Premises also represent a distinct thing. Finally, it is typical to include the costs associated with technical observations in the budget for an R&D department. Participation in numerous professional or scientific activities, as well as journal subscriptions, might be included here. The R&D division can also produce specific goods in addition to paying for these costs, such as proceeds, grant funds, or royalties from the dedication of patents. Additionally, they might charge outside parties for their services.

Aside from its contributions to management control in the strict sense, the budget also serves two additional purposes: during the budget cycle, it is a motivation and dispute resolution tool. The budget is also a decision and simulation support tool during the design phase. The final element is crucial in a task where the demands of technical progress may conflict with the goals of financial stability.

10.3 Project Management for Innovation

Two layers of project management may be considered: the project manager, who controls the project "internally," and the overall organization, which typically maintains a project portfolio. Initiatives are the main focus of project portfolio management to mediate and select projects.

10.3.1 Project Economic Evaluation: The Two Methods

In project management and financial analysis, evaluating a project's economic viability is crucial. Two primary methods are commonly employed: the Net Present Value (NPV) and the Internal Rate of Return (IRR). These methods assess the financial feasibility of a project by considering the time value of money and the potential returns on investment. Let's delve into these two methods in more detail.

Net Present Value (NPV) NPV is a widely used method for economic evaluation that measures the present value of a project's future cash flows. It calculates the difference between the present value of cash inflows and outflows over the project's lifespan. The formula for NPV is as follows:

$$NPV = \sum \left[Ct / (1+r)^t \right] - C0$$

- Ct represents the net cash flow at time t (inflow-outflow).
- r is the discount rate or rate of return representing the project's capital cost.
- t stands for the period.

- C0 signifies the initial investment or cash outlay.

If the NPV is positive, it suggests that the project is expected to generate a profit, and the potential return exceeds the initial investment. A positive NPV is generally considered a favorable outcome.

Internal Rate of Return (IRR) IRR is another essential method that calculates the discount rate at which the project's NPV equals zero. In other words, it identifies the rate of return at which the present value of inflows equals the present value of outflows. The formula for IRR is:

$$0 = \sum \left[C_t / (1+\text{IRR})^t \right] - C0$$

- C_t represents the net cash flow at time t.
- IRR is the internal rate of return.

The IRR is the project's break-even discount rate. If the project's expected rate of return exceeds the IRR, it is considered economically viable.

Both NPV and IRR have their advantages and limitations. NPV provides a clear monetary value and accounts for the time value of money, making it well-suited for comparing projects with different scales and timeframes. On the other hand, IRR expresses the project's return as a percentage, which can be easier to interpret but may lead to multiple solutions in certain scenarios.

In practice, using both methods in project economic evaluation is common. By considering both the NPV and IRR, project managers and investors can understand a project's financial feasibility comprehensively, enabling more informed decision-making regarding resource allocation and investment.

Project evaluation and selection remain a top issue for businesses despite the generality of project management because of the high uncertainty and failure rate that prevents the use of control methods and conventional assessment. This common practice applies to the design activity sphere as well. Time and cost abuses occur often in this area, partly because these tasks are less foreseeable and programmed than other firm activities. Methods have been created to stop this activity, and they fall into two major categories, according to the study by Garel [13]: profitability approaches and cost.

Cost Evaluation

This topic is extensively covered in project management literature, separating several cost ideas. The terms used by Garel [13] study state that "design to value," "design to cost," and "design to target" are the three primary categories of project cost evaluation. In "design to value" methodologies, the project starts with understanding a client's demand. Technical specifications are obtained from a specification, and a cost estimate based on cost accounting is completed.

In "design to cost" methodologies, which are a component of competitiveness optimizing techniques, calculating cost is no longer dependent on the specifications but rather on the profit needs of the business, in line with the client's expectations. Cost control is essential in this situation and applies to every product's life cycle stage, including the design phase (target costing). Companies increasingly often utilize this strategy, which also tries to shorten design times. Some extremely static design-to-cost methodology approaches have drawn criticism from authors. The argument made by Gautier and Giard [14] focused on the method's flaw of estimating a cost at a certain point without considering changes in the product life cycle. Plans that are "designed to target" answer this objection.

In project management, the intricacies of cost evaluation play a pivotal role in any venture's success and economic feasibility. Many strategies and methodologies have been developed to address the multifaceted aspects of cost management. The work of Garel [13] significantly contributes to this field, categorizing cost evaluation into three distinct but interconnected concepts: "design to value," "design to cost," and "design to target."

Design to Value: "Design to value" is a strategic approach that starts with a profound understanding of the client's needs and desires. It hinges on the meticulous analysis of these requirements, meticulously deciphered to extract the essential technical specifications. These specifications serve as the guiding stars for the project, leading to the creation of a robust and reliable cost estimate based on comprehensive cost accounting. The "design to value" approach aligns the project's value proposition with the client's expectations, emphasizing creating a product or service that provides optimal value for the investment.

Design to Cost: In stark contrast, "design to cost" methodologies shift the cost evaluation paradigm from specification-driven to profitability-oriented. Here, cost calculation is intrinsically tied to the financial objectives of the business, and it aligns with the client's financial expectations. Central to this approach is stringent cost control, which permeates every stage of the product's life cycle, including the design phase—a concept known as "target costing." Companies are increasingly adopting this methodology to enhance their competitiveness and shorten product development cycles. However, it's not without its detractors. Some critics argue that excessively rigid "design to cost" approaches can fall short of accommodating the dynamic changes occurring throughout a product's life cycle.

Design to Target: In response to the criticisms of the static nature of "design to cost" methodologies, the "design to target" approach emerges as a refined solution. This methodology emphasizes adapting to changes and unexpected developments that may unfold during a product's life cycle. By allowing for more flexibility in cost evaluation and adjusting the cost targets as necessary, "design to target" approaches seek to overcome the rigidity observed in the traditional "design to cost" methods. This adaptability ensures that cost evaluations remain relevant and aligned with the evolving dynamics of the project.

In conclusion, cost evaluation in project management is a multifaceted endeavor, with different methodologies suited to diverse contexts and objectives. "Design to value," "design to cost," and "design to target" are not mutually exclusive but rather complementary approaches. Project managers and organizations must carefully consider the nature of their projects, the expectations of their clients, and the fluidity of their product life cycles when choosing the most suitable cost evaluation strategy to optimize their project's economic feasibility.

Financial Profitability Evaluation

Financial profitability techniques were initially created when deciding whether to begin a project. However, these strategies can still be used after the project has started as long as cost reductions and improvements can be made. To increase profitability, some Japanese research has also underlined the significance of postponing certain choices as long as feasible. The first strategy, which is the most common, bases its evaluation of projects on cash-flow capitalization, on the net present value, and views them as investments that should bear fruit. A project is approved and funded if its net current value (NPV) of positive cash flows is higher than its net present value of negative cash flows and is larger than that of another project with whom it is vying for resources. Although an R&D project may be linked to a succession of costs followed by revenues, in inventive design, there is substantial uncertainty regarding the numbers on which expenditure assessment is based, let alone for prospective payments, concerning both their quantities and their implementation timetable; so, calculating rationality looks fairly unreal.

Ultimately, "an investment project might be undertaken despite a negative net present value (NPV) since the senior management thinks it is strategic. On the other hand, a project with an intriguing NPV can be turned down if it does not fit with the company's overall strategy [15]. The second scenario only sometimes results in the eclipse of concepts that can result in a spin-off, which is the term for establishing a new business that will guarantee the financial valuation of know-how and research findings. Making an activity regarded as non-strategic independent will give project owners a framework that is more receptive to the growth of their ideas and a framework that makes financial reports easier for investors to understand.

The second, the more contemporary method, is based on the options theory created for financial markets in the early 1970s. Remember that an option is a right to sell (put option) or purchase (call option) an asset (currencies, shares, bonds, etc.) at a given price at any time within a predetermined period. This theory has been used to evaluate nonfinancial assets like R&D projects, industrial units, or capital goods. It tries to optimize strategic decision-making by examining, over time, its possibilities and connected risks. This is known as a "genuine option." Estimated profitability assessments are used to evaluate opportunity and risk.

Real option assessment has frequently been suggested as a good tool for evaluating high-risk investment initiatives, as may be true for technical innovation projects with extremely high failure rates. The real option technique was used in pharmaceutical [16] and biotechnology [17]. The key problem is assigning money to the greatest research projects in these two industries, where a company's performance heavily depends on R&D.

Only Profitability and Cost Are Two of the Decision-Making Factors

Project evaluation within a corporation is a multifaceted process that extends beyond the mere consideration of projected benefits and costs. While these financial aspects are crucial, they are not the sole determinants of whether a project gets approved or rejected. Berland and de Rongé [18] shed light on several other factors that weigh heavily in the decision-making process. These factors help the organization prioritize projects and allocate resources effectively.

Integration with Current Manufacturing Processes: One critical consideration is how well a project aligns with the organization's existing manufacturing processes. This integration can impact the ease of implementation, potential disruptions, and the project's overall feasibility within the current operational framework.

Internal Competencies: The company's internal competencies and expertise are pivotal in project selection. Assessing whether the organization possesses the necessary skills, knowledge, and resources to execute the project is vital. In some cases, the project might be innovative but require skills the organization needs to gain.

Compatibility with Current Technology: The project's compatibility with the technology landscape is another essential factor. A project that seamlessly fits with the company's existing technology infrastructure is more likely to be approved, as it can minimize technological disruptions and costs associated with transitioning to new systems.

Conformity to the Business's Plan: Projects must conform to the business's plan and objectives to maintain strategic alignment. Projects that support the overarching strategic direction of the organization are often given higher priority.

Patentability and Innovation Protection: For organizations that rely on innovation, the potential to patent the product resulting from the project is a valuable consideration. Protecting intellectual property can be a competitive advantage and an important factor in project approval.

While these factors contribute significantly to the decision-making process, it's essential to emphasize that the qualitative aspects of R&D are equally crucial. These factors extend beyond quantitative measures and cannot be entirely captured by algorithms or numerical evaluations. Ultimately, the correct project choice involves a nuanced assessment that considers the projected financial returns and the strategic alignment, internal capabilities, and potential for innovation protection. Project evaluation is a multifaceted endeavor, and a balanced approach that blends quantitative and qualitative elements is essential for making informed and strategic decisions in the realm of R&D.

10.4 Enhancing Team Collaboration for R&D Success

In the realm of R&D projects, success often hinges on the ability of cross-functional teams to work together seamlessly, harnessing collective intelligence and creativity to drive innovation. Effective team collaboration is a critical factor in achieving R&D goals, as it can lead to increased productivity, higher-quality outcomes, and faster project completion. R&D projects often involve professionals from diverse backgrounds, including scientists, engineers, designers, and market analysts. To enhance team collaboration, creating an environment where these different expertise areas can converge effectively is essential. Encourage open communication, shared objectives, and mutual respect among team members. This can be achieved through regular meetings, cross-training sessions, and team-building activities that promote understanding and cooperation. Effective communication is the lifeblood of successful collaboration. R&D teams should establish clear and efficient communication channels, ensuring information flows seamlessly among team members. Utilize digital collaboration tools, project management software, and reporting systems to facilitate real-time data sharing, ideas, and project updates. By reducing communication barriers, teams can make informed decisions and respond to changes swiftly. A clear leadership structure is essential to ensure that everyone understands their roles and responsibilities. Appoint a project leader or manager who can guide the team, set priorities, and provide direction.

Moreover, accountability should be emphasized, with each team member taking ownership of their contributions to the project. Regular progress checks and milestones help maintain a sense of responsibility within the team. R&D projects thrive in environments where innovation is encouraged and rewarded. Foster a culture of experimentation, where team members are encouraged to take calculated risks and explore unconventional approaches. This approach can lead to breakthrough solutions and motivate team members. Collaborative decision-making ensures that all team members have a say in important project choices. Encourage open discussions, brainstorming sessions, and the collection of diverse perspectives when making critical decisions. This approach results in better-informed choices and increases team buy-in and commitment.

Conflicts are natural in any collaborative endeavor, and it's crucial to have mechanisms to address them constructively. Train team members in conflict resolution strategies and provide a safe space to express concerns. A well-managed conflict can lead to better solutions and stronger team relationships. R&D projects are dynamic, and adaptability is key. Encourage a culture of learning from both successes and failures. Regularly review project performance, identify areas for improvement, and adjust strategies accordingly. This continuous improvement mindset ensures that the team evolves and grows over time. By prioritizing collaboration and implementing these strategies, R&D teams can unlock their full potential, fostering innovation, efficiency, and success in their endeavors.

Figure 10.3 illustrates the intricate relationship between key elements in enhancing collaboration for successful R&D outcomes. It is structured around four main

10.4 Enhancing Team Collaboration for R&D Success

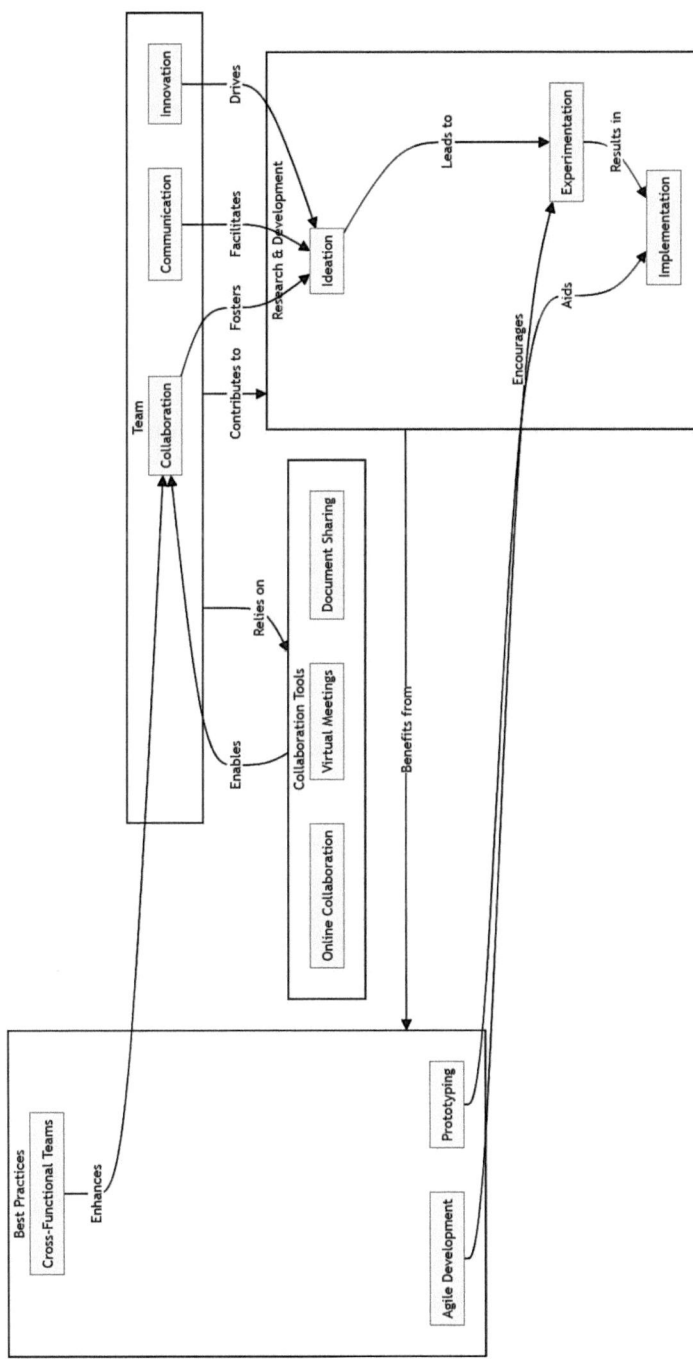

Fig. 10.3 Enhancing team collaboration for R&D success

components: "Team," "Research and development," "Collaboration Tools," and "Best Practices." The diagram showcases how effective collaboration within a team leads to the generation of innovative ideas ("Ideation"), which, in turn, guides the experimentation phase and culminates in the implementation of successful R&D projects. Additionally, it highlights the vital role played by collaboration tools such as online collaboration, virtual meetings, and document sharing in supporting team collaboration. Best practices such as agile development, cross-functional teams, and prototyping further enhance the R&D process, contributing to its overall success.

10.5 Leveraging Technology for Efficient R&D Performance

R&D is the lifeblood of innovation and progress in today's fast-paced, technology-driven world. Organizations across various industries invest heavily in R&D to stay competitive and drive growth. To ensure efficient R&D performance, leveraging technology has become a critical imperative. In this ever-evolving landscape, harnessing the power of technology is instrumental in accelerating research, improving collaboration, and maximizing the return on investment.

One of the primary ways to leverage technology in R&D is through data analytics and big data. The exponential growth of data can be managed and analyzed using advanced tools, providing valuable insights and shortening research cycles. High-Performance Computing (HPC) is crucial for computationally intensive tasks, such as simulations, molecular modeling, and data processing. HPC clusters and cloud-based solutions reduce the time required for experimentation.

Collaboration is essential in R&D, often involving geographically dispersed teams. Technology tools like video conferencing and cloud-based document-sharing platforms facilitate seamless collaboration, making virtual environments and augmented reality instrumental in enhancing remote teamwork. Machine Learning and Artificial Intelligence (AI) automate repetitive tasks, analyze complex datasets, and predict outcomes, reducing the need for manual intervention. Digital twins, virtual replicas of physical objects or processes, enable researchers to conduct experiments in a virtual environment, reducing the need for physical prototypes. In fields like health care and manufacturing, IoT and sensor technology provide real-time data to optimize processes and drive R&D. Blockchain technology is used to secure intellectual property and data integrity.

Automation and robotics in R&D laboratories increase productivity by allowing robots and automated systems to perform repetitive tasks. 3D printing and rapid prototyping technologies enable the quick and cost-effective creation of physical prototypes in product development and materials science.

Table 10.1 summarizes the key technologies and their applications in enhancing R&D performance to provide a comprehensive overview.

10.5 Leveraging Technology for Efficient R&D Performance

Table 10.1 Catalyzing R&D excellence: technologies and their impact on research and development

Technology	Description	Application in R&D	Benefits in R&D
Data Analytics and Big Data	Utilizes advanced analytics and big data platforms to process and analyze large datasets quickly	Data processing, trend identification, and data-driven decision-making	Shortened research cycles, improved decision-making, and enhanced data quality
High-Performance Computing	Employs powerful computing clusters and cloud-based solutions for complex simulations, molecular modeling, and data processing	Accelerating simulations and data analysis in various R&D domains	Faster experimentation, more accurate results, and increased efficiency
Collaboration Tools	Utilizes video conferencing, project management software, and cloud-based document-sharing platforms to enhance collaboration among geographically dispersed teams	Facilitating seamless remote collaboration, project management, and document sharing	Improved teamwork, enhanced knowledge sharing, and streamlined project management
Machine Learning and AI	Leverages AI and machine learning algorithms to automate tasks, analyze complex data, and make predictions in R&D processes	Automation, data analysis, and predictive modeling	Reduced manual work, enhanced data insights, and faster decision-making
Digital Twins	Creates virtual replicas of physical objects, systems, or processes to conduct experiments in a virtual environment	Virtual experimentation, prototype reduction, and process optimization	Cost-effective testing, reduced need for physical prototypes, and improved design processes
IoT and Sensor Technology	The Internet of Things (IoT) and sensor technology are used to provide real-time data for optimizing processes and monitoring various R&D activities	Real-time data collection, process optimization, and environmental monitoring	Improved real-time decision-making, process optimization, and data-driven insights
Blockchain	Utilizes blockchain technology to secure intellectual property, protect research findings, and ensure data integrity in R&D	Intellectual property protection, data integrity, and secure data storage	Enhanced data security, intellectual property protection, and data integrity assurance
Automation and Robotics	Integrates automation and robotics into R&D laboratories to perform repetitive tasks and increase productivity	Automation of repetitive tasks, laboratory work, and data collection	Increased productivity, reduced human error, and more efficient research processes
3D Printing and Rapid Prototyping	Utilizes 3D printing and rapid prototyping technologies for quick and cost-effective creation of physical prototypes in R&D activities	Rapid prototyping, product development, and material testing	Faster prototype development, iterative design processes, and cost-effective testing

References

1. A. Bourguignon, *Peut-on definir la performance? [Can We Define Performance?]* (Revue Française de Comptabilité, 1995)
2. P. Lorino, *Méthodes et pratiques de la performance: le pilotage par les processus et les compétences* (Ed. d'organisation, 2003)
3. S. Hooge, S. Roland, *Performance de la R&D et de l'innovation-Du contrôle de gestion à la gestion contrôlée* (2016)
4. B. Nixon, Research, and development performance measurement: A case study. Manag. Account. Res. **9**(3), 329–355 (1998)
5. F. Gautier, *Les systèmes de contrôle de gestion des projets de conception et de développement de produits nouveaux: une analyse empirique* (Institut d'administration des entreprises de Paris, GREGOR, 2002)
6. Farjaudon, A.-L., E.-P. Gallié, and C. Kuszla, Les indicateurs de la R&D et l'innovation. 2010
7. S. Fugier-Garrel, R. Masson, J. Heine, *Votre R&D est-elle performante* (Etude KL-UTC, 2016)
8. R. Kaplan, D. Norton, *The Balanced Scorecard: Translating Strategy into Action* (Harvard Business School Press, Boston, 1996)
9. W.G. Bremser, N.P. Barsky, Utilizing the balanced scorecard for R&D performance measurement. R&D Manag. **34**(3), 229–238 (2004)
10. D. Agostino et al., Developing a performance measurement system for public research centres. Int. J. Bus. Sci. Appl. Manag. (IJBSAM) **7**(1), 43–60 (2012)
11. R. Simons, *The Role of Management Control Systems in Creating Competitive Advantage: New Perspectives, Readings in Accounting for Management Control* (Springer US, Boston, 1990)
12. J. Hagedoorn, M. Cloodt, Measuring innovative performance: Is there an advantage in using multiple indicators? Res. Policy **32**(8), 1365–1379 (2003)
13. G. Garel, Pour une histoire de la gestion de projet. Gérer et comprendre **74**(1), 77–89 (2003)
14. F. Gautier, V. Giard, Vers une meilleure maîtrise des coûts engagés sur le cycle de vie, lors de la conception de produits nouveaux. Comptab. Contrôl. Audit **6**(2), 43–75 (2000)
15. G. Garel, *Le Management de projet* (La découverte, 2012)
16. M. Hartmann, A. Hassan, Application of real options analysis for pharmaceutical R&D project valuation—Empirical results from a survey. Res. Policy **35**(3), 343–354 (2006)
17. T. Miloud, W. Azan, *Evaluation de projets de R&D: Elaboration d'une méthodologie par les options réelles dans l'industrie des biotechnologies* (Journée AIMS-BETA, University of Strasbourg, 2015)
18. N. Berland, Y. De Rongé, Le contrôle de la R&D et de l'innovation, in *Contrôle de gestion: Perspectives stratégiques et managériales*, (Pearson, Paris, 2016)

Chapter 11
Success Evaluation in R&D

Understanding and assessing success is pivotal within the dynamic research and development (R&D) realm. This chapter delves into the multifaceted world of success evaluation, aiming to provide a comprehensive perspective on key concepts and methodologies. Defining success in R&D is a complex and nuanced undertaking, explored in the first section of this chapter. Whether it is groundbreaking innovations or incremental progress, the criteria for success vary, and it is crucial to establish clear definitions and benchmarks. Success in R&D is not an absolute, one-size-fits-all concept but requires adaptable and context-specific definitions. To accurately gauge R&D success, one must build a strong foundation. The chapter delves into the critical factors underpinning R&D activities, such as well-defined objectives, efficient resource allocation, and fostering collaboration. Various measurement types, including quantitative and qualitative methods, are introduced to capture the multifaceted dimensions of success in R&D. The chapter also discusses techniques for analyzing variance and trends in project outcomes, enabling the identification of patterns and areas for improvement in R&D performance. It further emphasizes the importance of tailoring measurement approaches to different project types involving basic research, applied development, or disruptive innovations. Finally, the chapter addresses R&D performance measurement, focusing on the role of Key Performance Indicators (KPIs) and balanced scorecards in assessing the overall effectiveness of R&D efforts.

11.1 Success Definition

For a variety of factors, success should be measured. However, first, it is important to clarify what success means:

> Meeting the requirements of the budget, timetable, and scope and ensuring that the stakeholders are satisfied with the results, outcomes, and deliverables are examples of success.

> **Learning Objectives**
> - Describe the major obstacles that hinder R&D success
> - Describe several methods for determining if R&D initiatives are successful and how to select them
> - Recognize the value of good governance in R&D projects
> - Explain how to include measuring into R&D projects
> - To manage an R&D project successfully, identify the key project management concepts
> - Acquire the ability to recognize and evaluate trends while assessing the performance of an R&D project

The practice of defining a range of success and designating an action as "successful" if it fits within the range is extremely frequent. However, it is common to discover that a project needs a clearer finish if stakeholders agree on what constitutes a good result. This is not to suggest that it has no end date on the calendar but as a project gets closer to that point; it is harder and harder to bring all activities to a conclusion that can be deemed complete. The overall success of a project can only be determined if it can be completed.

Success in project management is characterized by the performance of the customer's expectations in terms of scope, time, and budget. Success is believed to have been attained if the project was finished within the boundaries established at its inception. A variety of metrics can be used to ensure that these characteristics are satisfied. However, things are more complex. Even if a project met these criteria but crucial integration points were overlooked, resulting in other activities or projects failing or suffering, the product or process's quality was lower than anticipated, or the project's final product or process was unusable, it could still be considered unsuccessful.

When planning a project, it is crucial to specify the results, outcomes, and deliverables that may be anticipated if the scope, timeline, and budget are fulfilled and then acquire buy-in from the stakeholders on what constitutes success. It may be more complex than it first appears because it is frequently simpler to define failure. However, if this phase is finished, measures may be taken to validate the path to success and its eventual realization. Applying intermediate metrics throughout the project increases the likelihood of good outcomes. Even if the specific outcomes of the experiments do not support an idea or hypothesis, a project that aims to advance knowledge in a particular field and successfully conducts several experiments that result in a deeper comprehension of the field of interest can be considered successful in the context of R&D. The factor that contributes to a project's effective completion is frequently the learning that was attained during the experimentation.

11.2 Crucial Foundations

There are a few foundational issues to solve, even if all the other components are implemented and expertly managed. If these particular areas are properly handled, there is a much higher chance of success. Selecting a trained and experienced project manager is especially important for R&D projects, which can gain greatly from employing unconventional project management techniques. The communications from each stakeholder should be properly addressed. The project should then be baselined, and any modification requests should first be thoroughly evaluated for their potential effects before being implemented.

11.2.1 Project Manager

How successful the project is can depend on the project manager who is selected. The project manager creates the atmosphere and tone for the project, secures commitments from staff and stakeholders, assembles the team that will see it through to completion, and aids the company in appreciating its advantages. The project manager is also in charge of team morale and creates a learning environment by investing in employee training and mentorship so they can work as effectively as possible.

A weakness in the principal researcher or the team might also hinder success. This is especially true for initiatives involving R&D and innovation, which frequently depend on a skilled individual or group. Despite all efforts, if the team's talent mix is insufficient, meeting project requirements may not be possible unless additional team members with the required skills are added, education or training is provided, or other circumstances that increase the cost and potentially affect a project's schedule.

11.2.2 Communications

Communication is a crucial foundation activity in addition to having a capable project manager who can lead a team to success. A project might suffer devastating effects if communication is not done correctly. Effective communication is required between project members, stakeholders, and the official client. According to research, a project manager's duties might be made up of up to 90% of communications-related tasks. The requirement for communication is more pressing in R&D projects since it is common for those initiatives to need more effective communication techniques. Formal communications, like those seen in reports and records, may also be unofficial and informal.

There may be natural hurdles that should be overcome for successful communication, such as cultural, linguistic, socioeconomic, or geographic barriers since only some stakeholders or clients have the same communication needs. Departmental challenges are also possible, with distinct acronyms, abbreviations, and terminologies. Because teams are frequently small, results are largely unproven, and change is frequent, communications in R&D initiatives are frequently informal. A communications matrix is a technique that can help determine and document the various stakeholder preferences for informational delivery on the project. A communication matrix is illustrated in Table 11.1.

Understanding the stakeholder's position on supporting the project is crucial when determining the communications approach for each one of them. Stakeholders who are uninformed or hostile to the project require various forms and intensities of communication. An enthusiastic and keenly engaged stakeholder in the project's development may require a different communication medium. The demands of each stakeholder should be fully comprehended before being compared to what the project manager feels are necessary to solve any project support challenges. Even if it goes against his or her preferences, a stakeholder who is uninformed of the project but where there are significant interdependencies between it and their area of responsibility would need more disciplined and frequent communication. An essential duty of the stakeholder and the project manager is to negotiate the appropriate communications channel. A timeline and strategy for execution should be created once the communication strategies have been decided upon.

11.2.3 Basis for the Project

To guarantee that the project has an adequate definition and that success in fulfilling the project's outcomes means the same thing to the project group and the client, a project foundation should be developed at the beginning of the project. When the client and the project team agree on the project's budget, timetable, scope parameters, and goals, a baseline for the project is established.

Table 11.1 Matrix of communications

Stakeholder	Support feeling level	Interdependency	Project frequency choice	Communication choice by stakeholder	Frequency choice by stakeholder
Person 1	Neutral	None	As required	Casual conversations	As required
Person 2	Unaware	Significant	Weekly	Web page updates	As required
Person 3	Supporter	Significant	As required	Social media updates	As required
Person 4	Neutral	Insignificant	On occasion	Meetings	Monthly
Person 5	Resistant	Significant	Weekly	Information bulletins	Weekly
Person 6	Unaware	None	On occasion	Activity reports	On occasion
Person 7	Advocate	None	As required	E-mail	As required

11.2 Crucial Foundations

When there are financial problems, it can occasionally happen in scheduling that a project is started without a baseline, dealt with as a "soft start," or is executed gradually. However, a project can be impacted and frequently derailed from the outset with this kind of start. Even if the schedule alterations are properly executed, the perspective of successful attainment of results will be impacted by incorrectly estimating time frames for work activities or overall project duration. A negative performance impression will result from failing to shut on the final few unfinished project activities and enabling them to continue after the baseline project's stated end date. The earlier in the life cycle the R&D occurs, the more this is particularly true. R&D and related creative endeavors are sometimes challenging to schedule. Consequently, care should be given to selecting intermediate phases and tests that will provide measurable results for the project to be successful. To avoid wasting time and resources by continuing down an unproductive route, the correct branch points that enable an organization to cease working or entirely shift course should be found.

Having a baseline plan in the budget area shows that managing, monitoring, and controlling expenses are crucial from both an organizational and stakeholder standpoint. When a budget adjustment request is made, a proper evaluation of the effect on the timeline and scope may be done. Cost, scheduling, and scope are closely related to one another. Even if a considerable amount of change was done on the project, there is a higher likelihood that the project will be viewed as successful if all the stakeholders agree to the baseline scope, schedule, and budget parameters, and change is closely controlled throughout the project. The likelihood that a project will be able to fulfill its goals is decreased, and the possibility that it will eventually not be seen as successful is increased whenever ideas start to lead to uncontrolled revisions of requirements, scope, or any other plan components.

11.2.4 Success Definitions in Various Fields

Success is described in many different ways throughout the many disciplines. The project's life cycle location affects how the term "outcome" is defined. Each of the examples below may involve projects ranging from R&D to operational activities; hence, the range of possible project results is similarly wide. These are only a few examples of the different project results that may be found per discipline (Fig. 11.1):

Humanities. The development of the English language dictionary, finishing a work of art, and releasing a new play, film, or book are examples of successful results in this field.

Government. A government project's success may be determined by whether it succeeded in changing community behavior or had a beneficial influence on economic growth.

Fig. 11.1 Fundamental categories of each level of incremental measurement

Social sciences. Successful results in this field may result from ethnographic research and initiatives, novel psychological procedures, the understanding, development, or modification of laws, or any combination of these.

Infrastructure. Creating a brand-new idea, a new telecommunications tower, a building renovation, or a new bridge for a future transportation system are examples of effective infrastructure project results.

Higher education. The installation of a new student enrollment system, the publication of academic papers, or the creation of prototypes are all examples of successful project results in higher education.

Livestock and agriculture. Successful project results in these fields include innovations in forest management, mining, or veterinary treatment.

Water and power. Successful initiatives can include the creation of a new water processing plant or power generator, a new method to obtain water from an undiscovered source, innovative designs for alternative energy sources, or procedures to ensure that water is pure.

Natural sciences. These project results might include fruitful scientific inquiries into innovative methods for addressing environmental problems like pollution.

Technology. Successful technological project outputs include investigating new biology, inventing new mechanical or electronic devices, or producing new materials.

Information. In the information discipline, examples of effective results include putting new procedures into place, creating new networks, and creating new systems, hardware, or software.

Medical sciences. The best-known scientific discoveries in this field result from advances in procedures, medical devices, medications, and research into illness and health.

Business. Suppose a new or different product or service has a beneficial influence on current or potential new markets or successfully modifies customer behavior. In that case, the conclusion of a business initiative can be considered successful. A successful business initiative may also result in increased productivity or favorable financial results.

The number of possible projects for each of these fields is limitless. Each project may be classified as operations, production, development, research, etc., and will fit within the life cycle continuum. Additionally, the project management strategy utilized to produce good results might vary from adaptable to traditional depending on wherever it is on the continuum.

11.3 Measurement Types

The best strategy for ensuring the greatest chance of getting successful results is to:

- Managed change reduces any adverse effects on the project
- Make sure of the systematic system development by the project plan
- Ensure the implementation of measures that aid in making smart decisions throughout the process
- Ensure that a process, product, or system satisfies the requirements and is prepared to move forward in the life cycle.
- Ensure that the final product or procedure is appropriate for the intended use

Documenting the expected outcomes, deliverables, requirements, and intended scope inside a project plan that is baselined is the first step in using measurements to assure successful outcomes in any project, at any point of the life cycle, and in any discipline. No matter where the activity falls during the life cycle, a strategy should be created to determine where it is in the cycle and what is anticipated of the effort. Any attempts to evaluate performance and results become very challenging, if possible, with this documentation.

The baseline paper should then be cited to make clear which intermediate metrics the stakeholders want and desire. Both the project management strategy used on the project and the project life cycle phase should support this. The following questions should be raised, clarified, and then documented:

- Who needs what information, and why?
- When is the data required?
- In what location will the data be kept?
- How will access to the information be made?
- Who will supply the details?

Each level of incremental measurement is covered in the sections that follow. The fundamental categories can be seen in Fig. 11.1.

11.3.1 Metrics Versus Measures

Metrics, measures, or any mix of both are frequently used to make measurements. The more objective and quantitative types of measures are typically referred to as measurements. Measures are certain acts, such as counting the amount of help desk tickets, hours allotted or consumed, or transactions. A metric is a collection of data that indicates how well something is doing compared to an intended objective. These are frequently qualitative and arbitrary evaluations of performance on a scale. Common metrics include adherence to process integrity, a baseline, and accomplishment of goals. Choosing the appropriate measurements and benchmarks requires both creativity and teamwork. The effective and beneficial ones can be chosen by a project manager with experience managing successful projects.

Metrics should track how well the project is doing in terms of sticking to its budget, timeline, and scope limits. The most effective indicators for R&D and innovation projects display advancement along the trajectory and highlight how closely test and experiment results align with the project's plan. These metrics can show performance along a single measure or be a part of a collection of metrics that show success across numerous measures. Metrics are created based on industry standards and are an essential tool for interacting with stakeholders about the project. Therefore, the metrics must accurately reflect the development of that crucial information. Metrics are often additive because more will be employed based on the strategy taken and where the project is in the life cycle. Although not a perfect phase function since there are many instances when some of the stricter techniques will be used sooner in the life cycle, it is a good representation of the cumulative aspect of the metrics. As the formalism of the project rises, metrics can be introduced. The addition of measurements along the diagram's continuum is not intended to indicate

11.3 Measurement Types

that they cannot be utilized for research, the development of novel ideas or concepts, or in situations where only minimal degrees of control are required for informal initiatives. However, it shows that each of these procedures would be in place by the time a project is in production, is a formal project, and needs strict controls.

The most typical metrics used for a project that is in the middle of its life cycle are those that assess how far along the project is from the baseline for scope performance, schedule, cost, risk management, and control of change and configuration. The number of speaking engagements, publications of papers and other documents, partnerships, etc., is frequently counted when calculating measurements for R&D initiatives. Metrics are sometimes used to evaluate a trajectory's development, the outcomes of trials and tests, and accomplishments in achieving production-level capabilities. Simple metrics include the quantity of social media visits or followers, crowdfunding activity, and crowdsourcing success.

Ensuring measures deliver correct data and information is crucial when picking measurements. Many initiatives fail despite the analytics showing no serious problems. This may be due to imprecise data that causes inaccurate reporting, an unfavorable climate for reporting negative trends, or incorrect metrics being calculated. If individuals rigorously monitor the precise measures, watching the incorrect metrics can guarantee that the genuine issues and hazards are not brought up and handled and may have unintended repercussions. A measure that prioritizes completion over customer happiness is an excellent illustration of this. This could result in rushing a client through the procedure without thinking about how to handle any problems that arise. This is when a helpdesk rushes to "close" cases by meticulously arranging the definition but neglects to confirm that the proposed remedy addresses the customer's demands. Inaccurate metrics reporting can result in quality problems and the improper allocation of resources to an organization's strategic goals.

Key performance indicators (KPIs), also known as key success indicators, can be developed and tracked to ascertain the best metrics for a business. KPIs aid in defining the primary performance indicators for a certain project. They assist in reducing the amount of information that is potentially significant but not crucial for obtaining desired results. Attempting to measure too much or incorrectly leads to not assessing anything since the measures may not result in desirable behavior or wise decision-making. KPIs serve as a public record of the important metrics for the project's success. They serve as a project goal statement by outlining the how, where, when, and what aspects of the project will offer the most value and involve the most risk in the event of success. Indicators might be predicting or leading, retrospective or trailing, quantitative or qualitative, like efficiency or quality. They can be absolute, as when a minimum standard should be reached, or directional and represent a trend.

The first stage in establishing and implementing relevant KPIs is to talk with the stakeholders to define the crucial processes, requirements, and expected outcomes. The second step is determining the best performance metrics in a confined period. Analyzing measures that are too distant in the future is impossible. KPIs are defined as industry- and organization-specific metrics. Any KPIs selected might be as many or as few as necessary to offer a reliable foundation for analysis and decision-making. Following their selection and implementation, they should undergo regular evaluations to confirm their validity or that the data they collect is relevant and helpful for decision-making. KPIs that do not provide actionable data should be decommissioned, and new metrics should be created in collaboration with stakeholders. Instead of focusing on deadlines and milestones, R&D metrics should consider experiment and test outcomes and progress along the trajectory.

11.4 The Analysis of Variance and Trend

Once a project is underway, responsible management involves regular, established performance analysis against the baseline. At some point, every project component is being watched (Fig. 11.2).

When deciding on the performance measurement procedures and reporting schedule, it is important to consider the project scope, contractual requirements, complexity, risk, and cost. If scope, money, and time are not important considerations, even tiny informal initiatives need some performance evaluation. Variance and trend analysis produced from the precise cost, schedule, and scope data is typically included in performance assessments.

Variance analysis compares current performance to the initial plan. Periodically, the project's performance is compared to where it should be according to the baseline plan, and the results are reported as a positive or negative variance. Positive variances show good performance, while negative variances show bad performance.

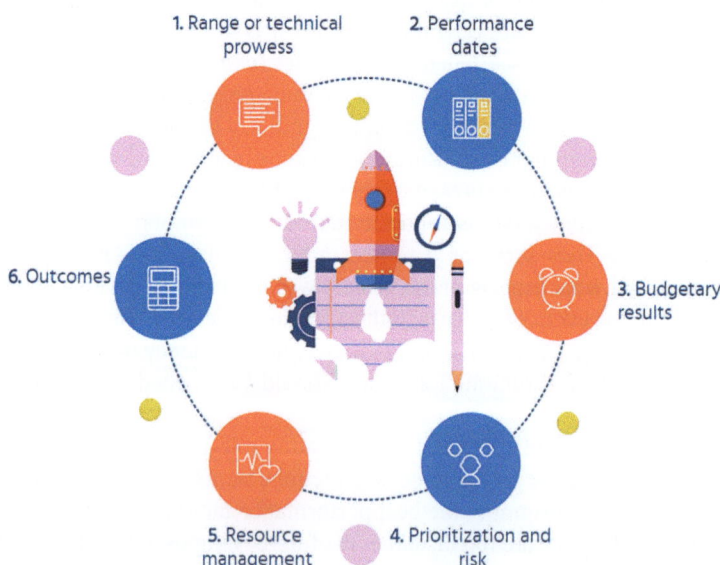

Fig. 11.2 Components of the project

11.4 The Analysis of Variance and Trend

To predict future performance expectations, trend analysis quantitatively assesses the project's historical pattern of actual performance. It is also helpful to determine how much the trend can shift while a project is still on schedule or how much it needs to shift to get it back on track. The choice of performance measurements might need to be modified as the project develops and new circumstances arise. The ultimate aim of performance monitoring and analysis is to enable efficient decision-making for the project's remaining phases and raise the likelihood of a successful conclusion. With this in mind, certain performance metrics are selected and applied. Due to their close ties, these project components should be considered concurrently.

11.4.1 Technical Performance

Organizations frequently prioritize managing costs and schedules while ignoring scope management. However, its scope should be evaluated first because the cost and schedule depend on the job scope. Technical progress may be disregarded at the R&D stage of a project when it is seen to be hard to qualify or measure, or it may be disregarded if there are just a few people with technical competence. Project managers should depend only on the technical leaders' assertions that progress is achieved without a technical performance evaluation. However, demonstrable incremental development toward the standards is required to assure the maximum possibility of project success. A project's cost and schedule will undoubtedly increase if the criteria are not satisfied when it is finished. As a result, it is critical to define and put in place technical progress measurements at the project's inception. Techniques of documenting scope success for less formal projects might be as straightforward as visual displays, charts, diagrams, metrics, etc. The technical plan and gradual progress toward that objective should be communicated explicitly for more formal or sophisticated projects. The technical progress may be assessed using actual technical advancement and forecasting probability toward reaching the expected performance, similar to how procedures are used to monitor schedule and cost.

By tracking the alignment of requirements to deliverables, a requirement traceability matrix may assist in guaranteeing that the technical components of the project remain connected with the rest of the project. A systems engineering management plan (SEMP) contains information that may be utilized to clearly define the characteristics and requirements that need management and monitoring when a project has a more substantial technical component [1]. This document comprises a collection of measurements produced to satisfy project criteria and usually depicts the completely integrated technical effort for the project. It may be used to assess the project's overall performance in attaining its intended goals. Additionally, it distinguishes clearly between customer-driven needs and the produced and suggested solutions.

The systems engineering master schedule (SEMS) is another component of the SEMP. The SEMS records the events that led to the needed outcomes, including technical events, demonstrations, reviews, decision points, and other significant verification activities. This schedule gives a broad overview of the systems engineering development process. The SEMS is event-driven, not date-driven, like the project-integrated master schedule (IMS); however, some events could have corresponding goal dates. It is not time-limited, in contrast to the IMS. To make it easier to track measurement accomplishments about the plan, each item in the SEMS is given a distinct number. Conversely, the IMS is organized around the deliverable good or service that results from hitting milestones. The project IMS and the SEMS must be consistent and aligned [2].

The project's customer-driven scope is the source of the technical specifications and metrics specified in an SEMP. How successfully the accomplished technical requirements are anticipated to fulfill the stakeholder's intent for usage in an operational setting is defined by measures of effectiveness (MOE). In other words, it queries if the stakeholders anticipated and envisaged behavior of the installed technology would occur. The key performance parameters (KPPs) and performance metrics are defined with the assistance of the MOEs. The company can employ KPPs to improve performance measures (MOP), which can subsequently be used to improve technical performance measures. If not addressed, KPPs are the functional requirements that would necessitate rebasing the project's technical portion. Similar to verification operations, technical performance measures (TPMs) verify that the process, product, or system complies with the criteria and specifications as defined. TPMs are the physical or functional characteristics of the technology. The crucial factors tracked during the TPM process are MOPs, which translate the MOEs into important TPMs.

TPMs give project managers control over crucial technical parameters and a mechanism to objectively and statistically evaluate performance. They identify technological dangers as well. Usually, TPMs are selected from the MOPs and MOEs. A thorough selection of important measurable characteristics is necessary to apply TPMs in technical management. The TPMs should contain any technical items that the risk registry has detected. Included should be anything that will be examined formally [3]. TPMs assess these things:

- Milestones that show how the specification performed after a technical review. The technical parameter's forecast value (or current estimate), anticipated to be fulfilled within the project baseline, comprises these escalating intended values
- Tolerance levels or parameters to offer high and low-performance characteristics and accommodate for estimation flaws
- Contractual performance benchmark, requirement, or criteria
- Current standing of the feat thus far
- The deviation from the present estimate in terms of performance

11.4.2 Schedule Performance

A timetable is an action-oriented strategy for meeting project obligations. It is determined by the project's scope and main deliverables. The project's major scope, requirements, and deliverables are often defined in documents like the contract and

11.4 The Analysis of Variance and Trend

statement of work (SOW). Identifying interim deliverables for project control and management objectives is also possible. Other management plans that outline the project's execution and control also contribute to developing the schedule. Program management plans (PMPs), integrated master plans (IMPs), SEMPs, make/buy plans, risk management plans, and resource management plans are a few examples of these [4]. The scope is broken down into doable and quantifiable activities that should be completed to produce the deliverables. An overarching strategy for carrying out the project is created by giving the jobs durations and rationally arranging them about one another. When determining the work durations and deadlines, resources and additional potential limitations should be taken into account. It is also possible to use historical information from previous projects with a comparable scope, size, length, and complexity as the schedule's source data. The schedule is a benchmark against which future performance is judged after being reviewed and approved by the task stakeholders and owners. The value of the activity indicated by each job on the schedule should be determined, as should the performance monitoring method to be applied. The status of each job is tracked as the project moves along, including the exact start and completion dates, amount of work performed, and percentage. The time needed to finish the remaining work on activities in progress is shown as an anticipated completion date. Numerous analyses are carried out using this essential data, and metrics are given to task management and owners.

Many performance-measuring measures are based on baseline dates. The baseline variance of a job or milestone is the discrepancy between the baseline date and the actual or anticipated date. Baseline hit/miss metrics compare the number of tasks completed on or before the baseline date (a hit) to the number of tasks completed after the baseline date (a miss). The baseline execution index (BEI) calculates the ratio of the total number of tasks performed during the measurement period to the total number of tasks baselined to be completed during that time.

Instead of comparing the anticipated dates to the baseline, some performance evaluations relate them to the deadlines for important milestones or project completion. A "forward pass" determines the earliest dates the residual tasks may be completed in a logically connected critical path method (CPM) scheduling network based on the tasks' recorded status, activity durations, stated constraints, and logical linkages. The phrase "critical path" refers to the jobs that comprise the project's longest path to completion. According to the existing plan, the critical path shows the earliest date the project is anticipated to be completed, along with the tasks that are contributing to that projection [5].

To guarantee that the overall project end date is fulfilled and to assess the amount of schedule risk developing as the project moves forward, it is crucial to compare the progress and deviations along the critical route. The project schedule's flexibility is referred to as float or slack. To ascertain if a slip in the schedule will transfer another activity into the line of the critical route, it is vital to comprehend the amount of float, both total and free float, in noncritical path jobs and evaluate performance in these areas. Activities that may be completed concurrently or overlap with other tasks can slip a bit in the schedule without affecting the project's overall schedule. Proper change control action is necessary if the project schedule runs beyond the official finish date. The project schedule must be continually reviewed and modified as a dynamic tool.

The schedule may be managed, and project activities prioritized effectively using CPM to assess total float and track changes to the vital route. Project managers can take remedial measures to keep their projects on schedule by anticipating any delays in advance. The project activities should be established with precise, logical linkages and proper task durations. A "reverse pass" is also calculated, which involves scheduling tasks backward from the project's conclusion or the deadline for an intermediate milestone and then figuring out the latest dates at which they may be completed while still meeting the deadline. The early dates are subtracted from the late dates to get the total float values. Total float is a measurement of schedule latitude that indicates how long work may be postponed without impacting the project's conclusion. A common management perspective is a diagram, plot, or visual representation of baseline job completions over time in comparison to actual and predicted completions over time. Another would be to show the predicted labor hour allocation for the planned work and the actual labor hours charged. Another option would be plotting the anticipated milestone completion dates against the milestones attained.

11.4.3 Budgetary Performance

Different estimation techniques are used to create budgets. When creating a budget, it is crucial to record the BOE (basis of estimate) so that a detailed history of how the expenses were calculated and the budget was created is accessible for future reference. When it is acceptable to have a fairly correct budget, analogous estimating may be used as a basis for estimation using historical data from previous projects with comparable complexity, duration, size, and scope. Bottom-up estimating is the most frequently utilized approach when creating a budget. The work breakdown structure (WBS) at the top level is then reached after this procedure has thoroughly reviewed each job package. The most popular approach also frequently leads to a greater budget since people automatically include contingencies at every stage of the calculation. If this kind or amount of estimating is required, it is occasionally possible to calculate a range of estimations, from the most optimistic to the most pessimistic. The expected budget plus a contingency, or a sum of money set aside to cover the project's known risks, will be included in the total project budget that will be baselined. The work package contingencies and budgets are contained in the individual WBS budgets, also known as control accounts, which are the level above the work package. Other approaches to obtaining more precise estimates include parametric estimating, which derives relationship information from previous data and currently understood parameters to provide an estimate.

As the project moves forward and the project's budget becomes known and baselined, the percentage of the budget spent will be recorded so that the project's progress status may be evaluated. Similar to schedule updates, budget updates can also

11.4 The Analysis of Variance and Trend

be made manually by the project manager, who gathers information on expenses from accounting and other systems that offer details on labor, material, and service bills. If a mechanism for directly tracking spending data is available, it can also be done electronically. No matter how the initial expenditure data are assembled, additional processes are needed for trend and variance analysis. The project manager should comprehend the significance of the variances and data to take appropriate action and successfully manage the project.

Several techniques may be used to evaluate trends and deviations once the initial project has been updated, often with the most recent cost performance data. The same status reports and work performance papers used to convey status, problems, and issues for schedule performance are also frequently utilized to describe cost performance. At the baseline meeting, how formally and quickly they will be generated and evaluated will be decided. Performance in the budgeting will be evaluated in the same way, independent of the project technique, as with schedule data; the only difference may be in the pace, regularity, and frequency. An essential metric is a visual representation, map, or diagram showing the budgetary difference between what should be spent and what has been spent. Another would be a comparison of the predicted cost allocation to the actual labor, material, and service expenses.

When evaluating budget performance, it is crucial to examine unexplained spending spikes that might result from incorrect labor charges or unforeseen costs for supplies or services. Spikes can sometimes happen when supplies are bought in bulk rather than incrementally throughout the project. In these cases, what would initially appear to be an anomaly or error can simply be validated as a legitimate charge. It is possible to determine whether the project budget is secure quickly. To ensure that the total project budget is met and not exceeded, and to evaluate and address the amount of budget risk that develops for the project, comparing the budget progress and variances is crucial.

A certain portion of the money is reserved for management reserve and contingency in well-run projects. Project risk is taken into account using contingency. Based on a preliminary project risk evaluation, the project manager chooses the amount to be set aside. These contingency funds are available for use in mitigating high-priority risks that would significantly negatively affect a project or have a significant possibility of occurring. Based on an evaluation of the degree of buffer required by the project management and organization, a managing reserve, or a portion of the total amount of cash put aside from the project budget, can be given. Throughout the project, it is closely controlled. The depletion rate may be quickly determined by comparing the proportion of projects completed to the amount of leadership reserve utilized thus far. This will give the chance to replenish the reserve, if necessary.

Although not frequently used on R&D projects, EVM can be utilized to control cost information when a project manager wants a formal approach to budget expenditure and allocation. When EV is utilized for scheduling, budgeting will follow. Utilizing EV offers a comprehensive view of the cost and scheduling aspects. To get a consistent understanding of the three main components of project management—scope, time, and cost performance—project managers will employ EV in conjunction with TPMs.

Budgets should be allocated to control account managers (CAMs) in a regulated manner and directly tied to the specified scope of work to comply with EVM. Distribution of the budget and the work authorization document for every control account (CA) inside a WBS. Before a WAD is in place, work cannot begin, and charge numbers cannot be made accessible. According to the project schedule baseline, the budget is time-phased, and each CA's period of performance (POP) is established from this. Each CA's work scope may be further divided into several short-term work packages (WPs) and long-term planning packages. A work package, planning package, and CA can only be a part of one WBS or CA at a time. The project management foundation is the total of all permitted expenditures in the CAs.

The degrees of control accounts used by various organizations vary; some collect charges at the top levels, while others drive down to several lower-level, chargeable accounts. The choice of the suitable level is often made by the organization's leadership and the project manager.

The cost variation (CV) is the discrepancy between the actual cost incurred (AC) and the value of work completed (EV) at any given time. The CV is the difference between the AC and the BAC at the project's conclusion. This useful measure is defined by the formula $CV = EV - AC$. When determining expected real project timelines and costs, converting the CV and SV values to efficiency indicators is important. Additionally, it offers a mechanism to quantitatively compare cost and schedule differences across different projects in a program. Each project's cost performance index (CPI) is determined by dividing EV by AC. CPI is calculated as EV/AC. Effective cost performance is indicated by a number greater than 1.0. In other words, more than an hour's worth of activity was completed for every hour spent on it. A score less than 1.0 similarly implies that more time was needed to complete an hour's worth of work.

11.4.4 Risk Management

Whether acknowledged or not, the risk is there in every endeavor. The first step in addressing these risks is to recognize and quantify them. Risk evaluation considers probabilities of occurrence in addition to qualitative and quantitative indicators. Quantitative metrics quantify the risk's size, whereas qualitative measures highlight the risk causes and dangers. These actions may generate information for cost and schedule risk evaluation, which aid the project manager in managing project risks effectively.

Technical risk typically shows itself as a failure to fulfill specifications. The sooner a project is in its life cycle, the greater the technical risk. High levels of risk are often acceptable for R&D initiatives, but managing and recording the risk enables reflection and superior decision-making about whether to refocus or end an effort. Assessing the degree of risk and possible risk mitigation measures is useful in risk management. An assessment of the likelihood and effect of that event is made, and a risk management strategy is put in place.

11.4 The Analysis of Variance and Trend

11.4.5 Resources Allocation

The project manager is committed to the success of the project. Correct project resource allocation is a crucial component of success. The distribution of increased resources is more complex, however. People frequently work on many projects concurrently, which adds complexity and necessitates management and tracking to assure success. Weak critical thinking abilities are common. It is crucial to plan the timing and distribution of work, cope with competing deadlines and overbooking of essential talents, and value the spare time to apply for new employment. These tasks need a thorough grasp of resource allocation data, which is sometimes challenging without putting in much effort to gather the information from systems, specific supervisors, or individuals.

This can be difficult for a company if the resource data is not electronically accessible. It is necessary to manually gather information across several divisions, displaying labor charge information by project and name. A program-level rollup of this data enables the identification of staff oversubscription and subsequent resolution. Project managers can communicate with one another on a more basic level to obtain the employee abilities required to make the project successful. Conflicts with work schedules can be found in various ways, but they frequently require discussion to be resolved. The most difficult problem arises when tracking labor charges. Once the team has been assembled, the project has started, and it is time to begin. It is typical to find that allotted time is not being charged in situations where projects compete with personnel who divide their work effort between operations and R&D projects because people frequently tend toward fulfilling operational chores, which are tangible and easier to judge performance against. To guarantee that the staff is moving forward with the plan, checking charges against the R&D projects is vital. This is where precise billing to the appropriate project codes is crucial, especially when using earned value, which necessitates a precise assessment of the progress that has been accomplished on projects thus far.

11.4.6 Outcomes

The project manager's goal is to guarantee that the project's results are achieved and that they satisfy or even surpass the demands of the stakeholders. The project outputs should be easily traceable via the criteria to monitor performance trends and variations. If they are perfectly matched, the project schedule's flawless execution and disciplined change management will lead to accomplishing the outcomes. Stakeholder satisfaction will be ensured both during and after the project via regular communication.

The work status may then be monitored using the techniques outlined in Schedule Performance and at the program level with the assistance of stoplight charts, which list all projects and display a red/yellow/green indicator of the progress to scope, schedule, and cost parameters. Combining the stoplight chart with a foursquare summary graphic is occasionally done. For simple reading and debate, the four-square chart presents each risk's baseline and current status, scope, schedule, and budget on a single page. These graphs can be manually constructed or electronically created by project management software. The organizational leadership, the client, and the project manager have different priorities, affecting how quickly compilation proceeds.

11.5 Various Project Kinds' Measurements

Measurements are necessary for any project, regardless of where within the life cycle it is. The measurements for the subsequent projects are listed below for review:

- Applied and basic research
- Development
- Innovation
- Production

This section outlines a technique for identifying the most relevant measurements at different points in the project's life cycle. The contrast between innovation, development, and applied and basic research is examined before being compared to production metrics. The choice of measures will also depend on the project's life cycle location and other elements, including complexity, money, and formality.

11.5.1 Measures for Applied and Basic Research

The best metrics for these kinds of projects consist of a limited number of focused reviews that are set up when the study has reached a crucial turning point, such as a manuscript that is prepared for publishing, a speech that will be made at a symposium, or a significant result that has to be validated. Measures can also include prototypes, simulations, models, demonstrations, publications, presentations, and articles. More than outputs and outcomes, progress is what is being measured.

The systematic and well-coordinated process of implementing changes is called change management. The present configuration is always understood thanks to configuration management techniques that make sure design changes are monitored and implemented carefully. Version management for documentation is the main emphasis of change management in fundamental research, which is purely theoretical. Tracking configuration changes is particularly required as modeling, simulation, and product prototyping are added after the study transitions into applied research.

A research notebook is frequently kept when conducting R&D. It is simple to follow what is developed and in what sequence because this material is presented sequentially. However, maintaining version control and order might become

11.5 Various Project Kinds' Measurements

increasingly challenging as documentation shifts to an electronic format. Document version management during the research phase is easily performed by assigning special identification numbers to the work products.

The meticulous documenting of what was changed and how it was modified is required for configuration control on components, systems, parts, and other products being produced. The current configuration will always be known whenever a modification is integrated, thanks to version control, which may take the form of special numbering or another type of distinctive identification. This is crucial in R&D since it allows for modifying experiments and testing results by small adjustments.

The client's and other stakeholders' preferences will determine which variance and trend reporting are most appropriate for the project. The specifications for measurements in this field are frequently stricter if an external funding agency is involved than if the project is domestically sponsored. Technical metrics, internal schedules, and budgets often start with maintaining a baseline and offering evaluations after certain intervals, such as every quarter. However, an external financing client could need a formal budget with precise, ongoing quantitative assessments that show effective change and time management.

It cannot be easy to manage a full project timeline in applied and basic research, particularly if a novel technology is being created. A collection of subordinate schedules can be systematically constructed in three-month increments or less, depending on the pace of the experiments, if the project scope and trajectory are properly documented and a defined set of anticipated outcomes, experiments, and tests have been established. The subordinate schedule may end at the experiment or test and not proceed further, with a branch or decision point at that moment. As there is typically no way to predict what the next phase will look like until the test or experiment is complete, enabling the experiment or test set to be the terminator of the project timeline up to that point with a go-no-go decision adds a sense of reality to the forecasting process. If the customer is uncomfortable with the uncertainty, using this strategy on an internal project is simpler than on an external one.

A risk assessment should be included in the schedule, regardless of whether it is created gradually to coincide with the experiments and tests or is finished and tracked throughout the project. The risk evaluation and mitigation plan for fundamental and applied research is rather challenging because the mitigating is frequently merely to stop and proceed in a different route. A different course of action would be chosen if the experiment does not yield the desired results. As a result, assessing risk fundamentally differs from how it would be on a project, including manufacturing.

Resource evaluation is crucial for both fundamental and applied research. These initiatives often have a key staff person who manages the project. The quantity of time the project manager devotes to it and the degree to which it distracts him or her from other activities frequently determine whether the project succeeds or fails. The time allotted vs distraction for the other essential staff members, who possess the crucial abilities required to conduct the experiments and tests, will also influence the likelihood of project success. Finally, if there is a need for such resources at that time, it may be challenging to ensure that any supplies, equipment, or services are available. Without knowing and being able to estimate the availability of the two essential staff members needed for the project, it becomes more difficult to meet the planned goals, even if the project has been allocated two FTEs (full-time equivalents), a budget of $50,000, and a year to create a prototype of new technology. The solution to this issue is simple. These factors can be taken into account by creating a baseline, resource-loaded timetable and then gauging performance against it.

Validation, verification, and testing are crucial for applied and basic research. Maintaining strict control over the procedures and outcomes cannot be emphasized whether a prototype is being built or a theory is being tested. It is critical to explain how one concluded, have the process repeatable, and provide the same outcomes. Generally speaking, regulated actual or simulated exercises with the capacity to communicate the procedure and outcomes properly suffice.

The need for an impartial examination of such procedures and outcomes depends on the client. For instance, an independent evaluation is required if the fundamental study results in a dissertation or peer-reviewed publication. If the project is internal, appraisals for fundamental and applied research will likely be informal, casual, and frequent as experiments and tests are finished, and results are needed to guide subsequent actions. Reviews for external projects are frequently less frequent but more formal and frequently involve presentations of the methods and findings.

The project manager is accountable for the quality of applied and basic research initiatives. Although it should be woven into the project's design, it is frequently a minor factor. The degree of quality can end there as long as the procedures and results are repeatable. This approach differs from those used in manufacturing or building projects.

Controlled progression may be a helpful assessment technique for applied and basic research initiatives. Once more, the client will determine the degree of approval required through an independent assessment and provide clearance to continue from one step to the next. Internal fundamental and applied research projects often rely on the subject matter expert and project manager to determine whether the project has reached a decision gate or a significant branch point and how to move forward.

Figure 11.3 illustrates a systematic R&D project management process. It begins with identifying the "Best Metrics" for such projects, emphasizing the importance of measuring progress rather than just outcomes. "Change Management" follows, outlining the need for well-coordinated implementation of changes. "Research Notebook" highlights the importance of documenting developments and maintaining version control. "Configuration Control" is crucial for tracking changes in components and products. "Measurement Specifications" stress the need for specific measurement criteria. "Project Timelines" explain how to structure project schedules, including risk assessments. "Resource Evaluation" addresses allocating staff and resources, and "Validation and Testing" focuses on maintaining control over procedures and outcomes. "Project Manager" takes responsibility for quality, while "Controlled Progression" allows for decision gates and branch points. The flowchart provides a comprehensive overview of the key elements in managing R&D projects, emphasizing systematic and well-documented processes, measurement, and quality control.

11.5 Various Project Kinds' Measurements

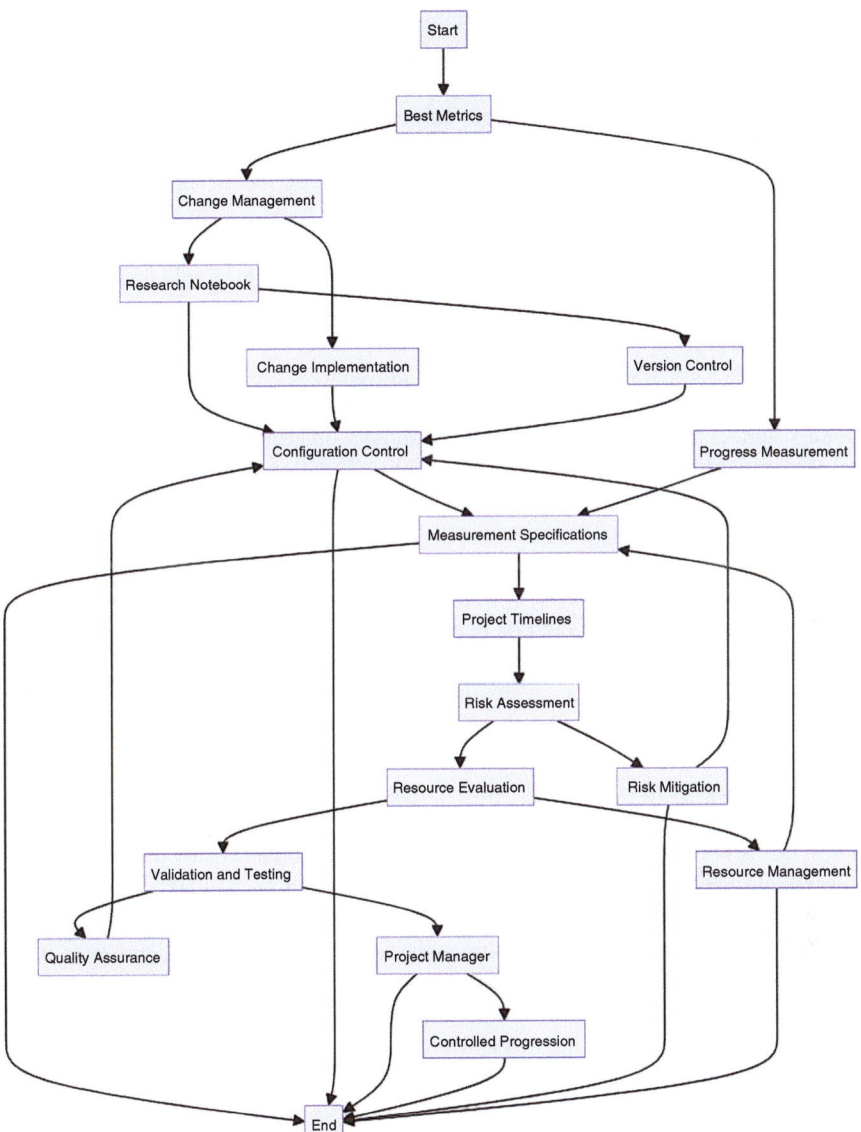

Fig. 11.3 Systematic management of R&D projects: a visual guide

11.5.2 Development Measures

Since more resources are allocated to the project, more discipline in implementing measures becomes crucial. The management of metrics, as well as modification and configuration control, presents the initial difficulty. Again, several decision-makers, like project management, client, or other stakeholders, should agree on how much additional discipline is needed. It becomes increasingly crucial to have more specific measurement needs and frequent evaluations once a project enters the development phase. Using a formal change control process using a CCB (change control board) is becoming more widespread, and configuration control of the developing process or product design is also becoming more significant. Controlling the design's constituent pieces is crucial since they develop independently. The configuration of each component section of the design would be locked down to whatever setting it is when the design has sufficiently matured and is prepared for implementation or production. Any modifications made after that point should undergo a thorough review since they will significantly impact the overall process, product, and individual components. The design configuration is crucial regardless of whether the project is for a new process, service, or product.

Although there is still an obvious requirement to align with the testing and experiments and continue using shorter subordinate timelines, timetables and result assessments often demand additional formality. Development efforts often ask for results that unequivocally show the value provided, necessitating semiform or formal discipline. In contrast to research activities, despite how closely they are related to them, development is concentrated on improving the current state of anything by modifying or creating a service, system, process, or product. To add additional discipline to the timetable, it could be necessary to do a qualitative evaluation of how the plan is performing. This goal may be achieved by looking back on each individual's performance thus far, evaluating the percentage completed, and comparing it to the anticipated percent completed.

When results are presented to a larger audience at regularly scheduled meetings with the management or customer, such as quarterly reviews, the structure of reviews often becomes more formal. In reality, evaluations can range from casual to official. Thus, this is not a set norm. Once more, this is heavily influenced by the project manager's and stakeholders' preferences. In a development project, quality metrics could still go largely unchecked. However, the project manager is expected to keep working to ensure that the processes and results can be replicated and verified. However, if the customer or management wants root cause analysis to explain the outcomes of tests and trials, the necessity for quality control procedures may rise. This analysis may be especially necessary if specific results were anticipated but not realized.

The importance of having a capable project manager in place increases as the project moves into the development phase. Because development typically has an ultimate aim, objective, or conclusion of becoming an established method, product, or service, these abilities are crucial throughout this phase. Therefore, it is important to recognize when "good enough" is sufficient. It is common for scientists and researchers to desire to keep pushing the boundaries of what is possible. This is preferred in both fundamental and applied research fields. The end state can only be reached with forwarding momentum and attention after a project has entered the development phase. This is accomplished by precisely describing the desired specification parameters and the tolerance range surrounding those values. For the project to succeed, it is crucial to ensure that the project manager has expertise in making these difficult choices.

11.5.3 Innovation Measures

In contrast to how R&D is traditionally associated with applied and basic research and product development, innovation encompasses various activities. Innovation involves interrelated actions that cut across many different process domains. The majority of innovation endeavors are inherently complicated. The links between processes inside and within organizations give complexity its name. Every process owner that the new capability will affect will be a stakeholder. To effectively manage all the interdependencies throughout R&D, communication becomes crucial. It may take more work to get the participation needed to adopt the new invention, and extensive two-way communication is needed. Innovation initiatives should be managed in a way that minimizes bad effects and maximizes good effects since they are typically more complicated and contain more significant quantities of outputs and inputs, interfaces, and risks.

With a properly defined compliance matrix that outlines what every stakeholder will get from the project and independently confirms that the capability is as expected, the innovation project may succeed in test, verification, and validation (TV&V). A compliance matrix may be created by utilizing a spreadsheet to list all the requirements along the rows, the details of the testing that will be run down the columns, and then verifying the performance within the associated boxes. It is simple for a project manager to make the mistake of claiming that ability has been provided before confirming that it meets the customer's or stakeholder's expectations. To make sure that all stakeholders are on board with the delivery being in line with the intended objectives, it is crucial to implement a TV&V process for innovation initiatives.

Additionally, the relevance of the quality processes rises. Diverse stakeholders have various viewpoints on quality. Additionally, if a thorough examination of the interdependencies is not methodically carried out before each modification while conducting a big innovation project, it is simpler to damage an existing capacity. The measuring procedure includes checks and balances to ensure nothing was overlooked before and after the modification. Controlled progression ensures that all implementation components, both in the process and final deliverables, are prepared to move forward.

11.5.4 Measures for Production

At this stage of its development, a project needs to have clear objectives. The setup needs to be finished. Any remaining issues or issues raised by the client or other stakeholders ought to have been addressed by TV&V. Risks need to have been eliminated. The procedure or item is prepared for manufacture or construction, indicating that the design is reliable and repeatable and has undergone validation.

The moment has come to organize official, ongoing evaluations. They will guarantee that the manufacturing or construction operations proceed as scheduled and remain on track. Control over configuration is essential. Any modifications to a system, component, or part should be thoroughly reviewed and managed to guarantee no negative influence on the anticipated design. Depending on the customer's and management's preferences, schedule, budget, and technical evaluations may be more formal or less formal. Technical performance measurements, earned schedules (ES), and earned value management (EVM) are examples of more formal implementation. Quantitative assessments that incorporate stoplight charts and four-square diagrams are quite valuable in situations like this.

Risk management is crucial at this project stage because contingency draws must be carefully controlled to prevent the total budget expenditures (budget plus management reserve plus contingency) from exceeding the allotted budget. Contingency is particularly important to meet project risks. Otherwise, management reserves are kept as a safety net against unforeseen events that arise for the project.

Although it is crucial to keep an eye on things, this stage of the project often calls for expertise sets that are simpler to come by. This is because personnel levels may influence the timetable. Usually, the design has been translated into replicable pieces. Thus, the crucial skills are less heavily involved. As a result, the criteria for staff management shift from guaranteeing that essential talent is accessible and that the task is completed as planned to manage a workforce around common difficulties, such as vacation time, sick leave, etc.

TV&V now has a significant new function. It is necessary to validate that the procedure or product produced by the project complies with the specification or falls within the tolerance range. Confirming that the finished product lives up to client expectations is crucial. Typically, TV&V assessments take on a more official tone, with presentations on a performance given throughout the regular reporting cycles. The customer's and management's preferences would determine how measures in these areas, such as a verification review and compliance matrix, independent panel approval that satisfies requirements, or other reviews of project management performance, are used.

In these projects, both quality measurements and controls are used more frequently. Strategies to improve production or construction processes, such as Lean, TQM (total quality management), etc., can be included in metrics. They frequently incorporate audits or other reviews to evaluate the quality of the service, process, or product. These evaluations result in action plans to address any problem areas. Quality controls become crucial to guarantee the project's outputs are high enough caliber and examine and explain the circumstances behind instances when things did not go according to plan. The controls are essential for identifying the underlying causes of issues and, consequently, for addressing them to prevent them from occurring again.

Having a competent manager in a position is more crucial for a project at this stage than having a skilled leader. This is an oversimplification since certain difficult initiatives need both skill sets to maximize the likelihood of success. The project manager should be able to oversee the entire process from beginning to end and make sure that the project's objectives are met. A good manager is needed to make the hard decisions that frequently arise in the last 5% of a project, such as firing or reallocating project workers and deciding how to allocate the last money. Completing a project is frequently more difficult than planned.

11.6 R&D Performance Measurement

R&D performance measurement is a fundamental practice within organizations, offering a systematic approach to evaluating the efficiency and effectiveness of research and development activities. This process is a compass guiding decision-makers in pursuing innovation and growth. R&D performance measurement encompasses various Key Performance Indicators (KPIs) tailored to an organization's specific goals and objectives, such as patent filings, project timelines, innovation costs, and product launch success rates. Striking a balance between short-term and long-term objectives is a central consideration in R&D performance measurement. While meeting immediate project deadlines and budgets is essential, it is equally important to assess the long-term impact of R&D efforts on an organization's competitiveness and market positioning. This necessitates focusing on the quality, not just the quantity, of innovations produced. Continuous improvement is also a core principle, as R&D performance measurement is an ongoing process that must adapt to changing market dynamics and evolving business strategies. Benchmarking against industry standards and peers is a valuable tool for evaluating competitiveness and setting realistic performance targets. Accurate data collection, analysis, and transparency are essential components of this practice, providing organizations with the insights needed to make informed decisions and enhance their R&D processes. Ultimately, R&D performance measurement should align with an organization's strategic goals, fostering a culture of accountability and collaboration among teams engaged in research and development. This comprehensive approach to performance measurement ensures that R&D efforts contribute to the organization's long-term success and innovation objectives.

11.6.1 Documentation

As a necessary consequence of R&D, process and outcome documentation are essential. Those fundamental components are necessary to accurately and repeatedly convey how it was finished, what was done, and the consequences. Additionally, most funding organizations demand a certain amount of reporting on how funds are used and comparisons between actual progress and what was projected or intended. Measurements of performance and results are challenging, if not impossible, without documentation. On an R&D project, the documentation process does not have to be time-consuming or significantly dependent on technology safeguards; it may instead be process-driven.

R&D activities can be recorded in a project book, shared board, research book, or another common document the team shares. The vision, strategy, trajectory, KPIs, in-depth notes describing trials and tests, outcomes, and following choices should all be included in one shared resource. There should be sufficient information to recreate the precise trials and tests and get the same findings at that time.

The choices related to the strategy for approaching the concept, issue, or hypothesis are the information that should be included when documenting the tests and experiments. It should include a list of the theories and methods that will be tested along the trajectory and confirm that there is a reasonable likelihood that the test or experiment will advance the trajectory by advancing comprehension of the theory or notion. Because continuing to pursue activities that do not promote learning along the trajectory would only save time and resources and influence the overall timetable, the experiment or test should be reformulated if there is no realistic probability.

11.6.2 Schedule and Budget

The project budget should account for the labor, ME&S, charges, and overheads related to the project. At the end of every financial reporting period, a brief analysis of actual billing will reveal expenditure trends and irregularities. The best tracking takes into account both commitments and actual booked charges. Keeping a running journal each time a promise is made or an invoice is filed is an easy method to do this. These will ultimately be deducted from the budget, but in the meantime, if they are monitored, the budget might be overcommitted. Even if estimations are available, there usually are no budgetary problems because the discrepancy between projections and actuals is only a few percentage points. However, failing to keep track of budget commitments might result in surges in real costs that the project manager might not have anticipated. A fast check can determine which charges have been processed and which are still outstanding after each financial reporting period.

The ideal project timeline, as previously noted, is a rolling wave of 3 months or fewer, and that is contained inside a larger, predetermined timetable period. It should include the trajectories, experiments, known tests that support those trajectories, and a clear start and finish date for the project. Although it does not establish milestones or a crucial path, it does indicate goal completion dates for the research, testing, and experiments. It can also employ TPMs, EVs, and SV (schedule variance) to measure performance, which will aid in decision-making at phase gates. It lists the personnel and ME&S resources required to execute such tasks. The project manager will be able to confirm that the employees assigned to the project are billing to it by checking which workers are charging against the account at the end of financial reporting periods.

11.6.3 Outcomes

As it focuses on the execution of activities that eventually result in the deliverables of the project, managing R&D advancement toward outcomes is crucial. Deliverables are a part of every R&D effort. They might be papers, prototypes, learning progress, etc. Design, research, development, experimentation, testing, verification, validation, and implementation are all steps that R&D initiatives in all fields take. The goal of outcome assessment is to understand how much progress is achieved at any given time.

The use of visibility to support decision-making at branch points is the most crucial component of putting this strategy into practice for R&D. To make choices about "go/no-go," "change course," or "stop," tests and experiments should be carefully specified. In initiatives where one is trying to see how far one can go, as well as experiments and tests that add definition to the realm of the possible within the context of a given set of criteria, clarity improves decision-making.

This is simple for projects requiring technical knowledge; at branch points, TPMs, KPPs, MOPs, and MOEs aid decision-making by explaining. TPMs can offer information on thresholds, tolerance levels, expected values, present estimates, and parameters for intended performance. It can be simple to decide to go back, make revisions, and retest using the information learned if the output is expected but has yet to be achieved.

The framework of MOEs may still be used as a guide to evaluating performance for non-technical projects like service, process, or basic research projects. A smart technique to gauge non-technical progress toward outcomes is to utilize the structure of MOEs to specify how effectively an implemented result would match the stakeholder's goal in the functional environment and then test along those parameters. This is true even though they are normally used to evaluate technical effectiveness.

For demonstrational reasons, the same documentary about eradicating world hunger was screened. The elimination of world hunger has the same strategic objective. The examples were divided into process-focused and technical-focused paths to illustrate the differences in applying TPMs, MOPs, KPPs, and MOEs. The first course of action is creating a system for distributing available food to needy places. The KPI is to confirm that food is being distributed to places with a need. The MOE checks to see that those in need of food are routinely fed. The measurements track those activities' progress in very specific ways.

A technological illustration of Trajectory #2 is creating a technology that employs a learning algorithm to recognize and indicate food overstocks. The KPI confirms that food overstocks have been reported. The MOE checks to see if surplus food is being provided. The MOP and KPP confirm that the algorithm successfully recognizes and reports only food overstocks. The KPP validates that the learning algorithm is acting appropriately. The TPM establishes the specification target. The measurements track particular actions.

11.6.4 Configuration and Change Management

For the team and the project manager to constantly be aware of where they are, what they have attempted, and in what order, using a systematic strategy to handle change on an R&D project is crucial. They should also be able to convey that at any moment to stakeholders. The data should be available whenever it is required. Change management should be a regular element of the project's daily operations for that level of situational knowledge on an R&D project.

A group project book is among the most successful organizational tools. Either a real book or an electronic file may be used. It is important to consider and comprehend any change's effects. All parties involved should be aware of the effects of the modifications and accept them. A user should always be able to access the project book location, understand the present configuration, and understand how the evolution occurred. There needs to be sufficient information to allow someone to repeat the experiment and get the same outcomes.

The client or other stakeholders may want more significant change control, including more formalized usage of TPMs, EVMs, etc., depending on the complexity and scale of the R&D project. However, in most R&D projects, the discipline in the modification process is done with the help of a brief review with a few important people.

Only when resources outside the project's boundaries are needed or when an experiment or test has forced a considerable rethinking of the project does it become a more serious exercise. Before execution, a more formal change review procedure should be used for substantial changes to a project plan.

11.6.5 Risk Management

Risk management for projects involving R&D and innovation differs from risk management for projects farther along in the life cycle. R&D risks often focus on implications from connected initiatives and resource risk (ME&S and FTE). Risk assessments can also be undertaken on the list of tests and experiments that will be carried out for the project. Suppose the experiment or test outcome differs significantly from projected, anticipated, expected, or theorized. In that case, R&D projects must evaluate within each trajectory whichever unknowns they seek to address have the largest influence on the overall project. Although it is frequently hard to estimate the likelihood of anything happening, it is crucial to arrange the experiments and testing to prioritize, starting with the ones with the most potential for effect.

The risk assessment should be completed within the same one- to three-month time limit as the project timetable. The experiments and tests conducted throughout that time range are subject to risk assessments. The detected and evaluated risk elements can be recorded in a straightforward spreadsheet. The spreadsheet may be kept current by periodically examining it and making the necessary corrections, such as when the subsequent incremental schedule is being created.

Using a visual matrix is another common systems engineering tool to guarantee that more complicated R&D initiatives and integrated innovation projects have properly addressed anticipated or actual implications. An N2 chart is a framework for identifying and tabulating physical and functional interfaces in technical projects with many interactions. The idea may be used to analyze various linkages, including intricate workflows, process or product road maps, etc. The method is equally effective when analyzing process flow diagrams to identify which departments actively engage in process activities and which do not. In defining stakeholder responsibilities for the communications plan, this activity also aids.

The project team would choose the physical or functional interfaces and position them diagonally on the N2 chart for R&D and innovation initiatives. The squares are then filled with inputs and outputs for the interfaces, leaving empty spaces without interaction. The chart will be complete once all functions have been analyzed and compared, often done by moving from function 1 to function 2 clockwise. At that point, a full set of interdependencies should be clear.

11.6.6 Reviews

Reviews for R&D projects should be as casual as is necessary for the project. They should offer discussion platforms that allow a good deal of time for question-and-answer sessions. Attendance from all interested parties is encouraged, especially those with related professions. An interdisciplinary review is typically the best option for fostering stimulating discussion on experiments and findings, particularly if subject matter specialists from various fields with overlapping criteria can be engaged.

The talks should, wherever possible, be unstructured. This report aims to give readers an update on the experiments and tests conducted, along with a summary of the findings and implications for further research. Less effort should be spent on programming information, such as budgets and schedules. The purpose of the review is to provide interested parties and individuals the chance to examine the results and methodology and to assist in exploring what they imply and the potential outcomes. Strong collegial communication should be used to spread awareness of the R&D project's activities throughout the company. There should be no agenda, and no outcomes should be documented. Project managers should schedule these evaluations after each scheduled period, whether monthly or quarterly. A more official evaluation would be conducted for gate reviews that required paperwork and recording.

Phase-gate and TRL reviews should be utilized as a more rigorous method for bigger, more complicated R&D projects. A group of impartial subject-matter experts is selected. Typically, senior management is in charge of this. During the review, notes are taken, and an agenda is established. The project manager ensures that the material is presented coherently and does not obstruct the panel's ability to evaluate and analyze it. The experiments and tests should be described, their dangers should be considered, and the outcomes should be known.

11.6.7 Quality

The amount of labor put into research, testing, and experimenting (R&D) projects determines their quality. An ongoing project that does not make any substantial progress wastes important resources. The project manager can delay critical decisions based on results that deviate from expectations. The longer time goes by, the less likely a project will continue to get funding. Assessing quality in these ongoing, multiyear programs that are not subject to independent evaluation is challenging.

Certain easy procedures may be used to guarantee the quality of R&D projects. Reviews, as was already noted, play a significant role in both keeping informed about the status and gaining opinions on the veracity of result claims. Branch points appear as testing and experiments are finished. These are the instances when the outcome of an experiment or test is drastically different from what was expected, raising concerns about the future's course. The choice should be made when a project reaches a fork, and it is obvious that continuing is no longer necessary. When this occurs, it is frequently a strategic shift for the company and can even call for abandoning a substantial past investment. To make those challenging choices, the project manager needs the support of management.

The project manager should also know when to justify continuing a project activity, especially if testing and trials indicate a lack of progress. Continued project development is often permissible as long as the project manager can demonstrate that the standard of learning via testing and experimentation is still consistent with the trajectory. It is important to show that forecasts are becoming more accurate, results are still of high quality despite not being what was anticipated, and risks are decreasing.

Bias can occasionally hurt the effectiveness of R&D efforts. Efforts should be made to guarantee that the normal forms of prejudice are eradicated and verified through debates and reviews. Even when all project processes are rigorous, bias can still occur due to overconfidence that the project is going much more smoothly than expected or is further advanced. Bias entails concluding without additional evidence based on a single, instantaneous result or utilizing familiar or straightforward justifications.

The project book should be a live record for a project to be considered high quality. The procedure for reviewing and controlling every step of the method should be established and followed. The development of R&D initiatives should be regularly reevaluated. The pace of learning determines how quickly project activities and documentation should be evaluated and amended.

11.6.8 Leadership

For R&D initiatives, leadership is essential. The project's advocate, the lead investigator, has a vested stake in the project's development and conclusion. He or she will contribute to the project's development and is frequently in charge of producing the scientific, technological, or other results necessary for the project's success. R&D project managers need to be capable leaders as well. They will be responsible for the results of their endeavor. If they are fortunate, they can select their team to include the critical competencies and personalities to guarantee the project's success. If they are less fortunate, a team is provided for them. Then, they should be able to recognize and utilize the abilities already in place, as well as be successful in managing a team with less-than-ideal skill sets for the task at hand.

The progress, fulfilling of the project's goals, and producing the results are the responsibilities of the R&D project managers. The difficult decisions that need to be taken will be their responsibility, and they will be held accountable. An organization should select a good choice for the R&D project manager and then be prepared to support the challenging decisions that must be made during the project. A skilled project manager can support the project at many stages of its life cycle.

11.6.9 Communications

It is not tough, yet communication is crucial while working on an R&D project. Building a communications matrix is a straightforward task that should be done. A spreadsheet may compile a list of all the project's stakeholders. The project manager categorizes the stakeholder's apparent degree of support as either positive, neutral, or negative. The interdependency with the stakeholder's area of responsibility should be recorded, and after that, the stakeholder should be questioned about the preferred form of communication and the frequency of updates. Even so, if the project manager's choice does not fit the stakeholder's desire, a dialog should occur, and they should agree; otherwise, the stakeholder's preference should prevail, provided the stakeholder genuinely has no meaningful stake in the project.

The project manager is in charge of organizing and holding the relevant reviews at the proper times during the project and developing the project's fundamental communication strategy. These should be held frequently rather than only on noteworthy occasions. These communication activities are necessary to create expectations for what the project would accomplish, clarify what it is not accomplishing, and gain the support of interested parties and other people who can sustain the initiative over time. It is important to recognize the value of effective communication.

References

1. A. Kossiakoff et al., *Systems Engineering Principles and Practice*, vol 83 (John Wiley & Sons, Hoboken, 2011)
2. E.P. Arnold, 9.1. 1 Systems engineering and project management intersects and confusion, in *INCOSE International Symposium*, (Wiley Online Library, 2012)
3. P. Ruiz-Minguela et al., Review of systems engineering (SE) methods and their application to wave energy technology development. J. Mar. Sci. Eng. **8**(10), 823 (2020)
4. K. Johnson, *Tailoring Systems Engineering Processes for Rapid Space Acquisitions* (Naval Postgraduate School, Monterey, 2010)
5. G.T. Edwards, *Project Management Fundamentals: A Practical Overview of the PMBOK* (Blue Crystal Press, Atlanta, 2013)

Chapter 12
Risk Management in R&D Projects

In the dynamic realm of research and development (R&D), risk management is a critical component that directly influences the success and innovation of projects. This chapter delves into the multifaceted world of risk management, with a specific focus on R&D projects. The first section, "Risk Management," provides a foundational understanding of risk, its various types, and the principles of risk management. This sets the stage for a comprehensive exploration of R&D-specific risk management strategies. Section 12.2, "R&D Risk Management," elucidates the unique challenges faced by R&D projects, such as the inherent uncertainty, long project timelines, and the ever-evolving technological landscape. It then presents a structured approach to identifying, assessing, and mitigating these risks, helping project teams navigate the intricacies of R&D endeavors effectively. Finally, "Reward Strategy and Risk" highlights the integral relationship between risk and reward in the context of R&D. It delves into the strategic aspect of risk-taking, where calculated risk is often essential for breakthrough innovations. This chapter outlines how organizations can align their reward strategies with risk management approaches to create a conducive environment for R&D success. This chapter equips R&D practitioners and decision-makers with valuable insights and practical tools to manage risk better, foster innovation, and maximize the potential rewards in the challenging research and development projects landscape.

> **Learning Objectives**
>
> - Describe typical risks in an R&D project.
> - Recognize essential risk management components in an R&D project.
> - Acknowledge the need to recognize, evaluate, and reduce risks in an R&D project.
> - Recognize the key papers for risk management in an R&D project.

12.1 Risk Management

Risk is defined as "the influence of uncertainty on objectives" by ISO 31000. Stakeholder viewpoints affect the project's aims. For instance, the goal of a company and its shareholders is probably to maximize the net value contributed. In contrast, the goal of its clients and consumers is to maximize the net result. The consumer, the primary value driver, also governs business objectives. The actual result, however, differs from the anticipated result, and project risks cause this disparity. Therefore, an improved risk management method will likely produce the desired result or even a greater return. Given the existing lack of confidence in a future occurrence or situation, as shown by chance distributions of the results of deliveries, uncertainty, and risk depend on this lack of certainty. These include risk going up (rewards) or down (losses). Even though a probability distribution looks quantitative, the information primarily depends on conjecture and, if available, a small amount of historical data. Risk analysis entails exposure to impact to identify the outcomes that could be significant to a topic. Risk assessment is, therefore, an individualized procedure that depends on how important the outcomes are. The known unknowns, unknown knowns, and unknown unknowns are the three main causes of risks and uncertainties [1].

Both systems engineering and project management techniques include risk management as a critical component. The procedures and equipment used to detect, analyze, minimize, monitor, and control risks are all a part of risk management [2]. Risk management is regarded by INCOSE, the International Council on Systems Engineering, as a standard instrument [3]. However, the project manager is often in charge of project-level risk evaluations and will eventually be held accountable for properly managing project risks. Either of these areas can take on the duty of risk management tasks.

There is danger whenever there is a planned change. A new idea's implementation, the creation of a new product, and the implementation of an updated service are all examples of planned change. Risk is frequently defined as probable occurrences that might affect a project's budget, timeline, or scope. Making deliberate choices on the best course of action is essential to managing risk. Project management and risk management must be separate and considered optional. While project management seeks to maximize the likelihood that a project will succeed overall by using controls and metrics, such as reviews, budget management, and earned value management (EVM), project risk management provides a framework to detect, analyze, and regulate the environment that was not originally planned for as well as to balance and limit the allocation of resources that are available for risk mitigation throughout the project. The process of deliberately lowering a risk's exposure and lowering the likelihood that it will materialize is known as risk mitigation.

Effective project management requires managing the project's environment and the opportunities and risks arising from it. These two procedures go hand in hand. Unforeseen needs will damage the project's parameters, or possibilities will only be

12.1.1 Processes of Risk Management

Typically, the project manager develops and oversees risk management procedures with the aid of a systems engineer. The stakeholders and the project manager, who comprise the project risk team, are responsible for carrying out the risk management procedures. Everyone with a stake in the project, including operational departments or divisions that interact with and support the project, should be involved in risk management since project risks typically impact the project's final success. It is crucial to participate actively. One frequent mistake in delegating risk management duties is to provide them to those who need more authority to carry out the essential tasks to address the risk.

The continual process of risk identification, impact and probability assessment with the necessary stakeholders, risk mitigation, control, and retirement, and updating the tasks' status in a central data repository are frequently overseen by a risk manager on big and complicated projects. Throughout a project, risk management actions are continual. Risk identification may occasionally be delicate, and the person who accepts the risk may occasionally come under unfair criticism. Those who raise risk issues must be fine, either by being given tasks they lack the power to complete or receiving negative feedback. At the project level, the crucial risk management procedures can be seen in Fig. 12.1.

At significant turning points or other significant moments, there should be a periodic evaluation of the risks currently being handled and a request for new risks. Even so, stakeholders should have a method to, at any time, present new hazards. Additionally, there has to be a procedure for adding new hazards and eliminating old ones. Beginning with the project's baseline, external and internal sources will pressure the project's budget, schedule, and requirements baselines. The performance of the timetable could be impacted if a requirement is not satisfied, particularly if a redesign is necessary. The budget may also be impacted if more employees should be engaged to make up the shortfall, a consulting agreement should be formed, a product should be acquired, or some other action should be taken to address the specification deficit.

Conversely, if a standard is not met, the project may re-evaluate it and select an alternative that exceeds the original specification. When appraising, one should consider both opportunities and hazards. When evaluating the outcomes of opportunities, risks and impacts should be considered.

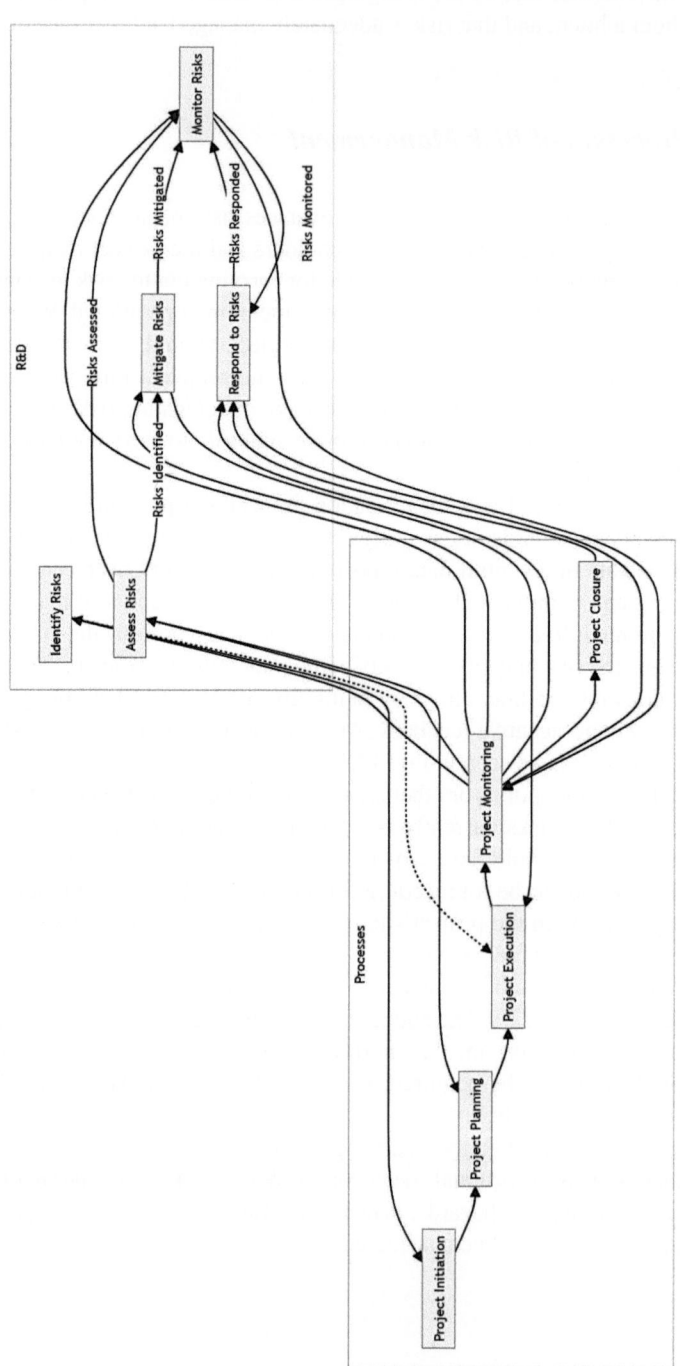

Fig. 12.1 Important risk management procedures

12.1.2 A Project's Level of Risk

The risk management process addresses risk at the project level. There are five steps in the procedure (Fig. 12.2).

Every project is prone to dangers that put the fulfillment of the specified objectives in jeopardy. Identifying and prioritizing risks enables the project manager to concentrate efforts on averting those that are both highly effective and highly

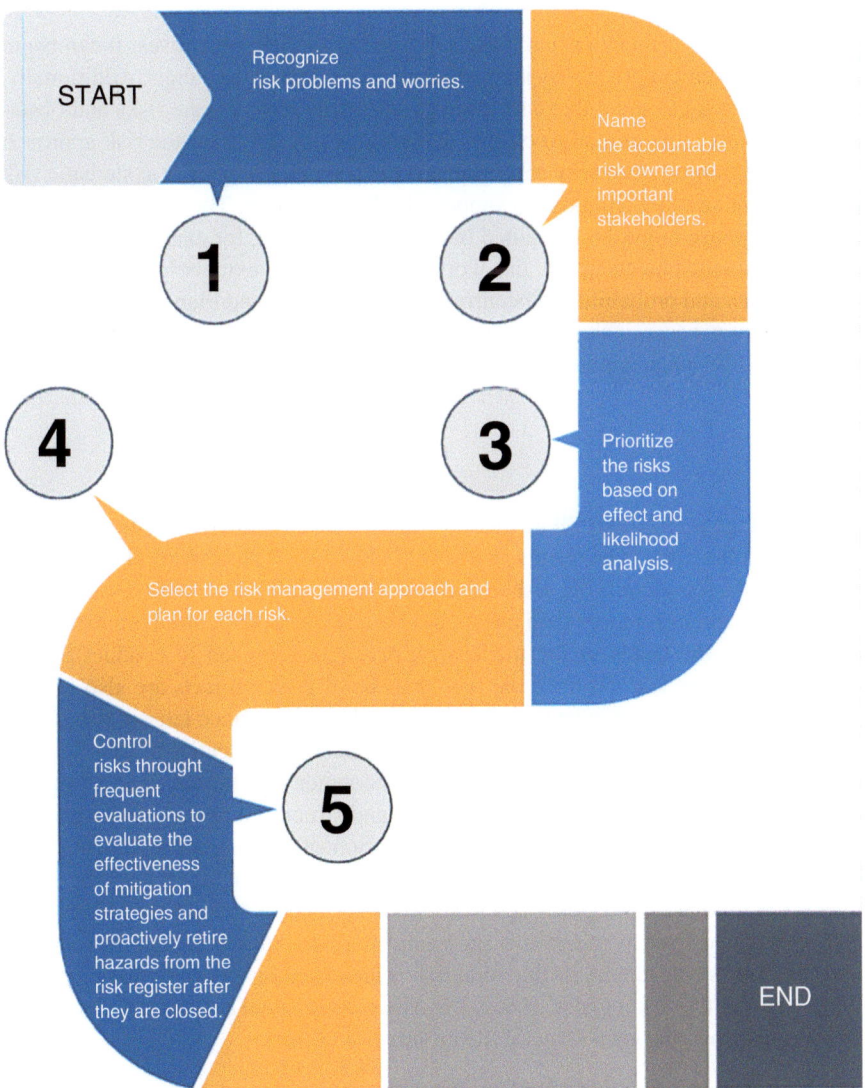

Fig. 12.2 Five steps in the risk management process

probable to occur. Once more, the time and resources spent managing it should be adjusted to the scope and difficulty of the project; still, each project should have, at the very least, a fundamental risk strategy and management procedure in place. Larger, more complicated projects will require specialized individuals or technologies to automate the process and heavily rely on risk management methods to provide information to the project manager.

Define and Plan

It is important to remember that risks only arise with clear objectives. It can begin the risk management process by identifying the goals that are in danger. Additionally, it is crucial to understand that each project will likely have unique risks and obstacles. Therefore, it is crucial to clearly define the risks and scale the risk approach from the beginning. At a very early point of the project life cycle, a systematic risk management strategy should also support the Define phase. The deliverables from the Define phase might be contained in a single document, for example, a risk register that lists the objects of risk in a worksheet along with extra columns for impact on objectives and probability of occurrence. At the Define and Plan phase of the risk management process, it should be decided how extensive and sophisticated the risk process will be when applied to the particular project.

Identify

The next stage is to further fill up the risk register and determine the risks that might impact the objectives. The results provide either a chance or a threat. There are several methods for identifying risks (such as SWOT analysis and key stakeholder interviews), and using more than one method is advisable. Structured brainstorming sessions with the heads of the main work packages are used to conduct SWOT analyses. Lessons learned from the risk analysis of prior projects are also helpful guidelines. Lists of known hazards and dangers that will surface later in the project are also included. Interviews with key stakeholders are essential to getting their viewpoints on any hazards the project could encounter. In addition to effecting and cause analysis, scenario analysis, flowcharts, and structured checklists are alternative methods for detecting hazards. The formation of the risk management group based on team members' expertise, knowledge, and judgment determines the effectiveness of all these strategies. Each risk that has been discovered has to have a risk owner assigned to it to mitigate it successfully. Every time a change in scope or specification is considered, its potential risk is identified by recording the modification objects in the risk register. That is why risk management is a dynamic and iterative process covering each stage of the project life cycle.

Analysis

Analyzing the order of risk items to determine the worst hazards and greatest benefits is important. The most common approach is to evaluate the effect and chance of incidence (as a percentage and on a scale of 1 to 10) and then calculate the risk factor. The evaluation of importance by risk register has limitations since the distribution of likelihood and effect size depends on how much the consequences matter to whom. However, it also provides a structured and systematic record with a single set of facts for management review. Failure mode and effects analysis (FMEA) is a more sophisticated variation of a risk register. The goal is to categorize all potential failures due to their impact as determined by the detection, likelihood, and gravity and then to identify ways to either eliminate or significantly reduce them.

Each function's probable failure modes are mentioned, followed by descriptions of each failure mode's implications, particularly those that the user would likely notice. After that, each failure mode's reasons are looked through and compiled. A probable failure mode's current controls are identified and evaluated. The potential risk to employees or the system is then rated on a scale of 1 to 10 for seriousness. The following stage determines how probable each failure is to occur, with the likelihoods ranging from exceedingly improbable (1) to most likely. On a scale of 1 to 10, the likelihood that the failure will be discovered is also evaluated. A risk priority number (RPN) is calculated for each failure as the product of the amounts estimated in the preceding three phases. The improvement activity should concentrate on the probable failure modes in descending order of RPN to reduce the chance of failure.

Even though the failure mode and effects analysis (FMEA) and risk register both employ statistics, these assessments prioritize risks based on subjective opinions. These are qualitative risk evaluation techniques that take each risk into account separately. However, it is frequently required to examine how hazards interact to affect the results of a project. Processes for quantitative risk assessment can be used in this situation. One of the most common quantitative methods is Monte Carlo simulation [4], which has software tools to support it and a history of statistical validity and dependability. The feasibility and implementation phases of the project life cycle are the main areas where the Monte Carlo approach is applied in project risk management. The ideal method for assessing a pro forma of investment prospects at the feasibility stage is simulation. Input at the beginning of a project needs to be revised because the pro forma is, by design, an estimate of performance.

Additionally, the Monte Carlo method's results will appropriately represent the given data's uncertainty. The Monte Carlo method assesses the total uncertainty of project completion timelines when a project is in the implementation stage. The method is also highly helpful for determining the likelihood and impact of a key risk in a risk register while a project is being implemented.

Respond

The next step after risk analysis is to consider the appropriate action. This may be the most important step in the entire risk management process. It is here that risk management plans and choices are significant. There are several ways to respond, but the majority of them involve one or more of the following options:

- Avoidance
- Deflection
- Reduction
- Acceptance

Avoidance aims to reduce or, if necessary, eliminate risk's primary sources or to alter the objective's parameters. This approach could promote creativity but frequently impacts the standards of excellence. Another party or supplier is used as part of the deflection approach to control the risk. The risk is often shifted to a third party with better expertise and understanding to address the problem. This tactic will probably have an impact on the completion time. By allocating more apropos resources, reduction aims to lessen risk's likelihood and effects. Altering the objective's scope might also result in a decrease. This tactic resembles avoidance, although to a lesser extent. When other tactics fail to work, acceptance is typically the last option. In this case, the solution is to provide cost flexibility with the anticipation that the project would proceed without material changes. Some risks are also protected by suitable insurance plans, particularly those that concern safety and health.

Control and Monitor

In addition to spanning every phase of the project life cycle, project risk management is also an adaptive process with regular evaluations and management control, as was already mentioned. The risk exposure also evolves. Allocating specialized personnel to record, analyze, and keep track of risk reports is a standard procedure. However, the group leader of the activity package where the risk is situated is ultimately responsible for implementing risk responses and corrective measures. A structured review procedure ensures that planned replies are delivered on time and, where necessary, create fresh responses. According to Turner's [5] study, the main goal of risk management is to communicate with every project participant. Therefore, the risk management strategy should be reviewed often (typically monthly) to verify that it applies to evolving conditions. The work scope, build technique, team members, suppliers, and regulations might all change. Additionally, it is a good idea to integrate the KPI into the main systems for project performance management, which are often based on a balanced scorecard and EFQM [6].

12.1 Risk Management

12.1.3 A Program's Level of Risk

Program-level risks consist of risks specific to a given project or program. Projects and other initiatives under its aegis interact, are connected, or depend on each other.

Risk management at the project level should also be active for program-level risk management to be effective. A list of prioritized risks and opportunities may then be structured to gather all the top priorities after the risks are gathered from every project or subordinate program. This approach is only slightly modified. This offers a cross-program perspective of all the greatest hazards as determined by the program.

12.1.4 A Portfolio's Level of Risk

Risks that affect a portfolio are shown in Fig. 12.3.

When activities taken as part of projects, programs, or operations are suspected of having an unanticipated effect on the organization, the risk is present at the portfolio level. Operations relating to the relationships between the activities in the portfolio might likewise be considered risks. If not recognized and reduced, these risks might impact strategy, finances, reputation, compliance, workforce, etc. Cross-organizational risks can be evaluated at the portfolio level using the same risk procedure. When the risks are gathered from each division, department, program, and project, prioritized risks and opportunities may be sorted so that all the top priorities are grouped, similar to how program risk management is done. This offers a cross-portfolio perspective of all the greatest risks as determined by the various groups.

It is essential to understand that portfolio risks frequently start far earlier than when projects are chosen and underway. The actions chosen should be the best options for the firm to minimize future risks at the portfolio level properly. Any new projects and initiatives should be carefully and methodically evaluated before being added to the portfolio. Generally speaking, reducing risk in this area involves ensuring that each new activity—whether an operational task, a project, or a program—supports the plan and has the resources necessary to be completed effectively.

Understanding what is being suggested and how these projects support the portfolio is often the first step in selecting the proper initiatives. Here is where the allocation of production and operational activities and radical and incremental R&D initiatives should be decided. Considerations for what to add to the portfolio include return on investment analysis, strategic alignment, the proper balance of short-term vs. long-term initiatives, and organizational-level resource planning. It is necessary to thoroughly evaluate a list of potential projects to ascertain which ones are most likely to result in the intended organizational results and which ones may truly result in the implementation of the strategy.

This is frequently difficult because initiatives at this level might have varying substance. It is critical to find a means to contrast completely unrelated projects, such as by contrasting the importance of hardware and software initiatives. This

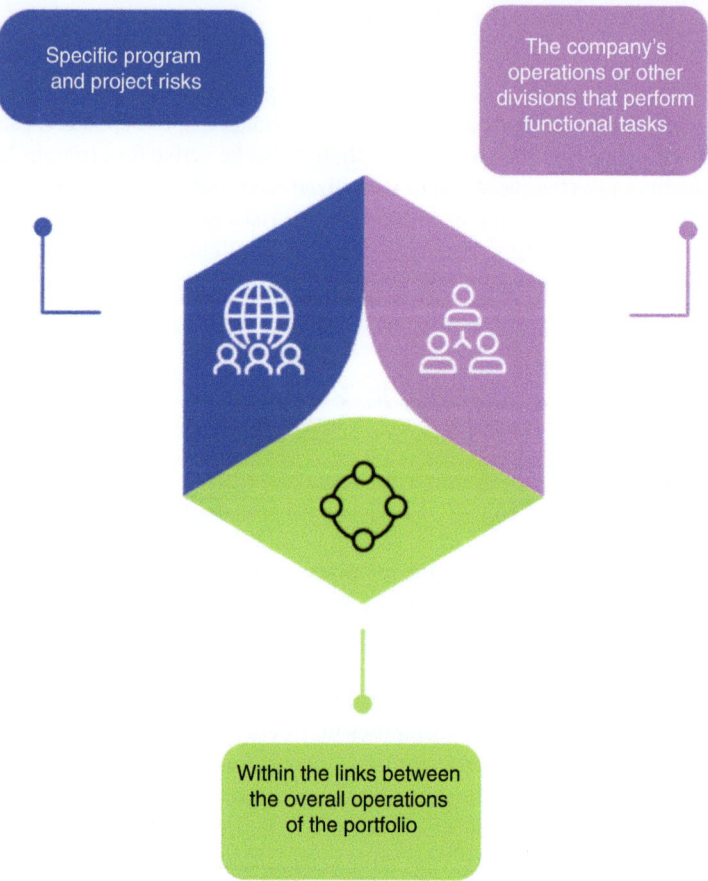

Fig. 12.3 Risk effect on a portfolio

section presents a strategy for gathering, evaluating, prioritizing, and picking a portfolio of programs and projects with varied content to minimize risk exposure. This technique relies largely on the availability of a well-defined strategy and set of trajectories, but it is particularly well suited to portfolio evaluation. If those conditions are met, the following essential components of risk management would not only guarantee the finest portfolio alignment but provide you the best chance to select the most prospective R&D projects.

In the portfolio risk evaluation, the following actions should be included:

– Putting together a list of prospective initiatives and programs.

- Determining the right balance of fundamental and applied R&D for the organization.
- The necessary resources.
- The proposer gives the precedence.

- How well they adhere to the corporate plan?
- What should the weighting criteria be?

– Combining the findings to create a collection that is prioritized and normalized
– Deciding on the best portfolio.

12.2 R&D Risk Management

R&D project features and goals are less definite than those of more conventional project categories [7]. The literature also makes the case that risk management may greatly increase the success of R&D initiatives. Let us now concentrate on the specific aspects of risk management in R&D after reviewing traditional projects' broad risk management methods. The origin of the risks in R&D is also diverse and more complicated than the elements alone. To deal with the added complexity of R&D projects, the integrated method of portfolio risk management is frequently used. Portfolio risk management aims to assist investors in controlling the risk involved with their financial decisions. It is primarily based on the notion that risk and profit are sometimes traded off. While portfolio risk management is often linked with finance and economics, high-risk R&D projects can also benefit from its integrated risk management strategy. The interaction between a project's strategy and scope, risk modeling, and risk reduction is known as portfolio risk management in project management [8].

What are these elements that make the complexity of risks in R&D greater? Despite the complexity and failure rates of risk factors in R&D projects being well publicized, notably in the pharmaceutical business, the present literature needs to be more developed in defining an organized strategy for classifying or identifying them [9]. Although Pass and Postle [10] tried to summarize the risks associated with R&D in the pharmaceutical business, more work has to be done. According to Saari's [9] study, the risks associated with R&D projects in general and drug development projects might be loosely classified as controllable and noncontrollable hazards since it shows four constructs—commercial risks, technological risks, regulatory risks, and organizational risks—as well as their reflected indicators.

12.2.1 Commercial Risks

The degree of financial risk that a company faces is the commercial risk of an R&D project. A company depends more on high returns, the bigger the business risk. A case can be made that the likelihood of rewards increases with increasing business risk. The product and its financial benefit on the market are the primary elements of commercial risks. A corporation can assume greater commercial risk if the product is anticipated to have a longer lifespan covered by a patent (such as a medicine product). Creating a new product should closely relate to an effective market intelligence procedure.

12.2.2 Technical Risks

An R&D project's vulnerability to loss from technical activities, including production, scientific procedures, testing, engineering, and design, is called technical risk. It may also be advantageous if an R&D team has technical expertise and experience, such as in platform technology. For instance, if a researcher creates a novel delivery system using a certain medicine and the optimal amount of excipients, other researchers can use the same excipients to create a new dosage form for a drug by just switching out the active component. In opposition to that, because it is impossible to forecast the process development risks associated with the fundamental characteristics of the compound, the only course of action available is to cancel the process or perhaps the project.

12.2.3 Regulatory Risks

The possibility that regulatory agencies would deny the product's registration or materially alter the law gives rise to the regulatory risk involved in an R&D project. Regulatory organizations are gradually altering their policies. A risk exists if you do not stay current with modifications. Understanding the many regulatory standards outside of the European Union is equally difficult.

12.2.4 Organizational Risks

A lack of resources, inadequate management, and knowledge gaps cause an R&D project's organizational risk. It could also result from shoddy collaboration agreements in joint R&D projects or contracts with significant suppliers. Implementing project management concepts effectively is the greatest way to control organizational risks. Due to the lengthy timeframe involved in launching new items to the market, commercial risks are significant and hard to foresee. The market may have completely disappeared, rivals may have introduced new items, etc.

Additionally, important is time-to-market. To be the first on the market or to replace existing items before their patents expire, products should be introduced within a specific timeframe. However, to acquire market share, a product typically has to outperform rival products in certain respects to be beneficial at all. Although external business risks are difficult to manage, excellent market research can increase a product launch's success. The dangers associated with the compound's intrinsic characteristics cannot be effectively reduced, and the only course of action can be to abandon the project if negative outcomes arise. Thus, potential "showstoppers" like safety and effectiveness concerns can be identified from the project's inception. Controlling these technological hazards is significantly more challenging. Regulatory risks may be managed by being current with changes in governance and regulations by the relevant regulatory organizations. Small businesses find it challenging, while bigger research units with enough resources can handle it.

Organizational risks, including scheduling, financial, resource, and knowledge-related hazards, are also in the controllable risks category. In theory, the dangers brought on by subpar project management are under your control. In development projects, budget overruns are typically less serious of an issue. Small businesses might suffer more from budget overruns than large industrial giants with enough means to cover them. Innovative businesses should decide whether to increase their commercial or technical risks. As technological risk rises, the corporation will increase its capacity for innovation and intellectual capital.

Conversely, by assuming more business risk, the corporation will aim to get a greater potential competitive edge in the market. Accordingly, from the study above, the most important risks to concentrate on in R&D projects are external risks, project management-related scheduling risks, and scope risks. A substantial number of risks in R&D projects may be under control if the sources of risk in the four categories above are used in a systematic approach to project risk management [11]. It is significant to remember that the risk management process should be included as a crucial part of project management.

12.3 R&D-Specific Risk Factors

R&D projects are unique, marked by innovation, experimentation, and exploration. Alongside any project's typical risks, R&D initiatives encounter a set of specific risk factors that demand careful consideration. Understanding and effectively managing these R&D-specific risks are critical for success.

12.3.1 Inherent Uncertainty in R&D

R&D is the breeding ground for innovation, fostering advancements across various industries. However, it is marked by a fundamental and inevitable uncertainty that challenges its progress. Several factors contribute to this uncertainty. First, the complexity of nature plays a pivotal role. R&D endeavors often seek to unravel intricate natural phenomena, be it exploring the cosmos or developing new drugs. The multifaceted aspects of these systems introduce uncertainty, as it is challenging to predict precisely how variables will interact. Second, human variables introduce another layer of uncertainty. The researchers, scientists, and engineers driving R&D projects are susceptible to cognitive biases and errors. This human element can influence the outcomes of research. Moreover, the financial and resource constraints inherent in R&D can affect the depth and breadth of investigations. Budgetary limitations and resource scarcity can result in incomplete or inconclusive results, magnifying uncertainty.

Market dynamics and competition add another dimension of uncertainty. Anticipating how a technology or product will perform in a constantly evolving market is challenging. Changing consumer preferences and market trends can derail the best-laid plans. Rapid technological advancements are another source of uncertainty, with cutting-edge developments quickly becoming obsolete. Keeping up with this pace of change is essential for the success of long-term R&D projects. Finally, unforeseen regulatory changes and ethical challenges can disrupt R&D endeavors, leading to uncertainty and delays. The need to comply with shifting regulations and ethical standards in fields like pharmaceuticals can be a significant hurdle. To navigate this inherent uncertainty effectively, R&D efforts require comprehensive planning, cross-disciplinary collaboration, continuous learning, an iterative approach, and robust risk management. These strategies help researchers and organizations adapt, mitigate risks, and embrace the ever-evolving innovation landscape. Embracing uncertainty as an integral part of the R&D journey is vital, as it is in this uncertain terrain that some of the most groundbreaking discoveries and innovations emerge.

12.3.2 Long Project Timelines

Long project timelines are a common feature of complex undertakings, spanning various domains such as construction, research, software development, and space exploration. While they bring advantages, they also present challenges. Advantages include the opportunity for comprehensive planning, allowing for a thorough consideration of all project aspects. Resource management benefits from an extended timeline, ensuring efficient allocation of materials, labor, and finances. Moreover, in-depth research and development, particularly in scientific and engineering fields, can be conducted methodically, reducing the chances of costly errors. Risk mitigation benefits from the additional time to identify and address potential issues, enhancing project resilience.

Furthermore, long timelines can be divided into phases or milestones, facilitating incremental progress and enabling frequent course corrections. However, challenges arise as well. Long project durations are often associated with higher overall costs due to increased labor, materials, and administration expenses. Maintaining a stable and motivated workforce over an extended period can take time and effort, with employee turnover and changing priorities potentially disrupting project continuity. Evolving project requirements and objectives can lead to scope creep, causing delays and cost escalation.

Additionally, adapting to rapidly advancing fields like technology can be problematic, as long timelines might hinder incorporating the latest advancements. Market dynamics in projects with commercial applications can shift unpredictably over time, impacting project relevance. Managing stakeholder patience and

12.3 R&D-Specific Risk Factors   261

expectations becomes crucial when dealing with prolonged projects. Addressing the challenges of long project timelines requires effective project management. Continuous monitoring, risk assessment, scope control, and adaptation to changing circumstances are essential. Clear communication with stakeholders and regular evaluation of project objectives and performance are critical to ensuring the project remains aligned with its goals.

Long project timelines offer benefits such as comprehensive planning, efficient resource allocation, and in-depth research and development. Nevertheless, they come with challenges, including potential cost overruns, resource allocation issues, scope creep, and difficulty adapting to evolving fields and market dynamics. Managing extended timelines requires careful project management, adaptability, and responsiveness to changing conditions in an ever-evolving world.

12.3.3 Rapid Technological Advancements

Rapid technological advancements define the contemporary era, ushering in a remarkable era of innovation and transformation across various facets of human life. These advancements offer many advantages, reshaping industries, enhancing productivity, and enriching our quality of life. In innovation and progress, technology constantly pushes boundaries, fostering the development of groundbreaking solutions to complex problems. Furthermore, it is a powerful engine of economic growth, creating new job opportunities and driving economic activity through increased efficiency and market expansion. Global connectivity is another significant advantage. Advancements in communication technology have made the world more interconnected than ever before, transcending geographical boundaries and fostering cross-cultural interactions. This has had profound effects on how societies operate and collaborate on a global scale. Technological advancements in the healthcare and medical sectors have resulted in substantial improvements in diagnostics, treatments, and patient care. These breakthroughs contribute to better healthcare outcomes and longer lifespans, making technology an indispensable asset in the medical field.

Moreover, technology plays a pivotal role in addressing environmental challenges. Advancements in renewable energy, sustainable agriculture, and waste management are all vital components of a more sustainable future. These innovations reduce our ecological footprint and promote responsible stewardship of the planet. Despite these advantages, the rapid pace of technological advancement brings challenges and considerations. One such challenge is the digital divide, a result of disparities in access to technology. Underserved and disadvantaged communities may find themselves without access to critical tools and information, exacerbating existing inequalities.

Privacy and security concerns also emerge in the wake of rapid technological advancement. The digital age has given rise to worries about data privacy and cybersecurity. As technology progresses, the need to protect sensitive information and critical infrastructure becomes increasingly vital. Furthermore, the potential for job disruption is a significant concern. Automation and artificial intelligence can transform traditional job markets, prompting concerns about unemployment and the necessity for workforce reskilling. Ethical dilemmas accompany technological progress. Advances in fields such as genetic engineering and artificial intelligence pose ethical questions about the limits of human intervention and the potential consequences of unchecked technological capabilities.

Additionally, there are environmental impacts to consider. While technology can contribute to environmental sustainability, it can also have adverse effects, including resource depletion, electronic waste, and increased energy consumption. Finally, there is a growing risk of overdependence on technology. Society's increasing reliance on technology has the potential to result in catastrophic consequences should critical technology systems fail. The ongoing management of the rapid pace of technological advancement necessitates a multifaceted approach. This approach includes implementing policies aimed at bridging the digital divide, regulations to safeguard privacy and security, educational and workforce development initiatives to address job disruption, and establishing ethical guidelines to ensure responsible innovation. In this dynamic landscape, businesses and individuals must adapt by engaging in continual learning and adopting a proactive, responsible approach to harnessing technology's power for humanity's betterment.

12.3.4 Resource Constraints

Resource constraints represent a ubiquitous daily challenge that individuals, organizations, and governments encounter. These constraints emerge due to the limited availability of essential resources, including financial capital, human resources, time, physical assets, and natural resources. Managing and mitigating these constraints is vital for achieving various objectives, whether running a business, executing a project, or addressing broader societal needs.

Several types of resource constraints are commonly encountered. Financial constraints are frequently among the most significant, necessitating careful budgeting and securing funds for operational and growth purposes. Limited financial resources can impact an entity's ability to invest in research and development, expand infrastructure, or attract and retain talented employees. Human resource constraints can arise from shortages in skilled personnel, high employee turnover, or a disconnect between the available workforce and the skills required, affecting productivity and organizational effectiveness. Time constraints are particularly critical in project management and goal attainment, as deadlines and timeframes must be adhered to meet objectives and deliver products and services.

In addition to these, physical resource constraints come into play in industries like manufacturing and construction, where access to essential equipment, machinery, and facilities is indispensable for operations. Natural resource constraints, often found in sectors relying on natural resources like agriculture and mining, can be

12.3 R&D-Specific Risk Factors 263

influenced by environmental factors, regulations, or geopolitical issues, impacting resource availability.

Resource constraints present several challenges. They can lead to missed opportunities for growth, innovation, or market expansion. Organizations may be compelled to cut corners, compromising the quality of their products or services and reducing their capacity for research and development. Time constraints can result in project delays, negatively impacting project timelines and potentially incurring penalties. In industries grappling with human resource constraints, attracting and retaining top talent becomes a significant challenge. Moreover, natural resource constraints can have broader economic and environmental repercussions, potentially leading to resource depletion and environmental degradation.

To effectively address resource constraints, individuals and organizations can employ various strategies. One of the fundamental approaches is effective resource allocation through careful planning and prioritization of projects and investments, optimizing the use of available resources. Efficiency and productivity improvements represent another vital tactic; by emphasizing streamlined processes and increased productivity, organizations can do more with fewer resources. Strategic partnerships with other organizations or the establishment of collaborations can provide access to additional resources, thereby alleviating constraints. Diversification, both in terms of resource sources and investment in resource exploration and development, can help reduce reliance on limited resources. Innovation plays a crucial role in addressing resource constraints, as finding alternative solutions and adopting sustainable practices can mitigate the impact of constraints.

In conclusion, resource constraints are an inherent component of most endeavors, and their effective management is pivotal to success. Acknowledging the challenges posed by limited resources and implementing strategies to optimize resource utilization empowers individuals and organizations to adapt to these constraints and continue thriving in their respective fields.

12.3.5 Intellectual Property Risks

Intellectual property risks in R&D are critical in today's innovation-driven landscape. Intellectual property comprises patents, copyrights, trademarks, and trade secrets, serving as a cornerstone of protecting innovative ideas and creations. However, the process of managing and safeguarding intellectual property within R&D activities comes with a set of inherent risks and challenges. One prevalent risk is IP infringement. R&D teams may inadvertently infringe upon existing patents, copyrights, or trademarks while developing innovations, potentially leading to costly legal disputes and damaging their reputation. Protecting trade secrets, including proprietary technologies or business processes, poses another risk. The leak of sensitive information can result in a loss of competitive advantage, particularly when R&D is a primary source of a company's value. Ownership disputes are a frequent issue, especially in collaborative R&D efforts. Complex projects may result in conflicts over intellectual property ownership, making it crucial to establish clear agreements and comprehensive documentation. Employee departures are another concern, as departing employees may take valuable intellectual property with them, potentially benefiting a competitor. To mitigate this risk, noncompete agreements and confidentiality clauses are essential. Finally, the lack of a coherent IP strategy can result in missed opportunities to protect valuable innovations, leaving them vulnerable to exploitation by competitors. To address these challenges, various strategies can be implemented. R&D teams should raise awareness and educate members about IP rights and potential risks. This awareness can help prevent accidental infringement and ensure team members appreciate the value of IP assets.

Additionally, thorough searches and assessments should be conducted before commencing an R&D project to identify any existing IP that might impact the project. Meticulous documentation of R&D processes and innovations is crucial for demonstrating ownership and defending against potential disputes. Clear agreements should be established when collaborating with external partners or employees to define ownership and usage rights and responsibilities.

Intellectual property risks in R&D are pervasive and require careful attention and proactive management. By fostering awareness, conducting comprehensive searches, maintaining meticulous documentation, and establishing clear agreements, individuals and organizations can better protect their intellectual property assets and navigate the intricate landscape of innovation with greater security and confidence.

12.4 Risk Assessment Tools and Techniques

Risk assessment tools and techniques are vital for evaluating and managing potential risks. These methods include SWOT analysis for identifying internal strengths and weaknesses and external opportunities and threats, risk matrices for visualizing risks based on likelihood and impact, Monte Carlo simulations for statistical analysis of uncertainty, and techniques such as fault tree analysis and failure mode and effects analysis (FMEA) to assess causes and consequences of failures systematically. Bowtie analysis, key risk indicators (KRIs), and scenario analysis help depict relationships between hazards, monitor progress, and prepare for future uncertainties. Sensitivity analysis evaluates the influence of variables, while brainstorming and the Delphi method facilitate creative idea generation and expert consensus on risk assessment. Finally, historical data analysis leverages records to identify recurring risks and patterns, aiding in data-driven risk evaluation. The selection of an appropriate method depends on the specific context and complexity of the risk being assessed, often involving a combination of these tools for a comprehensive approach to risk management.

12.4.1 Probability and Impact Analysis

Probability and impact analysis is a fundamental risk assessment technique extensively used in R&D. It enables R&D teams and organizations to systematically evaluate the likelihood and potential consequences of specific risks in their projects, innovations, or objectives. Probability analysis involves assessing the likelihood of risks materializing assigning numerical values or qualitative scales to categorize their probability. On the other hand, impact analysis evaluates the potential consequences or severity of risks if they occur, considering a wide range of impacts, from financial losses to safety hazards. The probability-impact matrix is a crucial tool in this process, offering a visual representation that categorizes risks based on their assessed probability and impact. High-probability, high-impact risks are the top priorities, as they pose the most significant threats, while low-probability, low-impact risks receive less immediate attention. R&D teams use this matrix to prioritize risks and allocate resources effectively for mitigation. This systematic approach helps make informed decisions and ensures that mitigation strategies are appropriately aligned with the potential risks. Continuous monitoring and adaptation are integral to this methodology, recognizing the dynamic nature of R&D projects. Risks should be continually evaluated throughout the project's lifecycle, and mitigation strategies should be adjusted as the environment evolves. Probability and impact analysis, by providing a structured framework for assessing and managing risks, enhances the resilience and success of R&D initiatives, allowing teams to make informed choices and allocate resources judiciously.

12.4.2 Risk Matrices

Risk matrices are powerful tools for assessing and managing risks across various domains. They visually represent risks based on two critical dimensions: the likelihood of a risk event occurring and the severity of its potential impact. This visual representation helps identify and prioritize risks, making allocating resources effectively and guiding risk mitigation strategies easier. A typical risk matrix consists of a grid with likelihood on one axis (usually the horizontal axis) and impact on the other (usually the vertical axis). Each cell in the grid represents a combination of a likelihood and impact category, which can be qualitative (e.g., low, medium, high) or numerical. The risk matrix provides a structured framework for assessing and categorizing risks based on their probability and potential consequences. Once risks are plotted on the risk matrix, they can be prioritized based on their position within the grid. Risks falling into the high-likelihood, high-impact category are considered the most critical and require immediate attention and thorough mitigation strategies.

Conversely, low-likelihood, low-impact risks may receive less immediate focus. The risk matrix guides decision-makers in effectively allocating resources for risk

Table 12.1 Risk matrix: assessing likelihood and impact

		Impact			
		Rare	Unlikely	Possible	Likely
Likelihood	Major	High	Extreme	Extreme	Extreme
	Moderate	Medium	High	Extreme	Extreme
	Minor	Low	Medium	High	Extreme
	Negligible	Low	Low	Medium	High

management, ensuring that the most significant threats are addressed promptly. Table 12.1 is a simplified example of a risk matrix.

12.4.3 Monte Carlo Simulation

Monte Carlo simulation is a potent computational technique extensively used in R&D. It offers a systematic and insightful approach to evaluating the impact of uncertainty and risk within R&D projects. This method is particularly valuable when project outcomes depend on numerous variables characterized by varying degrees of uncertainty. Key elements of Monte Carlo simulation include using random variables to represent uncertain project parameters, probability distributions that define the likelihood of these variables' values, iterative modeling through many simulations, and computational analysis to generate a probability distribution of potential project outcomes. The benefits of utilizing Monte Carlo simulation in R&D are multifaceted. It allows for quantifying project risk, offering clarity on the range of potential outcomes. Decision-makers can leverage this insight to make informed choices about resource allocation, project scheduling, and risk mitigation strategies. The technique excels in handling complex projects with intricate models and multiple interacting variables, contributing to more effective R&D management and a higher likelihood of project success, even in the presence of uncertainty.

12.4.4 Expert Judgment

Expert judgment is a cornerstone of R&D activities, offering specialized knowledge and experience in various fields. These experts possess subject matter expertise, often honed through years of practice and education, making them invaluable resources for organizations engaged in R&D endeavors. Whether the challenge is related to scientific research, technological innovation, product design, or industry-specific insights, experts provide critical guidance. They excel at problem-solving, diagnosing root causes, and proposing innovative solutions to complex technical challenges. Their ability to stay updated on the latest technological trends and advancements ensures that R&D strategies align with evolving market demands.

One of the significant roles of R&D experts is risk assessment. They evaluate the technical feasibility of projects, assess regulatory compliance, and gauge market acceptance, thereby aiding in identifying and mitigating potential risks. Their insights empower organizations to make well-informed decisions regarding project direction, resource allocation, and investments in new technologies. Furthermore, experts validate and verify the results of R&D work, assuring the quality and accuracy of research findings and technical solutions. This validation contributes to the credibility and reliability of R&D outcomes, which is crucial in scientific and technological pursuits. The benefits of expert judgment in R&D are multifaceted. They facilitate the transfer of specialized knowledge and best practices, nurturing the professional development of less-experienced team members. Expert judgment accelerates problem-solving when faced with complex technical challenges, reducing project delays and uncertainties. Experts also play a pivotal role in fostering innovation by introducing novel ideas and approaches that can lead to breakthroughs in R&D projects. Their guidance reduces the likelihood of project failures and costly setbacks by identifying and mitigating risks, making R&D efforts more robust and resilient. Expert advice supports informed strategic decisions, ensuring that R&D aligns with organizational goals and industry trends. Experts enhance the quality, efficiency, and strategic relevance of R&D activities, contributing to successful project outcomes and technological advancement.

12.4.5 Benchmarking

Benchmarking in R&D is a strategic practice that systematically compares an organization's R&D processes, practices, and performance metrics with those of industry leaders or best-in-class competitors. This approach provides valuable insights to R&D teams and organizations by helping them identify areas for improvement, establish performance targets, and drive innovation. The process of benchmarking in R&D typically involves several key elements.

First, organizations must select relevant performance metrics and key performance indicators (KPIs) specific to their R&D activities. These metrics include research productivity, time-to-market, innovation success rates, and project costs. Data is then collected from within the organization and from identified benchmarking partners. This data can include quantitative and qualitative information regarding R&D processes and outcomes, enabling organizations to perform a comprehensive analysis.

Once the data is collected, it is next to compare it to industry benchmarks, best-in-class organizations, or competitors. This comparison highlights areas of strength and areas in need of improvement. Through benchmarking, organizations can identify and assess best practices and processes that yield superior results in R&D,

providing a blueprint for improvement. The results of the benchmarking process are used to conduct a performance gap analysis, which informs the organization of the extent to which it lags or surpasses benchmarks. This analysis guides the setting of realistic performance targets and improvement goals.

Benchmarking in R&D offers several advantages. It fosters a culture of continuous improvement within R&D teams, enabling organizations to identify areas for enhancement and adopt best practices. By learning from industry leaders, organizations can enhance R&D efficiency and productivity, reducing time-to-market, lowering costs, and improving innovation success rates. Additionally, benchmarking provides valuable insights into emerging R&D trends and technologies, allowing organizations to adapt their strategies and remain at the forefront of innovation.

Moreover, benchmarking informs data-driven strategic decisions in R&D, including resource allocation, project prioritization, and process optimization. It helps organizations maintain competitiveness in the market by aligning their R&D efforts with industry standards or even surpassing them. This process aids in risk mitigation by identifying potential areas of risk or underperformance, allowing organizations to reduce the likelihood of R&D setbacks proactively.

In conclusion, benchmarking is a pivotal practice in R&D that offers a structured approach to assess and improve R&D activities. Organizations can enhance their R&D efficiency, innovation success rates, and overall competitiveness in a dynamic market by comparing their performance and processes to industry benchmarks and best practices. Benchmarking catalyzes innovation and performance improvement in R&D, leading to better outcomes and a more effective R&D strategy.

12.5 Reward Strategy and Risk

Big R&D initiatives are disliked because they are considered excessively risky, and the benefits will only materialize for a while. It is reasonable to recognize that the likelihood of failure increases significantly when a firm advances beyond incremental endeavors inside well-known areas. However, ignoring hazardous endeavors completely also stifles development. The answer is to take a methodical, disciplined approach that will more equitably disperse your ideas throughout the risk spectrum. A business must clearly understand where its initiatives are on the risk-reward spectrum to manage its innovation portfolio. The reward and risk matrix might aid a business in achieving this.

Strategic partnership-based R&D may also be taken into account. However, Stepping Stone is the only remaining choice when the capabilities and the market are still being determined. Small investments are considered in these experimental ventures to use the knowledge obtained as stepping stones. Exploring enabling technologies also makes use of collaboration with external R&D and institutions. The methodology helps focus on R&D portfolios, but it has a clear flaw in that it presumes that all portfolios would receive equivalent financial rewards. A helpful tool

Table 12.2 Five advantages of risk-taking businesses with a track record of success

Potential benefits	Description
Information advantage	Having superior information available in a crisis—especially early—can be extremely helpful
Speed advantage	A company can take advantage of possibilities that arise amid danger by acting swiftly (and responsibly)
Experience/knowledge advantage	Businesses (and management) who have previously encountered comparable situations can use what they have learned
Resource benefit	Having more resources than competitors might help a business survive a catastrophe that would destroy its rivals
Flexibility	When faced with danger, having the ability to alter directions rapidly might be advantageous

for choosing risk management tactics in the financial sector is the reward and risk matrix. Roggi [12] proposes five potential benefits that successful risk-taking organizations could harness to utilize the reward and risk approach better than their rivals (Table 12.2).

References

1. Z.W. Kundzewicz et al., Uncertainty in climate change impacts on water resources. Environ. Sci. Policy **79**, 1–8 (2018)
2. D. Crnković, M. Vukomanović, Comparison of trends in risk management theory and practices within the construction industry. Adv. Civil Archit. Eng. **7**(13), 1–11 (2016)
3. L. Yang, K. Cormican, M. Yu, Ontology-based systems engineering: A state-of-the-art review. Comput. Ind. **111**, 148–171 (2019)
4. M. Nabawy, L.M. Khodeir, A systematic review of quantitative risk analysis in construction of mega projects. Ain Shams Eng. J. **11**(4), 1403–1410 (2020)
5. J.R. Turner, J.R. Turner, T. Turner, *The Handbook of Project-Based Management: Improving the Processes for Achieving Strategic Objectives* (1999)
6. R. Basu, *Implementing Six Sigma and Lean* (Routledge, 2009)
7. J. Wang, W. Lin, Y.-H. Huang, A performance-oriented risk management framework for innovative R&D projects. Technovation **30**(11–12), 601–611 (2010)
8. S. Floricel, M. Ibanescu, Using R&D portfolio management to deal with dynamic risk. R&D Manag. **38**(5), 452–467 (2008)
9. H.-L. Saari, *Risk Management in Drug Development Projects* (2003)
10. D. Pass, M. Postle, Unlocking the value of R&D: Managing the risks. BioPharm **15**(6), 67–71 (2002)
11. J.F. Hair et al., Multivariate data analysis 6th Edition. Pearson Prentice Hall. New Jersey. Humans: Critique and reformulation. J. Abnorm. Psychol. **87**, 49–74 (2006)
12. O. Roggi, *The Corporate Risk. Risk, Value and Default* (2014)

Chapter 13
R&D Tools and Technologies

Innovation is not limited to the discoveries and breakthroughs it attempts to achieve; it is also present in the tools and technologies that propel the journey toward progress. The dynamic landscape of research and development (R&D) is evidence of the ever-changing relationship between scientific exploration and the instruments that make it possible. This chapter examines how laboratory equipment, computational tools, and communication technologies transform R&D by accelerating research, enhancing precision, and facilitating global collaboration. Emerging technologies promise to expand the boundaries of knowledge in the future, which we will also consider.

> **Learning Objectives**
>
> - Trace the historical development of R&D tools and technologies.
> - Categorize R&D tools into laboratory equipment, computational tools, and communication/collaboration tools.
> - Analyze how R&D tools accelerate research, enhance precision, support data-driven decisions, and promote global collaboration.
> - Explore cutting-edge innovations in R&D tools and their potential impact.
> - Consider future trends in R&D technology.

13.1 Evolution of R&D Tools and Technologies

For ages, human growth, invention, and technical progression have been fueled by R&D. Examining the evolution of R&D tools and technology reveals how closely these developments have been related to the growth and transformation of different industries.

13.1.1 Milestones and Breakthroughs in R&D Tools

R&D has always been at the forefront of invention and knowledge discovery. This section examines significant turning points and innovations in developing R&D tools and technologies (Fig. 13.1), emphasizing how they have influenced the current R&D environment.

The Microscope: A Glimpse into the Microscopic World

The development of the microscope was one of the early innovations in the field of R&D instruments. Inventing the straightforward single-lens microscope in the seventeenth century, the Dutch scientist Antonie van Leeuwenhoek was able to study microorganisms for the first time [1]. With the help of this discovery, biologists and microbiologists can now investigate the complicated and hitherto invisible world of cells, microbes, and tissues. Numerous scientific fields, including biology, medicine, and materials research, rely heavily on microscopes.

The Industrial Revolution and Analytical Chemistry

Significant improvements in analytical chemistry were made due to the Industrial Revolution, which started in the late eighteenth century [2]. Understanding the composition and characteristics of materials has undergone a revolutionary change due to the discovery and development of chemical analysis techniques like spectroscopy and chromatography. These methods, which developed over the nineteenth and twentieth centuries, have significantly impacted various fields, from environmental science to pharmaceuticals, allowing researchers to identify and study chemicals with astounding precision [3].

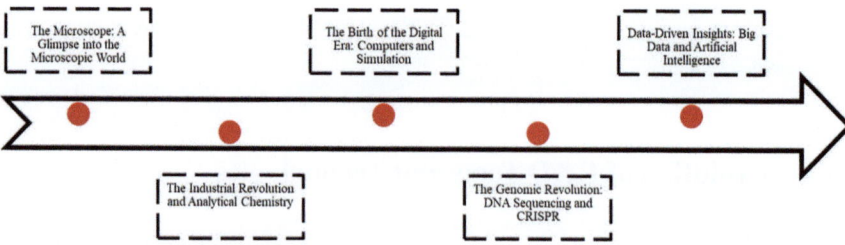

Fig. 13.1 R&D tool milestones and innovations

13.1 Evolution of R&D Tools and Technologies

The Birth of the Digital Era: Computers and Simulation

The development of digital computers in the middle of the twentieth century had a significant impact on R&D [4]. Computers substantially accelerated physics, chemistry, and engineering research by enabling scientists to simulate complicated systems and run mathematical models. R&D procedures became more productive and cost-effective thanks to the development of computational tools and software for scientific simulations. Researchers may now use previously impossible simulations thanks to supercomputers and high-performance computing clusters, such as modeling weather patterns and predicting chemical interactions in drug development.

The Genomic Revolution: DNA Sequencing and CRISPR

Due to advances in DNA sequencing and genome editing technology, genomics has made significant strides recently. The Human Genome Project's completion in 2003, which produced the first thorough map of the human genome, was a crucial turning point. Genetic research has also been transformed by the development of CRISPR-Cas9 gene editing technology, which allows precise and targeted changes to DNA sequences [5]. These developments have wide-ranging effects on everything from agricultural biotechnology to customized treatment.

Data-Driven Insights: Big Data and Artificial Intelligence

The explosion of data and the use of artificial intelligence (AI) have greatly impacted R&D in the twenty-first century [6]. Large-scale data collection, storage, and analysis have improved R&D decision-making procedures. AI and machine learning (ML) techniques are used to find trends, generate predictions, and improve studies. These instruments have sped up drug discovery, materials science, and renewable energy research.

These R&D tool milestones and breakthroughs are a small portion of the profound improvements over time. They emphasize the crucial part that cutting-edge instruments and technologies play in influencing the direction of scientific research and technological development. As this chapter progresses, we will analyze the current and prospective state of R&D tools and technologies and their impact on various fields and industries.

13.1.2 How R&D Tools Have Transformed over Time

In the dynamic field of R&D, progress depends on creativity and adaptation. To meet researchers' evolving needs and problems, the tools and technology employed in R&D have undergone a remarkable evolution. The future of innovation is still shaped by R&D tools in this constantly changing environment.

R&D, where tools develop to meet current needs, has long been known for its ability to adapt to contemporary challenges. Data collection, processing, and analysis have changed due to the digital revolution, giving academics access to high-performance computing, data analytics, and simulation tools. With the use of gene editing and genetic research instruments, biotechnology has significantly changed the fields of environmental science, agriculture, and medicine. Global teamwork is becoming more prevalent thanks to contemporary solutions like virtual conferencing and project management software. Data is fundamental to R&D's future, and predictive analytics and artificial intelligence are now essential for making data-driven decisions. Emerging technologies like nanotechnology, quantum computing, and 3D printing can revolutionize R&D by providing new tools, techniques, and possibilities for individuals prepared to accept change and harness the power of creativity.

13.2 Key Categories of R&D Tools and Technologies

The tools and technology used in R&D fall generally into three categories: laboratory equipment, computational tools, and communication and collaboration tools. These areas are essential for developing innovation and research in various industries.

13.2.1 Laboratory Equipment

Laboratory equipment is the foundation of experimental research and advancement in various scientific fields. These specialized and exact tools are essential for conducting experiments, collecting empirical data, and examining samples. It is possible to further divide the laboratory equipment category into several subcategories, each with a special collection of tools that address certain requirements and applications.

13.2 Key Categories of R&D Tools and Technologies

Analytical Instruments

For accurate measurements of physical and chemical properties, analytical instruments are essential. They are at the forefront of chemical analysis, materials testing, and quality control in R&D. They give researchers the necessary resources to take sensible actions and learn much about their samples. Popular analytical tools include mass spectrometers, which are essential for characterizing complex molecules in proteomics, metabolomics, and forensic science; spectrophotometers, which determine the concentration, purity, and chemical composition of substances; and chromatography systems, which separate and quantify components in complex mixtures.

Biotechnology Tools

Biotechnology tools are crucial in the life sciences and medical R&D because they allow scientists to work with genetic material, explore genomes, and discover novel therapeutic agents. They are at the cutting edge of biological research innovation. The most important biotechnology tools are DNA sequencers, which decipher organisms' genetic codes and revolutionize fields like genomics, personalized medicine, and evolutionary biology; polymerase chain reaction (PCR) machines, which are essential for DNA cloning, disease diagnosis, and genetic engineering; and CRISPR-Cas9 gene editing technologies, which enable precise DNA alterations in living organisms with significant implications for gene therapy, bioproduction, and genetic engineering.

Materials Testing Equipment

Understanding the qualities of various substances, from the microstructure to the macroscopic behavior, depends heavily on materials testing equipment. These tools are used by researchers in various fields, including materials science, aerospace, and the automotive industry. X-ray diffraction machines for examining the crystal structure of materials, aiding in materials characterization, mineralogy, and pharmaceuticals; tensile testing apparatus for evaluating the mechanical properties of materials; and electron microscopes like scanning electron microscopes (SEM) and transmission electron microscopes (TEM), which provide nanoscale imaging capabilities for in-depth surface and structural analysis. This information is crucial for the design and testing of materials for a variety of purposes, including manufacturing and construction.

13.2.2 Computational Tools

Computational tools have become an essential component of current R&D in the age of digital transformation. These tools greatly improve the R&D process by using computers to simulate complicated phenomena, analyze massive datasets, and create predictive models.

Simulation Software

To describe and mimic real-world processes, simulation software is a potent subset of computer tools used in R&D. These tools allow for the replication and analysis of complicated occurrences, resulting in new knowledge, predictions, and an improved understanding of how various systems act. Applications for simulation software are numerous in both industry and research. Simulation software in physics and engineering is essential for comprehending and improving physical systems. Engineers and scientists use these tools to tackle a variety of problems. Examples include determining the structural integrity of structures and infrastructure, designing cars aerodynamically, and forecasting the behavior of electromagnetic fields. These simulations help with design optimization, safety verification, and system behavior prediction under diverse circumstances.

Simulator software is used in environmental modeling to forecast and comprehend intricate natural processes. Researching and preventing environmental problems like climate change and natural disasters depend on it. Climate models use simulation software to forecast climatic trends and investigate human activity's effects on the environment. Intricate simulations of atmospheric variables are used to forecast short- and long-term weather patterns. Additionally, simulation software makes modeling and analyzing natural catastrophes like earthquakes, tsunamis, and hurricanes possible, which offers useful data for disaster response and preparation planning.

In the fields of pharmaceuticals and biotechnology, simulation software is essential for the development of new drugs. Biomolecular behavior, such as protein folding and drug binding, is studied using molecular dynamics simulations. The finding of prospective drug candidates is accelerated greatly by virtual screening using simulation software. Additionally, pharmacokinetic models help researchers better understand how medications are absorbed, distributed, metabolized, and eliminated from the body. These methods lessen the need for animal testing while assisting in developing safe and efficient medications. As simulation software develops, researchers can now perform virtual tests that would be expensive, time-consuming, or impossible to perform in the actual world. These tools will continue to be essential for R&D activities across numerous scientific and industrial fields as computational power and modeling methodologies advance.

13.2 Key Categories of R&D Tools and Technologies

Data Analytics Tools

In many businesses, R&D has come to rely heavily on data analytics technologies. These advanced technologies have been created to extract, process, and interpret the enormous and complex datasets at our disposal, efficiently converting raw data into useful insights. Tools for data analytics are the cornerstone of R&D, trend spotting, and hypothesis testing.

Big data analysis is one crucial area where data analytics technologies excel. Modern technologies are required to manage the enormous amount, velocity, and variety of data generated daily due to the big data boom. Researchers and organizations from various sectors use these tools to glean important insights from huge datasets. To enable predictive modeling, anomaly detection, and data classification, machine learning techniques, such as neural networks, decision trees, and clustering approaches, are essential for spotting patterns and relationships within the data. Businesses increasingly rely on real-time analytics for quick decision-making in fraud detection, network security, and dynamic pricing. In many fields, including healthcare, finance, and marketing, predictive analytics is invaluable because it uses historical data and statistical algorithms.

Data analytics technologies in market research and consumer insights are another crucial use. Businesses need to comprehend consumer behavior and market dynamics to develop products and marketing plans that appeal to their target market. Thanks to data analytics solutions, organizations may conduct extensive market research and customer insights. Tools for analyzing consumer behavior give businesses information on consumer preferences, purchasing behaviors, and buying trends that they may use to customize their products and marketing initiatives. Sentiment analysis tools analyze text data from social media posts and customer reviews to determine how people feel about particular goods, services, or brands. This helps organizations understand customer sentiment and respond quickly to new problems. These tools can also spot market trends and new business prospects, allowing companies to adjust their strategies in response to shifting customer demands.

Additionally, data analytics technologies have a significant influence on scientific research in a variety of fields. These technologies help genomics and bioinformatics assess DNA sequencing data, comprehend genetic variants, and pinpoint disease-linked genes. Bioinformaticians use data analytics technologies to evaluate biological data and create discoveries, particularly in the field of personalized medicine. Data analytics is used in astrophysics and environmental sciences to process and analyze massive datasets from environmental monitoring systems and space observatories, enabling the discovery of celestial events and the evaluation of environmental changes. Additionally, by utilizing statistical analysis, data visualization, and machine learning approaches, data analytics supports researchers in understanding the outcomes of experiments and observations and reaching relevant conclusions in various scientific fields. The primary uses of data analytics tools are outlined in Table 13.1, with use cases, related applications, tools used, and data sources in many fields highlighted.

Table 13.1 Applications of data analytics tools in various fields

Area	Key features	Applications	Tools used	Data sources
Big data analysis	Handle large datasets	Research, business, healthcare, finance	Machine learning, Real-time analytics, predictive analytics	Various data sources
Market research and consumer insights	Understand consumer behavior	Marketing, product development, customer insights	Consumer behavior analysis, sentiment analysis, market trend identification	Surveys, social media, sales data
Scientific research	Applied in scientific disciplines	Genomics, astrophysics, experimental research	Bioinformatics tools, Data analysis software, machine learning	Laboratory experiments, telescopes, environmental sensors

Artificial Intelligence and Machine Learning

R&D is changing due to the emergence of artificial intelligence (AI) and machine learning (ML) as revolutionary computing technologies. These innovations open new horizons in R&D by enabling robots to learn from data and carry out activities independently. AI and ML analyze large biological datasets in healthcare and drug discovery to uncover novel drug candidates, foresee drug interactions, and advance genomic medicine.

A further important area where AI and ML have substantially impacted is predictive analytics and forecasting. Applications for these technologies include demand forecasting, financial analysis, and weather and climate modeling. These technologies fuel predictive models that enable data-driven decision-making. The predictive powers of AI and ML improve catastrophe preparedness, direct investment choices, and optimize supply chains.

Additionally, manufacturing and laboratory procedures are being transformed by AI-driven robotics and automation. High-throughput screening, laboratory automation, and manufacturing optimization are all performed in R&D using automation and AI, greatly enhancing efficiency and precision. Additionally, researchers may extract knowledge from various textual and unstructured data sources, including scientific publications, patents, and different data sets, thanks to AI technologies focusing on natural language processing (NLP) and knowledge discovery. This makes it easier to find pertinent studies, analyze patents, and integrate data, all of which help foster innovation in various sectors.

13.2.3 Communication and Collaboration Tools

Successful R&D initiatives are built on strong communication and teamwork. Tools for collaboration and communication have considerably changed throughout time to serve academics, scientists, and professionals across a wide range of disciplines. Teams can collaborate easily across distances thanks to their facilitation of knowledge, ideas, and resource exchange.

Collaboration Platforms

Collaboration platforms are versatile, dynamic technologies that have revolutionized how professionals and researchers collaborate in the R&D industry. These platforms make communication, project management, and information exchange more efficient. As a result, a centralized area is created where teams can work together and efficiently coordinate their activities. This part will examine the value of collaboration platforms and the essential characteristics that make them so useful in R&D settings

Project Management Software

The tools category, project management software, is essential for improving communication and teamwork in the R&D industry. These software options offer an organized approach to project management and are professionally created to ease the planning, implementation, monitoring, and control of R&D projects. By doing this, they enable scientists, researchers, and project managers to effectively manage key project components, including resources, deadlines, and tasks, assuring the successful completion of R&D projects. These software options provide several essential characteristics that render them indispensable.

One such crucial aspect is the ability to manage and schedule work using tools like Gantt charts and Kanban boards, providing a clear view of job sequences, dependencies, and deadlines. Furthermore, project management software enables efficient resource allocation, ensuring personnel, capital, finance, and supplies are dispersed to particular project tasks or stages, avoiding resource shortages that impede project development. These technologies also include real-time progress tracking, making monitoring work completion easier and quickly identifying and addressing potential bottlenecks. Last but not least, collaborative planning tools enable project stakeholders to work together on project plans, encouraging communication and updates and ensuring that project objectives are met while utilizing the collective wisdom of the project team.

Virtual Conferencing Tools

Tools for virtual conferencing have become crucial parts of the landscape of R&D. Regardless of their physical locations; these tools allow knowledge sharing among researchers, scientists, and professionals while bridging geographic gaps, facilitating real-time communication, and facilitating collaboration. Many different virtual conferencing platforms exist, including video conferencing platforms, webinars, and online meeting tools.

Video conferencing, which enables users to interact face-to-face from several locations, is one of the main characteristics of virtual conferencing systems. In addition to improving communication, this feature allows sharing of visual information like data presentations, whiteboard drawings, and real-time experiments. For research teams working remotely, video conferencing is essential for promoting communication and real-time discussions.

The ability to share the screen is yet another crucial feature of virtual conferencing software. Researchers may easily display data, conduct live demonstrations, or present research findings thanks to the ability to share their screens. Sharing intricate data visualizations, experimental setups, or software demos can all benefit greatly from this functionality.

The ability to host webinars is a feature available on many virtual conferencing platforms. For distributing research findings to a larger audience, such as peers, stakeholders, and the general public, webinars are invaluable. The study results can be presented, followed by Q&A sessions and discussions. Researchers can interact with a wide audience and disseminate their expertise through webinars, which can be used as educational tools. These platforms frequently offer the opportunity to record meetings and webinars. This is crucial for saving research presentations and debates. These recordings can be listened to again by researchers for archival purposes or to communicate with colleagues who could not make the live session. Virtual conferencing systems with real-time chat and messaging capabilities enable text-based communication in addition to audio and video chats. This function helps exchange materials, links, and clarifications during meetings to ensure thorough and fruitful conversations.

13.3 Impact of Tools and Technologies on R&D

The impact of tools and technologies on R&D is multifaceted. They enable researchers to be more productive, innovative, and collaborative. However, it is essential to adapt to these technological advances, manage the associated challenges (data privacy, ethical considerations, etc.), and provide training to researchers to make the most of these tools for their R&D endeavors.

13.3 Impact of Tools and Technologies on R&D

13.3.1 Accelerating Research

R&D instrument and technology innovations have ushered in a new rapid innovation and discovery era. This section examines how these instruments have accelerated research in various fields. Experiments were typically time-consuming and significantly dependent on manual procedures in conventional research environments. Scientists and researchers meticulously conducted experiments and data collection, resulting in a lethargic rate of advancement in various fields. Incorporating cutting-edge instruments and technologies has revolutionized the rate of experimentation and data analysis.

Automation and robotics implementation in laboratories is one of the most significant contributors to accelerated research. Automation has enabled the precise execution of experiments, eradicating the possibility of human error. Robots can conduct repetitive tasks with extraordinary accuracy, and they can operate continuously without requiring breaks, drastically reducing the time required for experiments. In pharmaceuticals, materials science, and chemical engineering fields, robotic systems perform tasks including high-throughput screening, compound testing, and sample preparation. Not only do these automated systems expedite experiments, but they also improve their repeatability and dependability. Researchers can now conduct multiple experiments in a fraction of the time it used to take, enabling them to investigate a broader range of variables and test hypotheses faster.

High-throughput screening (HTS) is a game-changer in drug development, genomics, and proteomics. Using automation, this strategy involves rapidly testing thousands to millions of chemical compounds or biological samples. High-throughput screening requires automated liquid handling systems, sophisticated imaging technologies, and specialized data analysis instruments. In pharmaceutical research, for example, high-throughput screening platforms allow researchers to assess the potential for multiple drug candidates to interact with particular targets or pathways. By outsourcing the procedure and combining it with sophisticated data analysis, researchers can identify promising compounds and eliminate ineffective ones with greater efficiency. This not only accelerates the drug discovery process but also reduces costs significantly. These expedited research methods have a substantial effect. The ability to conduct experiments much faster has facilitated new avenues of investigation and innovation, such as the identification of materials with novel properties and applications, ranging from the development of new medications to the identification of materials with novel properties and applications. Researchers can now respond faster to new challenges and opportunities, fostering scientific advancements and discoveries that were previously impossible within the constraints of traditional research timelines.

13.3.2 Enhancing Precision and Accuracy

In many professions, the capacity to conduct experiments and collect data with the greatest precision can mean the difference between success and failure. Tools and technology have improved R&D's precision, accuracy, understanding of the world, and innovation ability. Integrating automation and robots into laboratory settings has been one of the most notable developments in R&D (Fig. 13.2). This advancement has given research that relied mainly on human participation a new level of accuracy and consistency. Automation lowers the possibility of human errors brought on by distraction, exhaustion, or variances in technique. Robots can work continuously, enabling unhindered experimentation and quickening research timelines. Furthermore, complicated investigations have been simplified by the perfect synchronization of numerous robotic arms and instruments in highly coordinated processes.

Precision is the key to success in genetic engineering and genomics. An era of unmatched precision and accuracy in modifying genetic information has arrived because of DNA sequencing and synthesis technology advances. Modern DNA sequencers have the speed and accuracy to read billions of DNA base pairs. This accuracy enables the identification of disease-causing genetic mutations, the comprehension of evolutionary links, and the tracking of genetic variation within populations. Precision bespoke DNA sequences can now be generated because of advances in DNA synthesis technology. This accuracy is crucial for gene editing and synthetic biology because it enables the creation of novel biomaterials, targeted medicines, and the engineering of organisms for particular uses.

Precision and accuracy have been greatly improved in R&D thanks to the integration of automation and robots and the development of cutting-edge genome sequencing and DNA synthesis technologies. These techniques have advanced areas like medicines, biotechnology, and materials research by increasing the validity of experimental results and the possible range. These tools' consistency and accuracy have opened up new opportunities and made it possible for breakthroughs that were previously thought to be out of reach.

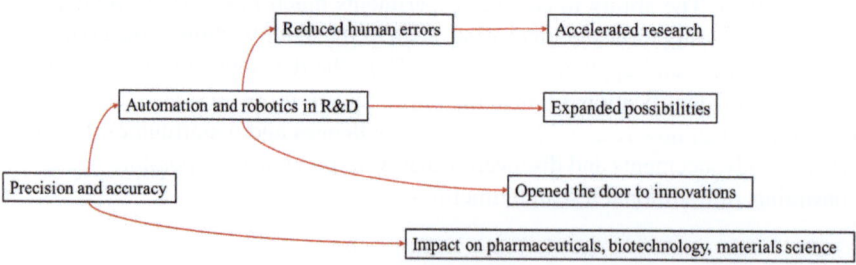

Fig. 13.2 The impact of automation and robotics in R&D

13.3.3 Data-Driven Decision-Making

Modern R&D is built on data-driven decision-making, ushering in a new era of innovation and effectiveness. The wide availability of data and powerful analytics tools has fundamentally transformed how research procedures are carried out, evaluated, and optimized. The idea of "big data," which includes enormous and complicated datasets produced by R&D tools and technologies, characterizes this paradigm shift. Modern analytics techniques and artificial intelligence (AI) combine and evaluate this information, ranging from climate sensor data to DNA sequences used in genomics.

Predictive modeling is made easier by advanced data analytics, which not only finds subtle patterns and correlations within enormous datasets. Models that predict future results based on historical data can be created by researchers, revolutionizing project prioritization and resource allocation. In this process, artificial intelligence (AI) is crucial, accelerating processes like materials development, individualized genetic treatment plans, and chemical compound screening in drug discovery. Data-driven decision-making, however, poses questions as it gains popularity. Data security and privacy are paramount, particularly in a time of rising cyber threats. Along with difficulties relating to the interoperability of diverse data sources and analytical tools, ethical issues like data ownership, informed permission, and algorithmic biases must also be addressed.

13.3.4 Facilitating Global Collaboration

Thanks to revolutionary tools and technology, international collaboration in R&D is now made possible on a scale never before possible. Modern technology has enabled the emergence of remote work and the creation of transnational R&D teams. Researcher collaboration across geographical borders is made possible by digital tools, including video conferencing platforms, cloud-based data storage, and other digital solutions. By utilizing global experience and varied cultural viewpoints, this accessibility not only improves data security and access but also widens the body of knowledge, ultimately encouraging creative responses to challenging issues.

Cross-disciplinary cooperation is increasingly at the forefront of R&D in addition to worldwide collaboration. The efficient coordination of researchers from many professions is made possible by specialized collaboration software and real-time data exchange systems. This strategy overcomes disciplinary and linguistic boundaries, allowing teams to collaborate on approach adaptation and communication. The resulting diversity of perspectives and knowledge has proven beneficial for addressing complex problems that necessitate novel solutions.

13.4 Innovations in R&D Tools

The persistent pursuit of innovation and advancement is a defining characteristic of the R&D industry. Recent years have seen an explosion of impressive R&D tools and technology advancements that promise to change the face of research in various fields. These developments are quickening the rate of discovery and broadening the range of potential outcomes.

13.4.1 Advanced Technologies on the Horizon

Emerging technologies that promise to alter how R&D operations are carried out fundamentally are poised to cause a significant upheaval in the R&D landscape. These innovations are poised to fundamentally alter the R&D industry and represent quantum leaps in human capability. Operating at the nanoscale, nanotechnology is becoming increasingly important in R&D. It entails tinkering with substances and structures with dimensions smaller than 100 nanometers, resulting in distinctive physical and chemical features [7]. Nanotechnology is enabling groundbreaking improvements in drug delivery in medicine. Drugs are transported directly to target cells or tissues using nanoparticles and nanocarriers to reduce side effects and improve therapeutic effectiveness. Additionally, for early illness detection and environmental monitoring, highly effective sensors utilizing nanomaterials are being deployed to provide data with previously unheard-of precision and sensitivity.

This technological revolution is being led by quantum computing, changing how R&D experts approach challenging challenges. Quantum computers use quantum physics to process information differently from conventional computers, allowing them to do complicated calculations and simulations with previously unheard-of speed and effectiveness. Quantum computers are modeling and recreating atomic and molecular behavior in materials research, greatly speeding up the creation of novel materials. Quantum computing accelerates drug development in pharmaceutical R&D by quickly evaluating chemical interactions and forecasting pharmacological features. It can be good to expect groundbreaking discoveries and previously unattainable solutions as these advanced technologies become more widely available and integrated into the R&D process. To use them to their greatest potential, however, they also provide a unique set of difficulties that call for careful consideration of moral, legal, and security considerations. Table 13.2 lists current trends, research areas, and advanced technology applications.

13.4 Innovations in R&D Tools

Table 13.2 Emerging technologies of R&D

Technology	Potential applications	Emerging trends	Research areas
Nanotechnology	1. Drug delivery at the cellular level	Miniaturization, targeted therapies	Drug design, nanomaterials, safety
	2. Advanced materials with unique properties	Sustainable materials, energy storage	Material science, energy, environment
	3. Targeted cancer therapies	Immunotherapy, personalized medicine	Oncology, immunology, genomics
	4. Environmental remediation	Nanosensors, green nanotechnology	Environmental science, chemistry
	5. Energy-efficient electronics	Quantum dots, nanowires	Electronics, materials, energy
Quantum computing	1. Cryptography and data security	Quantum cryptography, post-quantum crypto	Cryptography, cybersecurity, quantum physics
	2. Optimization problems in logistics and finance	Quantum algorithms, quantum optimization	Operations research, finance, algorithms
	3. Drug discovery and molecular modeling	Quantum simulations, drug screening	Chemistry, pharmacology, quantum physics
	4. Climate modeling and simulations	Quantum simulations, climate modeling	Climate science, atmospheric modeling, quantum physics
	5. Artificial intelligence acceleration	Quantum machine learning	Machine learning, artificial intelligence, quantum physics
3D Printing	1. Customized prosthetics and implants	Bioprinting, multi-material printing	Biomedical engineering, materials science, design
	2. Aerospace and automotive parts manufacturing	Large-scale printing, high-performance materials	Aerospace engineering, materials science, engineering
	3. Construction and architectural prototypes	3D-printed buildings, sustainable construction	Architecture, civil engineering, materials science
	4. Bioprinting for tissue and organ regeneration	Organ printing, regenerative medicine	Biomedical engineering, tissue engineering, biology
	5. Art and fashion design	Artistic 3D printing, sustainable fashion	Art, fashion design, materials science

(continued)

Table 13.2 (continued)

Technology	Potential applications	Emerging trends	Research areas
CRISPR and gene editing	1. Genetic disease treatment and prevention	Base editing, therapeutic applications	Genetics, molecular biology, medicine
	2. Crop improvement and agricultural innovation	Precision agriculture, gene-edited crops	Agriculture, genetics, biotechnology
	3. Gene therapy for inherited disorders	In vivo gene editing, gene therapy	Molecular biology, genetics, medicine
	4. Stem cell research and regenerative medicine	Stem cell editing, tissue regeneration	Regenerative medicine, stem cell biology, medicine
	5. Bioengineering for biopharmaceuticals	Synthetic biology, biopharmaceuticals	Synthetic biology, pharmaceuticals, biotechnology

13.4.2 Breakthroughs in Data Management and Storage

Data's importance in today's R&D environment cannot be emphasized. It has a significant impact on both the speed and caliber of study findings. The R&D industry has undergone revolutionary changes due to data management and storage innovations. This section examines two significant innovations: blockchain and edge computing for R&D data security.

In R&D settings, edge computing, a paradigm shift in data processing, is a game-changer. Processing data at or close to the source avoids the usual drawbacks of centralized cloud computing. This method has several important benefits in R&D. As a result of reducing data latency, it supports real-time analysis and decision-making for various applications, including monitoring scientific experiments and robotics and autonomous cars. By limiting the visibility of critical research data during transmission and maintaining it within local networks, edge computing also improves privacy and security. In sectors with strict requirements for data protection, this quality is priceless. Additional benefits of edge computing include scalability and cost-effectiveness because it allows for the strategic distribution of processing nodes that cater to the needs of certain R&D projects.

Blockchain technology is becoming more widely acknowledged as a guarantee for the validity and integrity of R&D data in data security. Documenting each modification in an immutable ledger assures data traceability and immutability. This promotes data integrity and guards against unlawful tampering. Such strong data integrity is essential in fields like clinical trials, where regulatory compliance depends on maintaining the confidentiality of patient data. The decentralized architecture of blockchains enables diverse parties to collaborate and share data securely. This function is crucial for international partnerships and preserving regulatory compliance. The impact of blockchain technology on the R&D industry is expected to increase as its use spreads, ensuring a safe and open environment for data management.

13.4 Innovations in R&D Tools

Data management and storage methods developments represent a sea change in R&D procedures. They enable researchers to operate more productively, safely, and cooperatively, pushing the limits of innovation and science. These improvements give R&D professionals better confidence and a larger capacity for innovation when they take on initiatives, whether by lowering data latency, boosting data security, or maintaining data integrity.

13.4.3 Novel Methods for Research and Experimentation

Innovative approaches to study and experimentation are constantly being sought after. These cutting-edge techniques are revolutionizing R&D and allowing scientists to delve into new waters.

3D Printing and Prototyping

Additive manufacturing, usually called 3D printing, has become a game-changing advancement in R&D. It has completely changed how R&D initiatives are conceived, created, and evaluated. The research design and testing phases have been expedited because of 3D printing's rapid prototyping capabilities, greatly cutting both time and expense. Digital designs may be quickly translated into physical prototypes by researchers, enabling the speedy examination and improvement of concepts. This agility fosters creativity by enabling the investigation of novel solutions and a deeper understanding of complicated issues. Furthermore, 3D printing has overcome the restrictions of conventional manufacturing processes, enabling the creation of elaborate and complicated geometries that would not otherwise be conceivable. This feature is especially useful for the medical industry's specialized implants, prosthetics, anatomical models, and industries like aerospace, where lightweight yet durable components are essential. It is anticipated that when technology develops further, it will have an even greater impact on R&D and open up new opportunities for advancement. However, it is crucial to consider quality control and material safety, especially in sectors with stringent rules and safety standards.

Beyond its use in engineering, 3D printing is setting the bar for improvements in bioprinting. In this area, scientists can use living-cell bioinks to construct three-dimensional biological structures, such as tissues and organs. Drug testing, organ transplantation, and regenerative medicine could all be drastically improved by this technology. Furthermore, by minimizing material waste, 3D printing is advancing sustainability. 3D printing is an additive procedure that only uses the material necessary to create the object, in contrast to conventional subtractive manufacturing techniques that produce significant waste. Additionally, materials that would otherwise be considered garbage can be recycled using 3D printing to create useful resources. As technology develops, it has the potential to push the limits of R&D, presenting fresh chances for creativity and discovery.

Incorporating 3D printing and prototyping in R&D promotes innovation and creativity, shortens the time needed for development, and opens up new avenues for research. Thanks to this revolutionary technique, researchers can now produce highly optimized components with complex geometries, which transcend limitations imposed by conventional manufacturing processes. It quickens the prototype procedure, enabling experimentation with multiple design iterations and a deeper comprehension of challenging issues. In addition to engineering, 3D printing is advancing in bioprinting and the construction of biological structures, opening up promising new avenues for regenerative medicine. Additionally, it has clear sustainability advantages, minimizing material waste and promoting environmental preservation. As 3D printing technology develops, it continues to transform the field of R&D, providing fresh opportunities for advancement while demanding attention to quality control and material safety by industry norms.

CRISPR and Gene Editing

Genetic research and gene editing have changed profoundly thanks to clustered, regularly interspaced short palindromic repeats (CRISPR) and the Cas9 technology. With the help of this groundbreaking technique, precise genetic manipulation is made possible, allowing for targeted genetic alterations. Researchers can replace or fix incorrect DNA sequences for future cures of genetic illnesses, selectively disable genes to explore their roles, and build disease models for medication development.

CRISPR-based technologies are changing agriculture in addition to the medical field. Crops with enhanced features, such as disease resistance and better yield, are being developed using technology. Researchers want to address food security issues and lessen agriculture's negative effects on the environment.

The therapeutic potential of CRISPR gene editing is enormous. Repairing broken genes offers the chance to heal hereditary illnesses at their source. It also paves the door for customized medicine, which adapts medical interventions to a person's genetic profile. However, the enormous potential of CRISPR creates ethical and legal problems, particularly in editing human germline, which has prompted continuing global discussions and the creation of regulatory frameworks to ensure responsible and advantageous use.

13.5 Role of R&D Tools in the Coming Years

The rapid evolution of technology will substantially impact how R&D tools are used in the future. It is predicted that machine learning and artificial intelligence (AI) will become widely used in R&D tools. These technologies will improve decision-making by providing data-driven insights, experiment optimization, and speeding up research procedures across multiple areas. AR and VR, which enable realistic data visualization and simulations, will find a role in R&D. In product design and structural engineering, researchers can interact with data in three-dimensional spaces, enabling a greater understanding of complex processes and enhancing collaboration.

With sophisticated laboratory robots and automated data gathering systems handling repetitive jobs with accuracy, automation and robotics are anticipated to play a larger role in society. The creative and strategic components of research will get more chance to shine. With open-access platforms enabling worldwide collaboration, increasing openness, and minimizing duplication of effort, collaborative research will experience a paradigm shift. In our increasingly linked world, improved data security methods, including the usage of blockchain, will be essential to safeguard the confidentiality and integrity of research data.

R&D processes will prioritize sustainability and environmental awareness in addition to the creation of tools. Particularly in disciplines like materials science and energy research, cutting-edge methods and technology will aid in reducing the environmental impact of research. Cross-disciplinary research will pick up steam as specialists from various fields and disciplines work together to address complicated, multifaceted problems. Additionally, personalized and precision medicine will become more prevalent in the healthcare industry. Genetic sequencing and data analytics techniques will make it possible to create customized treatments, vastly enhancing healthcare outcomes.

References

1. A. Berezow, et al., *The Next Plague and What Science Will Do To Stop It* (2018)
2. E. Homburg, The rise of analytical chemistry and its consequences for the development of the German chemical profession (1780–1860). Ambix **46**(1), 1–32 (1999)
3. S.A. Matlin, B.M. Abegaz, *Chemistry for Development. The Chemical Element: Chemistry's Contribution to Our Global Future* (2011), pp. 1–70
4. J. Van den Ende, R. Kemp, Technological transformations in history: How the computer regime grew out of existing computing regimes. Res. Policy **28**(8), 833–851 (1999)
5. A.L. Goldstein, *Denatured: Emergent Realities of Encyclopedic DNA Elements* (2017)
6. I.M. Cockburn, R. Henderson, S. Stern, The impact of artificial intelligence on innovation: An exploratory analysis, in *The Economics of Artificial Intelligence: An Agenda*, (University of Chicago Press, 2018), pp. 115–146
7. B. Rogers, J. Adams, S. Pennathur, *Nanotechnology: Understanding Small Systems* (CRC Press, 2014)

Chapter 14
R&D Intellectual Property and Copyright

Intellectual property (IP) and the research and development (R&D) industry, which is focused on innovation, meet in a complex and dynamic way that raises many legal, ethical, and strategic questions. A range of legal rights known as "intellectual property" are meant to safeguard the products of human ingenuity and incentivize innovation by protecting the works of human intellect. Among them, copyright legislation stands out as a pillar for upholding and fostering originality, giving authors exclusive rights to their creative works across various R&D fields. This chapter explores the complex interactions between R&D, intellectual property, and copyright and offers a thorough grasp of copyright law's foundations and applications in the field of R&D. It examines copyright ownership, protection, enforcement, and collaborative aspects while taking into account emerging trends.

> **Learning Objectives**
> - Define the importance of intellectual property (IP) in R&D.
> - Explain the core principles of copyright law and how they apply to R&D.
> - Analyze copyright ownership and licensing issues in the R&D process.
> - Understand the implications of collaborative agreements on IP in R&D.
> - Explore strategies for protecting R&D work through copyright and other means.
> - Discuss emerging challenges in R&D IP and copyright.

14.1 Understanding Intellectual Property in R&D

The foundation of innovation and advancement in many industries, including technology, science, health, and the arts, is R&D. IP is a key notion in the context of R&D essential for safeguarding and advancing innovation results [1]. Various legal rights and protections given to people and organizations for their creative and innovative work are referred to as IP. These rights are a vital guarantee for the worth of R&D investments, promoting additional study and economic growth. It is possible to think of IP as a collection of exclusive rights and governing laws that protect human intellect's works. Intangible assets produced by human intelligence and innovation are covered by it. These assets can include inventions, creative works, discoveries, and more in the field of R&D.

14.1.1 Types of IP

Patents

Inventors and artists are granted exclusive rights to their discoveries through patents, a legal protection for IP that prevents unauthorized parties from producing, using, selling, or importing the patented product. Particularly in the area of R&D, patents are essential for defending IP rights. Several important characteristics define patents. They demand that the innovation be novel— fresh and original—with no earlier public disclosure or patenting. Second, patents shield ideas that serve a real-world need, provide a beneficial service, or solve a real-world problem. Third, a patent requires that the invention be nonobvious, which means that it must not be obvious to subject matter experts and must mark a substantial departure from prior art or technology. Patents have a finite lifespan of approximately 20 years from the filing date [2]. The inventor is granted exclusive rights to the innovation during this time. The invention, however, enters the public domain after the patent expires (Fig. 14.1).

To safeguard various R&D innovations, numerous sorts of patents are available. The following categories are especially important: The most prevalent kind of patents, utility patents, which cover methods, apparatus, manufactures, and compositions of matter, are frequently requested for technological advancements such as new hardware, computer software, or chemical compounds. Plant patents, which are specific to breeding and cultivating unique and distinctive plant varieties, are applicable in R&D projects in agriculture and horticulture. Design patents, which protect the ornamental or aesthetic aspects of a product's design, are used in R&D to safeguard the distinctive appearance of products, including shape, surface, and decorative elements. The steps in the patent application process include submitting an application to the relevant patent office, such as the United States Patent and Trademark Office (USPTO), having it examined to see if it satisfies the requirements for patentability, and, if approved, having a patent issued, which gives the invention exclusive rights for the predetermined amount of time.

By encouraging innovation and defending the money invested in R&D, patents are essential to R&D. They give R&D organizations a legal framework to protect their technological innovations and, if wanted, to license or commercialize their

14.1 Understanding Intellectual Property in R&D

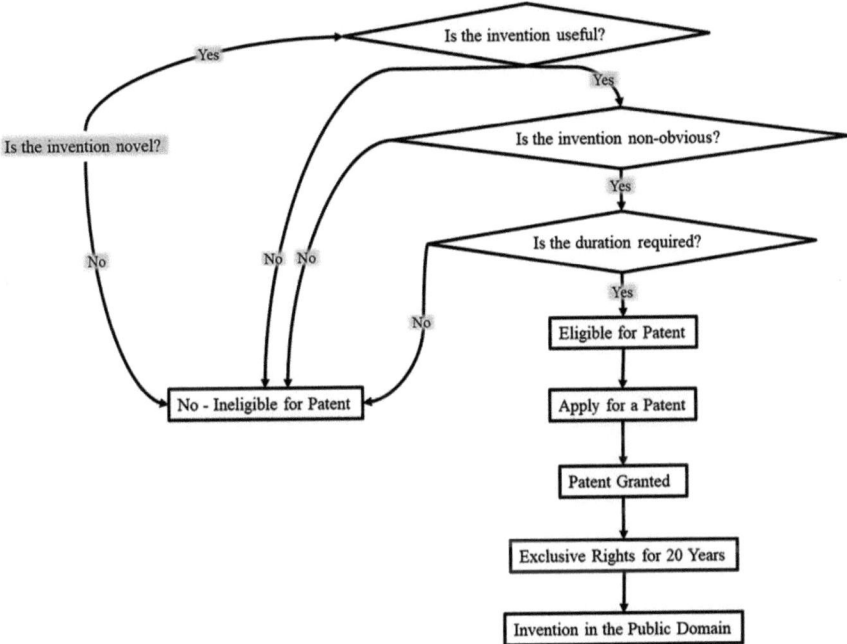

Fig. 14.1 Patent eligibility decision flowchart

creations. Due to its definition of ownership and rights about inventive discoveries, patents can also encourage collaborations and joint ventures.

Copyright

A legal safeguard known as copyright is given to those who create original works of authorship. It includes a wide variety of artistic, musical, literary, and other fixed works that are expressions of creativity and intellect. The owner or creator of the work has the only authority to decide how it is used, reproduced, distributed, customized, performed, and shown. This is known as copyright. Without the requirement for registration or other formalities, this protection kicks in as soon as the work is generated and established in a concrete form.

A broad range of artistic creations, including literary works like books and articles and creative products like paintings, sculptures, and pictures, are subject to copyright protection. Scripts and performance rights for dramatic works like plays and operas are also shielded by copyright, as are musical compositions and lyrics. The scope of protected content includes audiovisual works, such as movies and multimedia presentations, and software, which is subject to the same protections as literary works. Additionally, building architectural plans and drawings may be protected by copyright laws.

The privileges granted by copyright give owners and producers various exclusive rights. These include the ability to make copies of the work and the right to duplicate it. Copyright grants the power to share documents with the general public, facilitating wider dissemination. Further, it encourages creativity and innovation by enabling the author to produce derivative works by modifying or adapting the original. The duration of copyright protection depends on the jurisdiction and the nature of the work, with general guidelines as follows. Copyright protection affords the right to publicly perform the job for positions such as music and theatrical performances. At the same time, it also includes the right to publicly display the results of visual arts and photographs. Copyright normally lasts for the creator's lifetime and an extra 50 to 70 years after they pass for particular creations. Copyright protection typically lasts 95 years from the date of publication or 120 years from the date of creation for corporate works created by workers while acting in the course of their job, whichever is shorter. Copyright protection for anonymous or pseudonymous works can last for 120 years from the date of creation or 95 years from the date of publication [3]. These rules are the foundation for the framework that controls the length of copyright protection for different kinds of creative works.

Trademarks

IP law is fundamentally concerned with symbols, names, or other devices that identify and distinguish the source of products or services. These devices are known as trademarks. Trademarks may not be the main topic in the context of R&D, but they are essential to the branding and marketing of novel products. Trademarks are distinguishing characteristics that help build a distinct brand identity, such as logos, brand names, and slogans. While R&D focuses primarily on technical and scientific issues, trademarks are important because they help identify products, increase brand recognition, and make possible marketing and licensing agreements. Enforcement is crucial to protect the value and reputation of these marks, and trademark registration offers legal protection against rivals using identical patterns. Trademarks are essential in transitioning from R&D to commercialization and market entry because, unlike patents and copyrights, their safety can last indefinitely as long as the mark is used and its registration is upheld.

Trade Secrets

Trade secrets are a specialized form of IP protection that is especially important. These trade secrets include private and proprietary data, including calculations, procedures, client lists, and business plans, giving a company a competitive edge. Trade secrets are identified by the stringent confidentiality they require, their monetary value, and the measures taken to protect them. Trade secrets play a role in R&D when new technologies, inventive techniques, and groundbreaking procedures are developed. They provide long-term protection in place of patents. They enable businesses to keep a competitive edge, particularly in sectors like pharmaceuticals where the formula of a groundbreaking drug may be jealously guarded as a trade secret. Trade secrets can be selectively disclosed to partners in cooperative R&D projects through contracts, fostering creativity while safeguarding confidentiality. Trade secrets are legally protected by state or federal legislation, and unauthorized use is punishable by fines and restraining orders. However, maintaining confidentiality is the main difficulty, calling for strong security measures, nondisclosure agreements, and employee education to protect these priceless intellectual assets.

Licensing

In the field of IP, licensing is crucial, especially in the fast-paced world of R&D. The permitted use of IP assets in this context is referred to as licensing, frequently with specified terms and restrictions. It performs as a flexible and strategic tool with numerous advantages. It first provides access to external technology by utilizing outside knowledge, boosting R&D initiatives. Second, licensing can generate income by allowing IP owners to monetize their assets through licensing fees and royalties, which can then be put back into R&D or applied to new forms of innovation. Third, it promotes teamwork by permitting joint ventures and partnerships where several parties pool their resources and skills to undertake challenging R&D initiatives. Additionally, licensing can increase market penetration, reduce risks, promote technological transfer, and improve IP portfolio management.

Licensing agreements are essential in the R&D environment to promote collaboration, grow markets, and efficiently manage IP assets. License agreements need to include clear terms and limitations. These should specify the license's boundaries, length, geographical rights, fee schedules, and usage limitations. Effective licensing agreements are essential to ensuring the preservation of the interests of all parties. To ensure compliance with relevant laws and regulations, it is advised to get legal guidance when designing licensing agreements because they can be complex and subject to several legal restrictions.

14.2 Copyright Basics

The legal system of copyright encourages innovation and creativity by giving authors exclusive rights to their unique works. It covers a range of intellectual and artistic expressions, guaranteeing that authors may manage how their works are cited, copied, and shared. These exclusive rights are not perpetual; they have a set duration, and the results become part of the public domain. Through the provision of restricted monopolies and the acceptance of exceptions like fair use for activities like research and education, copyright tries to balance the interests of creators and the general public.

In addition to literary works (books, articles, and code), artistic works (paintings, sculptures, and pictures), musical compositions, theatrical works (plays and screenplays), architectural designs, and software programs, copyright protection also covers a broad range of other creative and intellectual works. By recognizing uniqueness as a requirement, it makes sure that only genuinely innovative expressions are protected. In the context of R&D, where creators and researchers frequently face copyright-related difficulties while accessing or producing copyrighted content, this balance between creator rights and the public's interest is vital to IP law and plays a crucial role.

14.2.1 What Can Be Copyrighted?

Original Works of Authorship

Being an original work of authorship is one of the primary requirements for a position to be eligible for copyright protection. According to copyright law, the work needs to be original, which means it needs to come from the creator and be creative in some way. The work needs not to be a replica or merely a derivation of an existing appointment, even though this threshold for inventiveness is typically relatively low. This requirement highlights human creativity's significance in producing copyrighted works. Works that are entirely computer-generated and lack significant human involvement might not be protected by copyright. However, they might qualify if human creativity was used in the generation process.

The exact expression or presentation of facts or ideas is protected by copyright, not the facts or opinions. This means that while "$E = mc^2$" cannot be covered by copyright, a book or essay that creatively and uniquely explains it can. Common knowledge and extremely succinct sentences or slogans are typically ineligible for copyright protection. Additionally, government-produced works are frequently not covered by copyright laws and are considered in the public domain in many places. However, some components of official works might remain under protection. Creators and researchers in the field of R&D need to be conscious of this need when developing content or producing IP since understanding the notion of original works of authorship is fundamental to identifying which works are eligible for copyright protection.

Fixation in a Tangible Medium

The fixation requirement is a crucial criterion for deciding whether a work is eligible for protection. According to this requirement, the position must be "fixed in a tangible medium of expression." Accordingly, the work must exist in a perceivable and stable form. Print media, such as books; digital media, such as computer files; analog media, such as vinyl records; and physical artworks are examples of tangible media. This fixation guarantees that the work can be viewed directly or with a tool, such as a computer, for digital data. It can be perceived across time while remaining stable and impervious to change or erasure.

For creators and researchers in R&D, the fixation requirement is a crucial factor to consider. It emphasizes the value of steady, tangible documentation and preservation of creative and intellectual works. Adhering to this condition guarantees that the results are eligible for copyright protection, protecting them from unlawful use or reproduction, whether producing written reports, digital content, or scientific research findings. However, it is important to remember that transient or ephemeral statements, such as uttered words in conversation, may only satisfy this condition if recorded or otherwise fixed in tangible media.

14.2.2 Rights Conferred by Copyright

A crucial legal principle, copyright protects authors' creative works and other IPs. It gives people exclusive rights to their original works of authorship, preventing unauthorized use, duplication, or distribution of those works. This protection covers various artistic and creative works, including literary works, musical compositions, and computer code. For copyright protection to be effective, a work needs to be "fixed in a tangible medium of expression," which refers to an observable and stable recording or storage method, such as a book written on paper, a digital music recording, or software code saved on a computer's hard drive.

The exclusive rights provided by copyright include several crucial rights for creators. The most fundamental of these rights is the right of "Reproduction," which enables owners of copyrights to make copies of their work in both physical and digital versions. The "Distribution" right allows creators to manage all aspects of how their work is made accessible to the public, including sales, rentals, and lending. The right to "Derivative Works"—including adaptations, translations, or other transformative works—is also granted to creators of their original positions. The final option allows authors to control how their work is "Publicly Performed and Displayed," including offline demonstrations, public exhibitions, or internet displays. However, these rights are not absolute and may be subject to limitations. For example, the "fair use" doctrine allows for the restricted use of copyrighted material in some situations, such as criticism, commentary, or study.

14.2.3 Duration of Copyright Protection

Copyright protection is a finite right that expires after a certain period. Copyright protection's duration varies from country to country and is based on the nature of the work and particular conditions. Knowing how long a work is protected by copyright is crucial since it determines when it enters the public domain and can be used by anybody without restriction.

In most jurisdictions, the copyright for works produced by individual creators typically lasts for the creator's lifetime plus a predetermined period beyond their passing, which is often between 50 and 70 years. The lifespan of the final surviving author determines the longevity of works with joint authors. Copyright frequently expires after a certain number of years from the date of first publication for anonymous or pseudonymous works. Usually, the copyright conditions for pieces made for hire or corporate positions differ from those for employment made by individual producers.

When the copyright protection period expires, the work becomes public domain. The result is no longer covered by copyright when it is in the public domain, and anybody is free to use it however they see fit without asking permission or paying royalties. As they offer many publicly accessible resources for aspiring creators to build upon, works in the public domain play a crucial role in encouraging creativity and innovation. For producers, researchers, and anybody accessing copyrighted works, understanding the length of copyright protection is essential for striking a balance between preserving creators' rights and the accessibility of knowledge and culture to the general public.

14.3 Copyright in the R&D Process

Copyright in the R&D process is an important legal concept that protects the intellectual property created during the development of new ideas, inventions, and creative works. Copyright law primarily applies to original works of authorship that are fixed in a tangible medium, but its application in the R&D process can be somewhat limited.

14.3.1 Copyright Ownership in R&D

R&D copyright ownership is a complex topic requiring careful thought and adherence to institutional guidelines and contractual commitments. It is crucial to comprehend who owns the rights to creative works when discussing R&D. The fundamental rule of most copyright jurisdictions is that the original owner of a copyrighted work is who created it. This idea also applies to the field of R&D, where writers and researchers are frequently regarded as the creators and proprietors of their original works, which may include research papers, reports, software code, and other kinds of IP. However, difficulties arise when R&D operations occur within organizations or businesses, each of which has specific copyright ownership policies and regulations.

Academic and corporate institutions frequently set copyright policies to control IP produced by staff members or affiliates. Universities and research organizations are examples of educational entities that frequently have regulations defining copyright ownership. These regulations may vary, but they often state that professors, researchers, and students maintain ownership of their works' copyrights. Institutions, however, may be granted particular rights, such as the right to utilize the work for teaching or research. People engaged in R&D in academic environments should know their institution's copyright policies because they can differ greatly. When it comes to corporate R&D, the employer frequently asserts ownership rights over the creative work produced by employees, especially when the work falls under the job's purview. Employment contracts and agreements clearly define copyright ownership and exceptions.

R&D frequently involves collaboration, and when this occurs, copyright ownership issues can get complicated. If numerous parties—individuals, departments, or organizations—are involved, joint ownership may result. Subject to the approval of the other joint owners, each joint owner has the right to use and license the work. Collaboration agreements should specifically address copyright ownership and the distribution of rights to avoid potential issues. Additionally, copyright ownership usually depends on the terms of the contract in situations when R&D activities are contracted out to independent contractors or freelance researchers. In these circumstances, it is crucial to have written agreements that clearly define who will have the copyrights to the work produced and under what circumstances. Understanding institutional policies and contractual duties is essential for navigating the copyright landscape in R&D. To minimize disputes and protect their IP, researchers, particularly those working on collaborative or corporate R&D projects, should proactively address copyright ownership at the project's inception.

14.3.2 Work for Hire and Copyright

The idea of "work for hire" is fundamental to copyright law. It specifies who owns the copyright to creative works and names the commissioning party or employer as the original copyright owner. This doctrine applies to particularly commissioned works when both parties have agreed in writing and works produced by employees acting within the scope of their employment. The work-for-hire doctrine's application in R&D has important ramifications for copyright ownership.

For various reasons, researchers, universities, and organizations involved in R&D must comprehend the work-for-hire philosophy. When workers conduct R&D, it assures obvious copyright ownership, simplifying the division of rights and obligations. Clear agreements defining the endeavor's status as a work for hire or not become crucial when R&D projects involve independent contractors or freelancers. Determining copyright ownership is vital to collaboration agreements in collaborative R&D projects that frequently involve several contributors from various businesses. These agreements should cover the final work's rights, license, and prospective licensing.

It is important for anybody involved in R&D, including employees, contractors, and collaborators, to be aware of the effects work done for hire has on their IP. Employee researchers should know the copyright rules and employment agreements that may set forth specific rights and responsibilities. To prevent future disagreements, contracted researchers should carefully study contracts to establish whether the task is classified as work for hire. Participants in collaborative R&D need to create legal agreements that include copyright ownership, licensing, and the use of R&D results. Legal professionals should be consulted to clarify the rights and obligations of all parties involved.

14.3.3 Fair Use and Research

A key principle of copyright law is the notion of fair use, which permits the limited use of works protected by copyright without the owner's consent or payment of royalties. Fair use is particularly important for researchers since it offers freedom when incorporating copyrighted works into their projects, articles, and studies. Researchers frequently have to quote, cite, or reference copyrighted materials in their scholarly writings, such as theses, essays, and research papers. Fair use authorizes the restricted use of copyrighted content for these objectives if proper citation and acknowledgment are upheld.

Academics and researchers who offer analysis, remark, or critique can critically analyze copyrighted works. This could entail making legitimate use of copyrighted material, pictures, or other media to support claims or offer evidence. Researchers may use parody or satire in specific R&D projects, particularly in the humanities and social sciences, to make a point or communicate their findings. Copyrighted resources may be used in these creative creations under the terms of fair use. In R&D, text mining and data analysis are becoming increasingly crucial. Fair use may be applicable if software is used to process copyrighted resources for data analysis and the service is transformative and furthers research goals.

It can be challenging to tell whether a given usage falls under the definition of fair use because it depends on individual circumstances. When determining whether a use is fair for R&D, courts frequently take into account elements such as the intended use, the nature of the work, the amount used, and the impact on the original work's market value. Researchers need to understand that fair use is a complex and changing term. When using copyrighted materials for R&D, researchers and developers should seek to create something new and original that furthers their research objectives. They should also seek the advice of legal professionals in complex cases or when using copyrighted materials for commercial purposes. Although its implementation necessitates carefully evaluating the unique circumstances and respecting the legal principles and guidelines set within copyright law, fair use can be a useful tool for researchers.

14.3.4 Licensing and Permissions

Sometimes, using copyrighted content necessitates going beyond what is allowed by fair usage. To ensure legal and moral R&D processes, this section discusses the critical role of licensing and permissions. The framework for securing the right to utilize works protected by copyright while maintaining copyright rules is provided by licensing and permissions. A methodical strategy is used in the licensing and permissions process for copyrighted works in R&D. Identifying the copyright holder is the first step for researchers, who may need to contact publishers and creators or use copyright databases. A request for permission should be sent after confirming the identity of the copyright owners, outlining the planned use and any modifications needed. License agreements should be carefully examined to ensure they meet R&D demands before negotiating terms and costs.

In the R&D setting, various licensing agreement types are used. These include royalty-free licenses, which allow use without ongoing payments; nonexclusive charges, which would enable use by numerous parties; and exclusive privileges, which give the licensee exclusive rights to the work. Customized contracts can also be established to meet the needs of a given R&D project. Legal adherence depends on complying with licensing agreements and keeping records of proof of authorization in case of disagreements or audits. R&D organizations can safely and legally use copyrighted materials in their research projects while respecting the rights of copyright holders by being aware of these procedures and forms of license agreements.

14.3.5 Open Access and R&D

Open access has become a transformative movement that offers an alternate method for sharing knowledge and encouraging innovation. This section examines the idea of open access and how it relates to R&D, emphasizing how it affects collaboration, accessibility, and IP issues.

In contrast to traditional subscription-based journals and exclusive databases, open-access publishing models, which aim to enable unlimited public access to research findings and creative works, permit content to be openly accessed, downloaded, shared, and frequently repurposed.

The advantages of open access in R&D include wider dissemination because paywalls and subscription barriers are removed, increasing visibility and impact; improved collaboration across institutions and geographies because open-access materials can be shared without restrictions; accelerated innovation through the free exchange of knowledge, fostering the ability to build upon others' work without extensive licensing and permissions; and increased public awareness. Different publishing models can be used to achieve open access, such as gold open access, where authors or their institutions pay for publication costs; green open access, which

Table 14.1 Comparison of open access publishing models

Publishing model	Description	Funding source	Accessibility
Gold open access	The author or institution covers publication costs, making the work freely available upon publication	Author/institution-funded	Immediately open to the public
Green open access	Authors deposit preprints or postprints in repositories, accessible after an embargo period	Usually, no direct funding	After the embargo period, freely accessible
Hybrid open access	Researchers pay to make individual articles open access in subscription-based journals	Author/institution/research	Specific articles open, rest behind the paywall

involves self-archiving in repositories; and hybrid open access, which enables paid access within subscription-based journals (Table 14.1).

Regards for copyright and open access are crucial in the context of IP and knowledge sharing. Open access does not eliminate copyright; rather, it coexists with it, allowing authors and other creators to keep their rights while choosing open-access licenses to encourage sharing and reusing. These licenses cover a wide range, from the most lenient, like CC BY, to the most stringent, like CC BY-NC-ND. Some artists even go so far as to donate their works to the public domain, waiving all copyright and granting unrestricted use rights.

Open access and IP do, however, have a complex relationship. Open access promotes the sharing of knowledge, but it does not negate IP rights. Researchers need to understand the licensing conditions of open-access materials carefully. Contrarily, open-access publication can provide transparent access to the work and its sources, acting as a strategic instrument for protecting IP. The open-access movement will expand to include additional channels, including free access to data, software, and different research outputs. This movement is poised to influence the future of IP, accessibility, and information sharing in the field of R&D as R&D organizations adopt open access more frequently to encourage innovation and cooperation.

14.4 Intellectual Property and Collaboration in R&D

R&D is built on collaboration, which allows different stakeholders to combine their knowledge, resources, and creative ideas beneficially. However, the complicated environment of IP administration comes into play when engaging in joint R&D projects. Starting with developing collaborative agreements, providing transparency around IP ownership and its governance becomes essential. These agreements need to carefully specify important elements, such as IP ownership, individual contributions, publication rights, and commercialization methods. Such clear definitions act as the cornerstone for averting future conflicts and creating a conducive working environment for partners.

Establishing duties and decision-making procedures that are clearly defined is essential when joint IP ownership is involved. A management committee made up of members from each cooperating entity is one practical strategy. This committee is crucial in administering, licensing, and commercializing IP. A dispute resolution system that can effectively resolve any arising disputes needs to be included in the cooperation agreement. This mechanism may consist of choices like arbitration, mediation, or predetermined negotiating methods, all of which try to stop disagreements from worsening and endangering the partnership.

Open and honest communication is fundamental to preserving a successful and harmonious collaborative R&D environment. Keep all collaborators informed and in sync by holding regular meetings and providing continual progress reports to reduce misunderstandings and potential conflicts. Engaging the services of legal counsel knowledgeable in IP and collaboration agreements is also regarded as a wise move. To guarantee that collaborative contracts comply with applicable laws and regulations and to protect the rights and interests of all parties involved in the collaborative R&D endeavor, these legal professionals provide vital help.

14.5 Protecting R&D Through Copyright

To defend the original and creative works, it is crucial to copyright protect the R&D activities. Whether they take the shape of written texts, computer code, creative works, or other things, original manifestations of ideas are protected by copyright laws. Copyright can apply to various resources in the context of R&D, including reports, software, datasets, and even multimedia presentations.

14.5.1 Strategies for Copyright Protection

A legal system known as copyright was established to protect the original manifestations of ideas. It covers various works, including academic articles, software code, datasets, and multimedia presentations. Understanding the extent of copyright protection is essential because it safeguards the distinctive features of the R&D output, focusing on the particular language, organization, or presentation that makes the work stand out.

Copyright protection depends heavily on documentation and recordkeeping. In the event of a disagreement, meticulously keeping track of all stages of the creative process, including drafts, revisions, and creation dates, provides strong proof of ownership and originality. A clear timeline of the creative process strengthens the copyright claims. Copyright notices are another easy-to-use but powerful countermeasure. These notices normally contain the copyright sign (©), the creation year, and the name of the copyright owner. By clearly stating the rights, communicating the rights deters possible infringers while also announcing the IP ownership.

Training team members and collaborators within the context of collaborative R&D is essential. Sharing information about copyright and the value of preserving IP helps to guarantee that all project participants are aware of the various resources' copyright statuses and respect others' rights. When dealing with outside parties or sharing the R&D assets, licensing and permissions provide flexibility. Particular usage rights can be given through licensing while still protecting the copyright. This enables setting guidelines for how the materials may be used.

Additionally, open-access considerations are crucial for R&D projects wanting to spread knowledge freely. Choosing an appropriate open-access license can strike a balance between sharing knowledge and protecting the rights. Examples of open-access licenses include Creative Commons licenses, which offer the power to define how the work can be used, shared, and credited while maintaining copyright ownership.

14.5.2 Copyright Notice

A copyright notice is an easy-to-use but powerful instrument for establishing ownership and copyright protection over creative and original R&D works. It can operate as a deterrent against possible infringers and acts as a visual reminder of the rights. The copyright notice needs to be placed correctly and follow a specific structure to be effective. The copyright symbol (©), the year of production or publishing, and the name or the copyright owner's name make up most of it. This structure should be easy to read and understand, just like "20XX [The Name]."

R&D content is frequently disseminated online in recent times. Adding copyright disclaimers on online content is crucial when distributing the R&D files on websites or other digital platforms. These disclosures may be included in the footers of websites, on specific pages, or in the text that connects to additional content. Accessibility and security need to coexist in harmony. Even though copyright notices are crucial for defending legal rights, it is critical to avoid using them in a way that could deter cooperation and information sharing. Be prepared to negotiate permissions, clearly describe how the R&D materials can be used and shared, and consider open-access licensing choices to encourage cooperation while safeguarding the interests.

14.5.3 Digital Rights Management (DRM)

To protect IP, regulate access to R&D materials, and govern their distribution, digital rights management (DRM) has emerged as a crucial instrument. Research reports, software code, datasets, and multimedia content are just a few examples of the types of digital assets that can be protected with DRM from unauthorized access, distribution, and alteration. DRM plays a critical role by giving content owners and creators the tools to manage who can access, copy, edit, and share their digital R&D

materials. Encryption, access control, copy protection, and watermarking are examples of common DRM strategies. By protecting digital content from unwanted access, copying, and sharing, these methods help to maintain the confidentiality and integrity of R&D output.

Utilizing DRM in R&D requires striking a balance between security and usability. Overly restrictive DRM controls can make it difficult for research teams to collaborate and share knowledge. To regulate how R&D materials can be utilized, specific licensing and permission terms need to be established. When installing DRM in the R&D context, one needs to comply with legal considerations, particularly about copyright laws and the Digital Millennium Copyright Act (DMCA) in the United States.

R&D personnel can choose from various DRM tools and solutions, including open-source and commercial DRM software. The DRM system chosen should provide the required level of protection and control over digital assets and align with the R&D organization's unique needs and goals of the R&D organization. As technology advances, staying updated on changing best practices and legal considerations in the dynamic digital landscape is essential.

14.6 Intellectual Property Enforcement and Infringement

R&D-related innovations and creations must be protected, and IP rights are essential. Effective enforcement methods and the capacity to stop infringement are necessary for protecting these rights. Identification of copyright infringement is the initial step in this procedure, and it is accomplished using techniques including routine monitoring, plagiarism detection software, third-party reports, and comparative analysis.

When copyright violations are discovered, several legal actions can be taken to protect IP. The Digital Millennium Copyright Act (DMCA), litigation, cease-and-desist letters, alternative dispute resolution (ADR), and international enforcement procedures for cross-border issues are among the available remedies. Each remedy can be used to stop unauthorized usage, demand compensation, and reclaim IP ownership [4].

The best course of action is to completely avoid infringement by incorporating best practices into the R&D efforts. To ensure that the R&D outputs are free from potential infringement issues, the necessary permissions and licenses need to be obtained; third-party content should be properly attributed; team members need to be educated about IP rights and internal policies; guidelines should be implemented; and regular audits should be carried out. In conclusion, R&D's successful and moral advancement of R&D depends on a thorough grasp of protecting IP rights and combating infringement. Proactive efforts and awareness of legal choices can maintain the honesty and inventiveness of the contributions to the R&D field. Figure 14.2 illustrates how IP rights in R&D can be protected by identifying, responding to, and stopping violations while fostering moral behavior and preventative actions.

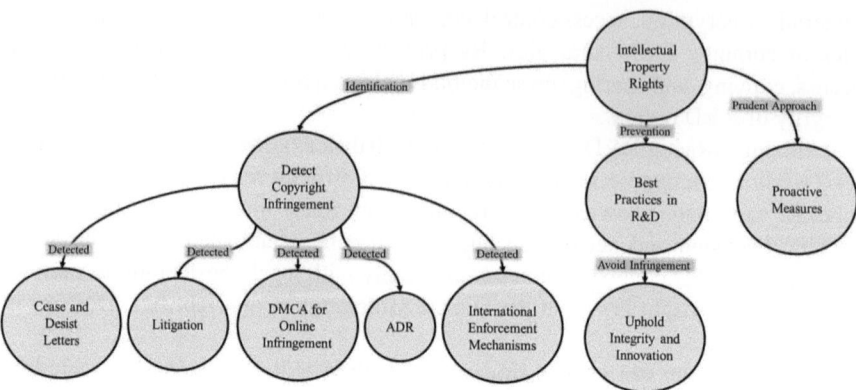

Fig. 14.2 Safeguarding IP rights in R&D

14.7 Emerging Trends and Challenges

Copyright in the R&D process is an important legal concept that protects the intellectual property created during the development of new ideas, inventions, and creative works. Copyright law primarily applies to original works of authorship that are fixed in a tangible medium, but its application in the R&D process can be somewhat limited. These trends and challenges are likely to continue evolving as new technologies and societal demands shape the R&D landscape. Staying adaptable and proactive is crucial for organizations and researchers in this field.

14.7.1 Digital and Online R&D

The digital revolution has significantly changed R&D in recent years, with the growing digitization of processes and the widespread use of online tools. The pace of invention has quickened due to this change, creating new opportunities and difficulties in copyright and IP. The ramifications of digital and online R&D need to be understood by researchers, organizations, and institutions.

The production and security of digital content are important factors. Research reports, publications, statistics, and software code are just a few of the many types of content that R&D operations in the digital age produce. All of these materials are protected by copyright. Researchers need to understand how copyright protections apply to these digital works and how to protect them against unlawful use, reproduction, and dissemination. Additionally, the emergence of online publishing platforms and data repositories has impacted writers' rights, licensing arrangements, and copyright duties, upending traditional publishing structures. It is critical to navigate the open-access publishing complications while balancing the desire to spread information with the necessity to protect IP rights.

14.7 Emerging Trends and Challenges

Additionally, the development of data sharing and collaborative research has changed how scholars work together across geographic distances. However, these techniques provide issues about access control, attribution, and IP ownership. In collaborative R&D, researchers need to define precise rules for data sharing and IP rights. Due to the increased prevalence of illicit digital content copying and modification, the internet and digital world also pose enforcement issues. Digital rights management (DRM) technologies and judicial procedures are frequently needed in this digital environment to protect IP rights. With breakthroughs like virtual and augmented reality, data analytics, and cloud computing primed to offer new considerations for IP and copyright, it is anticipated that digital and online R&D will continue to evolve in the future. In the constantly changing environment of digital and online R&D, researchers and organizations need to adapt proactively to new digital trends while ensuring that their rights are upheld.

14.7.2 AI and Machine-Generated Works

The cutting edge of contemporary R&D is machine learning and artificial intelligence (AI). These technologies can produce various intellectual products, from research papers and data analyses to original music, art, and even literature. The consequences of AI-generated works for copyright and IP are intricate and constantly changing.

AI as a Creative Force

AI has shown an impressive ability to produce content that can be mistaken for human-made, especially in the form of generative neural networks. This includes artificial intelligence (AI) music composers, deep learning networks that produce art, and natural language processing models that can write articles and reports. The distinction between authorship and ownership gets murkier as AI systems create content that formerly required human inventiveness. Given that AI-generated works may not include a human author who is actively involved in the creative process, questions about who might assert copyright in these works arise.

Copyright and Authorship

The author of an original work normally receives exclusive rights under copyright law. Reproduction, distribution, and the development of derivative works are all included under these rights. Traditional copyright cases identify the human author. AI-generated art, however, challenges this paradigm. There is disagreement regarding who should be regarded as the creator and owner of the copyright in AI-generated content—the person who created the AI model, the person who activates the AI, or the AI system itself. This problem still needs to be resolved, and different legal systems may have different ways of doing it.

Ownership and Licensing

Understanding ownership and license agreements for AI-generated works is crucial when navigating the IP ecosystem. Clear policies and contracts need to be established by groups and individuals engaged in AI R&D. AI system creators may assert ownership in some situations. In contrast, AI may be considered a tool without rights to others. When AI-generated content is shared, altered, or used for commercial purposes, licensing requirements can become more complicated. Managing the usage and dissemination of AI-generated works requires careful consideration of licensing agreements, open-source approaches, and collaborative agreements.

Legal and Ethical Considerations

There are ethical issues with AI-generated content in addition to the legal ones. Using AI to create content may raise concerns about responsibility, transparency, and cultural sensitivity. The public's impression of authorship, artistic expression, and the possible erasure of human creators in various sectors need to be considered by researchers and organizations. Ethical standards and procedures should cover these AI research, development, and implementation issues.

Ongoing Developments

The ethical and legal climate surrounding artificial intelligence and computer-generated art is continually developing. As AI-related copyright challenges arise, various jurisdictions are formulating their strategies. Legislators and legal experts are actively considering copyright law changes that could be made to account for AI-generated content. Staying current on these advancements, participating in interdisciplinary discussions, and feeling the potential effects on their work and IP strategies are crucial for R&D professionals and organizations. Future ideas for AI and computer-generated art in the framework of IP and copyright are presented in Table 14.2.

14.7.3 Open Source and Collaboration

Open-source development has gained popularity in R&D disciplines thanks to its dedication to open access, community cooperation, and transparency. The many contributions of the open-source community assist organizations and researchers,

14.7 Emerging Trends and Challenges

Table 14.2 AI and machine-generated works: future considerations

Future considerations	Key questions	Challenges	Proposed actions
Legal frameworks and legislation	What legal rights should AI systems have? How can they be enforced?	Varying national and international laws	Create AI-specific copyright legislation
International agreements	Can global agreements effectively address cross-border issues?	Complex international negotiation processes	Establish global standards for AI copyright
Defining authorship and ownership	Who is the author of AI-generated works, and who owns them?	Legal and philosophical challenges	Define AI authorship and ownership rights
Copyright duration and protection	How does the duration of AI copyrights compare to traditional copyrights?	Balancing protection with public interest	Review and adjust copyright duration rules
Ethical guidelines and bias mitigation	How can AI-generated content avoid perpetuating biases and harmful stereotypes?	Ethical dilemmas and cultural variations	Develop ethical AI content guidelines
Enforcement and detection technologies	What technologies can effectively identify and attribute AI content?	Rapidly evolving technology landscape	Invest in AI-powered detection tools
Creative collaboration and AI assistance	How should credit and ownership be shared in collaborative AI projects?	Legal, practical, and ethical complexities	Establish guidelines for collaborative AI projects
Public domain and fair use revisions	How does AI impact traditional concepts of public domain and fair use?	Balancing innovation and public access	Revise public domain and fair use criteria
Cultural and artistic preservation	How can AI aid in cultural preservation while respecting copyright and IP laws?	Maintaining authenticity and cultural integrity	Collaborate on AI-driven cultural preservation projects
Educational and research implications	How can AI-generated works be leveraged in education and research ethically?	Access to AI-generated content in educational settings	Develop guidelines for educational AI usage
Future-proofing copyright laws	How can copyright laws remain relevant in a rapidly evolving technological landscape?	Balancing innovation with legal consistency	Periodically review and update copyright laws

fostering innovation through group wisdom. Open source, however, poses a distinct set of problems for copyright and IP. The choice of open-source licenses, which specify the conditions for sharing IP, such as the MIT License and GNU General Public License, is crucial to this problem. To ensure compliance with the chosen license, researchers need to carefully select the suitable right that is in line with the aims and goals of their project. Complexity arises from the need to balance openness and the protection of trade secrets and unique techniques. Strategies to control this balance include dual license options, transparent documentation, and records of contributions.

The ethical and legal issues surrounding IP in R&D become more crucial as open-source techniques develop. Clear ethical standards are essential because conflicts might occur when proprietary software or research interacts with open-source projects. Furthermore, one area for improvement with open-source research is safeguarding private or sensitive data. To successfully integrate open-source techniques into R&D, careful consideration needs to be given to licensing decisions, ethical behavior, and adherence to community norms. This will maximize the benefits of collaboration while effectively managing any difficulties.

14.7.4 Future of IP in R&D

R&D use of IP will be determined in the future by a complex interaction of technological growth, changing legal landscapes, and shifting societal goals. As we look ahead, several significant trends and areas of interest emerge, offering perception into the future state of IP in R&D.

Blockchain and IP Management

Incorporating blockchain technology is one of the most exciting advancements in IP management. Blockchain provides a decentralized, secure, and open method for tracking and managing IP rights. Organizations can create immutable records of their IP assets using blockchain, enabling effective licensing, tracking, and enforcement. This technology may make innovative IP monetization techniques like microlicensing and real-time royalty distribution possible. The creation and acceptance of blockchain-based IP management systems will greatly impact how IP is used in R&D in the future.

IP and Sustainability

The importance of IP in tackling these challenges rises as the world struggles with urgent issues like climate change and sustainability. Strong IP protection is necessary for developing and commercializing green technology, renewable energy options, and sustainable agricultural methods. In the hunt for sustainable solutions, a rising movement supports open access and cooperative research. Future discussions regarding balancing IP rights protection with enabling universal access to vital technology for advancing society and the environment will likely continue.

International Harmonization of Copyright Laws

Since national regulations have historically impacted copyright laws, there are differences in cross-border protections and enforcement. Increased efforts to harmonize copyright rules internationally may be made in the future. In a globalized society, organizations, governments, and international organizations are looking at ways to simplify copyright compliance, making it simpler for scholars and producers to engage in international partnerships and knowledge sharing. The ability of countries to harmonize their copyright laws will determine the viability of such initiatives, which could substantially impact global R&D processes.

Ethical and Cultural Considerations

A growing understanding of the ethical and cultural ramifications of IP will impact how IP is used in R&D. Indigenous knowledge, customary ways of life, and biotechnological discoveries create difficult queries regarding benefit sharing, ownership, and access. The difficulty of preserving cultural history and encouraging innovation while upholding IP rights will always exist. Future legal and ethical frameworks are expected to include more frequent protections and considerations for these factors.

References

1. A. Gorbatyuk, G. Van Overwalle, E. Van Zimmeren, Intellectual property ownership in coupled open innovation processes. IIC-Int. Rev. Intellect. Propert. Competit. Law **47**, 262–302 (2016)
2. E. Salmerón-Manzano, F. Manzano-Agugliaro, *Low-Cost Inventions and Patents: Series II* (MDPI, 2023), p. 20
3. P.B. Hirtle, *Copyright Term and the Public Domain in the United States* (2019)
4. O. Bulayenko et al., *Cross Border Enforcement of Intellectual Property Rights in the EU*, vol 10 (Centre for International Intellectual Property Studies (CEIPI) Research Paper Forthcoming, Study for the Committee on Legal Affairs (JURI) of the European Parliament, 2022), p. 255094

Index

A
Applied research, 2, 3, 6, 7, 171, 175, 232–234, 236

B
Basic research, 2, 4–7, 11, 13, 20, 215, 232–235, 237, 241
Budgeting process, 75–76, 78, 80

C
Careers in R&D, 135, 136, 146, 167
Change management, 103, 132, 180, 191, 231, 232, 234, 242
Commercialization, 38, 43, 111, 125, 128–131, 294, 302
Competitive intelligence (CI), 55–56
Copyright, 19, 36, 38, 40–42, 56, 85, 86, 88, 90, 101, 108, 124, 175, 263, 291, 293–309, 311

D
Digital rights management (DRM), 304–305, 307

F
Feasibility assessment, 115–119
Funding sources, 21, 70, 71, 73–74, 92, 93, 193, 302

Future trends, 91–93
Future trends in R&D, 91–93

I
Innovation, 1, 2, 6–8, 13, 15, 16, 18–21, 23–27, 29–49, 53, 57, 59, 62, 63, 65, 66, 69–72, 74, 75, 85, 88–90, 93, 95, 98, 101, 104, 106, 107, 111–115, 119, 121, 124–128, 131, 133, 136, 140, 143, 145–148, 153, 155, 157, 161, 169–172, 179, 180, 184, 189, 193, 194, 197–202, 204–210, 212, 215, 217, 221, 222, 232, 237, 239, 242, 243, 247, 259–268, 271, 272, 274, 275, 278, 281–288, 291, 292, 294, 295, 298, 301, 302, 305, 309, 311
Innovations and R&D, 26, 27, 29, 30, 39, 40, 43, 49, 133, 157, 202, 264, 272, 278, 281, 284, 292, 295, 305

J
Jobs, 77, 112, 121, 135–138, 142, 145, 155–157, 159, 163, 167, 225, 227, 228, 261, 262, 279, 294, 299

M
Managing R&D, 16, 23, 36–39, 64, 81–84, 99, 102, 106, 136, 158, 168, 201, 234, 241

© The Editor(s) (if applicable) and The Author(s), under exclusive license to Springer Nature Switzerland AG 2024
H. Taherdoost, *Innovation Through Research and Development*, Signals and Communication Technology, https://doi.org/10.1007/978-3-031-52565-0

O

Open access R&D, 289, 301, 308

P

Patent, 19, 36–42, 55, 56, 85, 86, 88, 90, 101, 108, 124, 133, 141, 146, 148, 172, 175, 193, 194, 202, 205, 209, 239, 257, 258, 263, 278, 292–294

Performance management, 167, 197, 200–202, 254

Permissions, 38, 116, 124, 283, 298, 301, 304, 305

Personnel management, 168

Project life cycle, 167, 179, 191–193, 222, 252–254

Project management, 3, 10–12, 14–17, 20, 29, 36, 37, 57, 65, 80, 81, 88, 93, 95, 101–104, 107, 122, 126, 136, 147, 154, 155, 160, 161, 163, 176, 177, 179–187, 191, 193, 197, 205–210, 213, 216, 217, 221, 222, 229, 230, 232, 235, 238, 248, 257–259, 261, 262, 274, 279

Project management process, 10, 11, 15, 17, 20, 176, 181, 186, 193, 234

Project zones, 189

R

R&D budget, 57, 61, 74–76, 78–84

R&D employment, 148

R&D financial reporting, 82

R&D funding, 69–74, 78, 91–93, 146

R&D intellectual property, viii, 18, 23, 36–39, 41, 56, 98, 106, 133, 194, 263, 264, 291–311

R&D leadership, 17

R&D licensing, 56, 107, 301, 308

R&D measures, 40, 84

R&D metrics, 224

R&D performance, 23, 39–44, 75, 197, 198, 200, 203, 212–215, 239

R&D performance measurement, 215, 239–245

R&D planning, 52, 54, 58

R&D practices, 20, 37–39, 42, 44, 112, 135, 154, 163, 174, 197, 198, 239, 241, 267, 268, 305

R&D process, 7, 13, 14, 43, 44, 88, 98, 99, 111–113, 115, 116, 118, 119, 123, 125, 127, 128, 131–133, 171, 194, 212, 276, 284, 298–302, 306

R&D project evaluation, 104, 209

R&D project management, 95, 99, 104, 163, 179, 234

R&D project prioritization, 61

R&D projects, 17, 40, 56, 61, 65, 78, 85–88, 90, 92, 95, 96, 99–103, 105, 107–109, 118–122, 160–165, 167–176, 208, 239, 242, 243, 245, 257, 258, 264, 301

R&D roles, 18, 19, 136, 137

R&D strategy, 25, 27, 48–50, 53–57, 59, 64–66, 268

R&D success, 20, 41–43, 57, 121, 198, 210–212, 215, 247

R&D team, 19, 36, 37, 44, 50, 87, 95, 122, 143, 160–164, 170, 175, 176, 258

R&D technologies, 285

R&D tools, 272, 273

Research and development (R&D), 1, 23, 47, 69, 95, 111, 135, 153, 179, 197, 247, 271, 291

Research methods, 8–11, 113, 281

Resource allocation, 1, 27, 28, 31–38, 40–43, 48, 51, 54, 56–58, 62, 64, 66, 72, 76, 81–83, 85–87, 93, 97, 102, 104, 105, 107, 120–122, 127, 133, 179, 206, 215, 231, 261, 263, 266–268, 279, 283

Return of investment (ROI), 27, 33, 35–36, 40–42, 57, 60, 61, 69, 72, 76, 83, 85–88, 93, 117, 118, 147

Risk assessment, 46, 64, 66, 83, 87, 90, 93, 104, 105, 107, 118, 119, 122, 233, 234, 242, 248, 253, 261, 264–268

Risk management, 1, 19, 31, 34, 52, 64, 65, 79, 80, 87, 88, 101, 103, 105–106, 112, 116, 122, 127, 133, 183, 184, 187, 188, 223, 227, 230, 238, 242–243, 247–260, 264–266, 269

Risk management process, 64, 251, 252, 254

Risk of R&D, 99

S

SMART R&D, 47, 50–51, 58, 104, 105

Strategic R&D, 7

Strategy and risk, 247, 268–269

T

Team building, 108, 122, 153, 164, 166, 210

Team management, 36, 108

Trademarks, 19, 36–38, 40–42, 56, 85, 86, 88, 124, 263, 292, 294

MIX
Papier aus verantwortungsvollen Quellen
Paper from responsible sources
FSC® C105338

If you have any concerns about our products,
you can contact us on
ProductSafety@springernature.com

In case Publisher is established outside the EU,
the EU authorized representative is:
**Springer Nature Customer Service Center GmbH
Europaplatz 3, 69115 Heidelberg, Germany**

Printed by Libri Plureos GmbH
in Hamburg, Germany